D0930986

I TATTI STUDIES IN
ITALIAN RENAISSANCE HISTORY

Published in collaboration with I Tatti
The Harvard University Center for Italian Renaissance Studies
Florence, Italy

FORGOTTEN HEALERS

Women and the Pursuit of Health in Late Renaissance Italy

SHARON T. STROCCHIA

Harvard University Press

Cambridge, Massachusetts
London, England
2019

Printed in the United States of America

First printing

Library of Congress Cataloging-in-Publication Data
Names: Strocchia, Sharon T., 1951- author.
Title: Forgotten healers : women and the pursuit of health in Late Renaissance Italy / Sharon T. Strocchia.
Other titles: I Tatti studies in Italian Renaissance history.
Description: Cambridge, Massachusetts : Harvard University Press, 2019. | Series: I Tatti studies in Italian Renaissance history | Includes bibliographical references and index.
Identifiers: LCCN 2019012378 | ISBN 9780674241749
Subjects: LCSH: Women healers—Italy—History—16th century. | Women healers—Italy—History—17th century. | Women in medicine—Italy—History—16th century. | Women in medicine—Italy—History—17th century. | Medical care—Italy—History—16th century. | Medical care—Italy—History—17th century. | Medicine—Italy—History—16th century. | Medicine—Italy—History—17th century.
Classification: LCC R517 .S77 2019 | DDC 610.94509/031—dc23
LC record available at https://lccn.loc.gov/2019012378

In memory of

STELLA STROCCHIA

Mother, teacher, caregiver

CONTENTS

LIST OF FIGURES

A NOTE ON DATES AND CURRENCY

The Florentine new year began on March 25, the feast of the Annunciation. Dates given in the text have been modernized to correspond to the calendar year beginning January 1. When citing archival documents in the notes, I give both dating styles when appropriate to avoid confusion.

Renaissance Florentines used a complex monetary system that included silver and gold coins, as well as moneys of account. The lira was a money of account. One lira was divided into twenty soldi; each soldo was subdivided into twelve denari, making a lira equal to 240 denari. In the early sixteenth century, the most common gold coin in circulation was the florin. After the Medici dukes came to power in 1532, the florin was replaced by the scudo or ducato. One scudo was worth seven lire.

FORGOTTEN HEALERS

Introduction

ONE OF THE MOST STRIKING FEATURES to emerge from recent studies of Renaissance medicine is the sheer diversity of female practitioners who anchored a wider medical economy.[1] Thanks to a growing body of scholarship, we know that women from northern Europe to the Mediterranean basin permeated every aspect of healthcare services both within and beyond the home between 1400 and 1700. The explosive interest in Renaissance domestic medicine over the past decade has firmly established the household as the primary locus of care well into the eighteenth century, despite the proliferation of hospitals and other charitable institutions.[2] But it is clear that women from all walks of life contributed to health and healing outside the home as well. They served as paid and unpaid nurses, hospital administrators and community caregivers; made and marketed medicines as commercial pharmacists and estate managers; and managed the reproductive process as civic midwives and wet nurses in charitable institutions.[3] Women identified by their communities as "expert practitioners" testified in court and assessed the health status of domestic slaves in contract disputes. Female empirics worked as oculists or carried out minor surgeries, while wives and widows ran health-related establishments alongside their husbands or apprentices.[4] Gentlewomen wrote and occasionally published collections of medical recipes, but many more circulated healing knowledge through broad epistolary networks.[5] Pious women working singly or in groups offered important end-of-life services; they not only comforted the dying but bore witness to the very act of dying well. Innovative forms of poor relief springing up throughout the sixteenth century frequently repurposed women's experiential knowledge of the body. Elderly parish women who first acted as "keepers" of the sick soon became "searchers" of the dead who advanced public health schemes by relaying epidemiological information to civic authorities.[6] In short, Renaissance women performed much of the day-to-day work of healing and caregiving throughout this period.

Part of the difficulty in integrating these contributions into a broader narrative of early modern medicine stems from the nature of the evidence itself. Much of the documentation for women's medical activities is fragmentary and must be excavated through painstaking work in local archives and obscure print materials. In addition, sources are silent on everyday matters of importance. Birth attendants who worked without compensation or neighborhood women who washed the bodies of the dead have left few traces in the historical record. Because these types of bodywork were thoroughly naturalized as feminine care practices, they entered the historical record only when they rose to the level of visible economic exchange. Complicating matters of evidence still further were collective work-sharing arrangements among family, friends, and neighbors that often make it difficult to know "which person was doing exactly what work."[7]

Scholars face an even more vexing problem, however, in trying to conceptualize this diffuse body of evidence. Conventional histories of Renaissance medicine have tended to focus on the transmission of text-based knowledge and institutional authorization by guilds and universities, rather than on healing skills acquired experientially. Our understanding of the interplay and entanglements between female healers and professional medicine as a body of knowledge and practice remains partial at best. A persistent focus on official titles and occupational identities has led us to both undercount and undervalue the healthcare services Renaissance women provided to household and community.[8] Other conceptual challenges stem from coding certain care practices frequently performed by women—feeding the sick, dressing sores—as charitable work, which empties them of medical meaning. The relationship between preventive health measures and much-heralded curative skills also warrants further examination through the lens of gender.

By redefining what constitutes medical work, we can map a more complex terrain for early modern healthcare and women's place in it. New interpretive frameworks have situated women squarely within health promotion and illness management, from pharmacy and household medicine to emerging structures of public health. The terms "medical agents" and "agents of health" coined by Monica Green capture a wider range of practitioners active in settings where formal titles were rarely used. Montserrat Cabré has taken a different tack by exploring the overlapping semantic associations between the terms "women" and "healers" in late medieval society. The concept of "bodywork" introduced by Mary Fissell has begun to dismantle some of the "hierarchies of value" first created in the early modern period and reproduced by later generations.[9] This reorientation results in a more accurate, dynamic depiction of medical provisioning in early modern society while revealing its deeply gendered nature.

My book expands this critical reappraisal by examining the broad palette of women's medical work in late Renaissance Italy (1500–1630), a period best known for its striking transformations in academic medicine. Drawing on rich but often fragmentary archival records from Florence, Bologna, and Rome, this study demonstrates both the sweeping scope and the social importance of women's contributions to provisioning care. At the center of analysis is a rich matrix of practitioners—household caregivers, nun apothecaries, noble-born empirics, poor hospital nurses—whose health-related activities positioned them as important knowledge-makers and commercial innovators within a changing medical landscape. As agents of health, Italian urban women operated along a differentiated continuum of skill and knowledge that was produced and transmitted experientially.[10] I argue that increased demand for healthcare services, coupled with the growing interest in preventive health practices, opened new opportunities for Italian women to participate extensively in both the medical economy and emerging cultures of experimentation. In so doing, I take an integrative approach that moves beyond a single methodology or type of document in order to connect the study of health and medicine to broader developments in politics, society, economy, and religion.[11]

The link between medical skills and knowledge production is worth stressing in light of the recent material turn in early modern scholarship. The pursuit of health often provided the springboard for women with skilled hands and inquiring minds to experiment with remedies or put body knowledge to new uses. By and large, their goals were results-oriented. Pharmacy in particular offered a congenial arena for elite women to advance experimental interests and circulate proprietary remedies within broader epistolary networks.[12] As part and parcel of good household management, making medicines certified wifely virtues while providing a gateway for further inquiry.[13] Medical reading offered another avenue for practical knowledge-making. By extracting and reorganizing information found in printed herbals and regimens, and then interpolating new observations into these extracts, literate women not only consumed but also produced new knowledge.[14] Poorer widows and single women often had more explicit commercial objectives when making and marketing remedies within an economy of makeshift.[15] Similarly, the countless women living in custodial institutions across Italy—orphaned girls, abandoned wives, converted prostitutes—developed and circulated embodied knowledge about healing by tending the health needs of other residents within these constructed communities. Having grown up or grown old in an institutional setting, these women bore much of the responsibility for their own care.[16]

Both the geographical terrain under consideration and the relationships between knowledge and practice posited here have important implications for several reasons. Most studies of women's medical agency in this period have focused on northern Europe, especially England and Germany. Because the historiography of Italian Renaissance medicine has been dominated by developments in academic medicine—anatomy, dissection, humanistic debates—studies of women's entanglement with household medicine, care practices, and experimentation have not flourished to the same degree as elsewhere. To be sure, this extant body of scholarship offers valuable perspectives on transformations in academic life and medical thinking, but it has been less concerned with the lived experience of illness on the streets or the activities of frontline healers in households and hospitals. To date scholars have not fully capitalized on the abundant archival evidence that would support these body-centered inquiries. Available materials, many of which are used here, range from narrative sources such as letters, petitions, and chronicles to economic transactions registered in account books and administrative deliberations embodied in hospital statutes and guild records.

By placing Italian women healers squarely at the center of analysis, this study both recuperates their importance to local health markets and highlights significant differences in medical provisioning across Europe. Sandra Cavallo and David Gentilcore have stressed "the importance of the local cultural context in differentiating the opportunities and choices" that consumers could make among active medical agents.[17] Late Renaissance Italy stood out from its northern neighbors by virtue of its high urban density, commercialized economy, extensive print culture, precocious medical humanism, and well-developed court societies. Sixteenth-century Italian cities and courts claimed a high degree of health literacy, thanks in part to the popularity of vernacular printed health regimens that focused attention on the importance of the six "non-naturals": diet, air, sleep, evacuation, exercise, and emotional equilibrium. In Galenic medical thinking, these external factors helped regulate the internal balance of the humors that determined good health. Domestic spaces featured prominently in this growing culture of prevention, giving Italian female householders significant responsibilities as guardians of healthy living.[18]

Other historical contingencies influenced medical provisioning on the peninsula. Despite its cultural splendor, late Renaissance Italy was marked by hard times and social unraveling. Troop movements prompted by waves of foreign invasions, along with an entrenched culture of prostitution, amplified the new disease threat of syphilis after 1494. Deteriorating economic conditions and widespread unemployment in the textile trades worsened the effects of recurrent famines in the second half of the sixteenth century.[19] The widespread plague of 1575–1578—the most important plague episode of sixteenth-century Italy—ushered in new civic attempts to track the spread of disease, quarantine suspected goods, segregate the sick, and otherwise contain this contagion.[20] Catholic reform turned the spotlight on social welfare; papal

Jubilees brought flocks of pilgrims into Rome; religious persecution forced migrations into and out of the peninsula. All of these circumstances increased the demand for healthcare services across Italy—a need met in significant part by cheap or unpaid female labor.[21]

In providing a comprehensive look at women and the pursuit of health in late Renaissance Italy, my analysis breaks new ground in three main ways. First, I take a more body-centered approach to healthcare that focuses attention on what Renaissance Italians actually did to treat and prevent illness. One advantage of this approach is that it cuts across the widest possible range of activities and social strata. In the course of this study, we meet a sweeping panorama of healers who came from every point on the social spectrum: princely consorts concerned with keeping their heirs healthy; court ladies who circulated medical practices across political and linguistic boundaries; patrician women and merchant wives who trafficked in the healing power of relics; unwilling nuns who found intellectual outlets in medicine and the natural world, and their more devout sisters who excelled at making and marketing remedies; peasant women entrusted with nursing the Medici children; widows and single women who staffed hospital pharmacies; orphaned girls who tended poor syphilis patients in hospital. Taking an integrative, body-centered approach not only highlights the social range of women involved in healthcare, but also showcases the value and meanings contemporaries attached to these different services in a pluralistic medical marketplace.

Second, I argue that Italian women working outside traditional guild and academic settings became invested in empirical knowledge-making through the pursuit of health. In so doing, my study extends the reach of empirical culture into the female spaces of Renaissance courts, convents, and charitable institutions. One of the threads guiding my investigation is the notion that vernacular knowledge production in the early modern world involved different "ways of knowing." As part of a broader reassessment of knowledge production, scholars such as Pamela Smith, Harold Cook, and Steven Shapin, among others, have asked how empirical knowledge was generated and how these processes worked and who participated in them. Rather than emphasizing book learning and isolated experimental endeavors, they have foregrounded the importance of apprenticeship, embodied skills, and complex sensory repertoires to modes of inquiry. In sum, these scholars propose a more collaborative, skills-based view of early modern knowledge production that unified the work of mind and hand.[22] As Pamela Long has shown, commonalities between artisanal practices created important synergies among the arts, crafts, and sciences in the fifteenth and sixteenth centuries.[23] Taking cues from this important body of scholarship, my study moves away from a focus on science and technology to examine how forms of bodywork provided avenues for Italian women to participate in the culture of experimentation.

Finally, my study offers a pioneering look at the commercial dimensions of women's pharmacy work in late Renaissance Italy. Among the most prominent medical vendors of their day were religious women living in enclosed communities. Their numbers were substantial: Italy had the highest proportion of nuns anywhere in Europe between 1500 and 1650.[24] Heirs to a long healing tradition, Renaissance nuns commercialized their pharmacy work after 1500. In so doing, they capitalized on the burgeoning interest in medicinals sparked by early globalization, as well as on local social welfare initiatives and growing immiseration. By producing and marketing drugs to the public, nuns both augmented the medical resources available in Italian urban society and acquired roles of public significance beyond the spiritual realm.[25] Thanks to rich documentation, I present the first in-depth look at these female-run businesses, including the spaces, technologies, and embodied skills women needed to become commercial producers. Nun apothecaries worked at the nexus of market and laboratory as both medical artisans and entrepreneurs. In making and marketing remedies on a large scale, they became deeply immersed in commercial culture while developing broad technical expertise. Studying these businesses brings other important issues in Renaissance studies and medical history to the fore. The avenues by which nuns kept abreast of new technologies and market trends maps the complex social circuits through which information flowed in early modern cities. Exploring the world inside convent workshops also highlights women's engagement with the materiality of making medicines—its fierce heat, pungent smells, distinctive tools. An examination of attempts to regulate these businesses reveals both the limited reach of the Medici state and its self-interested role in mediating competition between purveyors.

Organization of the Book

The sheer diversity of women's medical activities poses an intriguing organizational challenge. This book uses the prism of class as an organizing tool for assembling the broad palette of health-related work by which women produced and circulated medical knowledge. My study opens with elite household practitioners at the early Medici court and ends with hospital nurses poor in financial and relational resources. Clustering discussions of different types of bodywork along vertical nodes defined by social hierarchy carries certain theoretical risks, since the very notion of class in the premodern era has been hotly debated. Nevertheless, this approach also presents clear advantages: it provides a strong organizational structure for handling disparate evidence; it permits the use of multiple scales of analysis that bring micro-level perspectives to bear on macro-level questions; and it generates fresh evidence for the relationship between experience-based medicine and professional practice in a variety of settings. The wide remit afforded by this organizational structure also allows me to examine women's health literacy across a broad social spectrum and to discern cross-class medical competencies. By health literacy, I mean the practical knowledge needed to make everyday remedies, recognize and treat common ailments, maintain preventive health regimens, and engage with new medical thinking. Throughout the book, I take an integrative approach to medical thinking and practice that embeds the pursuit of health in a thick historical context. Taking cues from the material turn mentioned earlier, I work from the premise that knowledge-making about the body was a networked social enterprise involving a rich tableau of participants.

Chapter 1 explores the politics of health at the early Medici court between 1530 and 1570. Despite the recent surge of interest in household medicine, we still know little about how it was practiced in a court setting. The medical skills that formed a conventional part of women's domestic duties acquired great political significance in Renaissance princely households, since health concerns affected dynastic interests and the stability of the state. Although European court women displayed high levels of health literacy, their participation in everyday care routines and medical decision-making at court remains poorly understood. Besides shedding new light on these practices, examining medical exchanges between noblewomen, court physicians, and other professionals problematizes healing approaches based on different ways of knowing. Without taking the interventions of medically informed court women into account, we risk misconstruing how Renaissance court medicine actually functioned and the ways in which courts operated as sites of knowledge exchange.

At the center of my analysis are two Medici women: Maria Salviati de' Medici (1499–1543), mother of Duke Cosimo I, and Eleonora of Toledo (1522–1562), his Spanish-born wife. Both women exerted real influence over preventive health measures, daily care routines, and critical decision-making processes as household practitioners. Tapping voluminous court correspondence, I analyze Salviati's medical expertise in relation to the emergence of pediatrics as a distinct body of knowledge and practice in Renaissance Europe; consider her impact on long-term medical provisioning at court; and assess her complex interactions with court physicians and other household caregivers. Rather than seeing household medicine and professional practice as inherently conflictual, I argue that, in this case, they operated as adjunctive approaches to healing that were tempered by considerations of power. The final portion of Chapter 1 shows how her daughter-in-law, Duchess Eleonora, put healthcare to sophisticated political uses as an instrument of patronage, diplomacy, and self-fashioning. Overall, the chapter advances our understanding of noblewomen's medical expertise while throwing medical decision-making practices at court into sharper relief.

Building on this discussion, Chapter 2 uses an examination of medical gift exchange to connect the health cultures of Italian courts and convents. Scholars have noted the many circuits of power that made these sites increasingly interdependent after 1500. Italian noblewomen colonized female religious communities as extensions of aristocratic living spaces, transformed these institutions into nodes of artistic and political patronage, and relied on nuns to implement medical charity. Part of this interdependency hinged on the many gifts of health that nuns and noblewomen exchanged between and among themselves, ranging from remedies and recipes to therapeutic advice and consoling words. I use two case studies to expand our understanding of women's medical competencies, especially in the realm of pharmacy, and to add new dimensions to the practice of household medicine at court. Together these cases bring macro-historical concerns about the circulation of new materia medica and microscale questions about the therapeutic value of consolation into the same analytical frame. In so doing, these cases lift up the multivalent nature of health-related gifts beyond their purely economic value.

The first case tracks the transnational commerce conducted between Queen Leonor of Portugal and the Florentine nuns of Le Murate from 1490 to 1565. At the heart of this exchange were new medicinal commodities drawn from Portuguese royal coffers, which the Murate nuns reciprocated with small handmade gifts meant to benefit the queen's spiritual health. I show how the Portuguese queen integrated distant religious women into her charitable strategies as both agents and recipients—a process that transformed these nuns into brokers of new medicinals in secondary distribution hubs. By contrast, the second case explores the more intimate medical exchanges between Archduchess Maria Maddalena de' Medici and the local Tuscan healer Sister Orsola Fontebuoni. Practicing between heaven and earth as both spiritual and naturalistic healer, Fontebuoni served as health consultant and spiritual adviser to the archduchess and other Medici court women in the 1610s and 1620s. Central to her reputation were healing eloquence that restored emotional equilibrium, along with the time-honored power of amulets, relics, and charms. Juxtaposing new and old remedies, material and immaterial gifts, local and transnational transactions, these two cases argue that the persistence of medical pluralism continued to create extensive spaces for female agency in a changing historical landscape.

Chapter 3 turns to the business of health pursued by Italian convent women. In major cities from Venice to Naples, nuns marketed a wide range of products to the public. Their apothecary shops served simultaneously as sales outlets, distribution points for medical advice, and production sites for making medicines. Focusing on Florence, the chapter opens by asking why nuns entered the medical marketplace around 1500 and what conditions fostered their continued involvement in the business of health. By examining the spatial distribution of pharmacies in the sixteenth century, I show that Florentine convents provided crucial resources to the urban periphery, home to dense concentrations of poor laborers and artisans. Renaissance nuns were commercial innovators who devised new products and marketing strategies. To sustain their businesses, they drew on the artisanal expertise, supply chains, and information networks they had developed through earlier craft endeavors, such as painting and textile work. Moreover, nuns engaged in horticultural experiments to facilitate low-cost sourcing of botanical materials used in making remedies. Confronting a volatile regulatory environment between 1500 and 1630, Florentine religious women successfully negotiated conflicts between reform ideals and market forces, on the one hand, and between powerful families and Medici state policies, on the other. My findings present new insights into the pharmaceutical trade in a city renowned for commercial and technological innovation.

Building on this evidence, Chapter 4 examines how Renaissance nuns acquired the skills needed to make medicines on a commercial scale. Convents developed systems of apprenticeship seen in other early modern crafts, in which expert practitioners taught younger assistants through hands-on instruction. This traditional model allowed Renaissance convents to sustain high levels of craftsmanship over multiple cohorts at no cost to the community. In addition, I ask how convents capitalized on new medical thinking by means of printed books and personal interactions—a process that transformed convent workshops, parlors and dispensaries into important sites of medical exchange. A close study of one Florentine convent pharmacy in the late sixteenth century takes us into the heart of the "kitchen," where available technologies conditioned strategies for production. This analysis also contrasts the experimental nature of the workshop with the performative retail space of the dispensary, where issues of social trust loomed large. The concluding section presents three case studies that document nuns' immersion in the culture of secrets—those hard-won bits of know-how about how to make everything from household products to industrial wares. Although scholars have argued that Italian women were less visible in this arena than were their northern counterparts, I demonstrate instead that nuns were deeply conversant with the language of secrets.

Chapter 5 considers the Renaissance hospital as an epistemic space that translated new strands of medical thinking into everyday practice. Hospital nurses who entered lifetime service at the institution possessed a large, varied portfolio of healing and administrative skills. These included examining the sick prior to admission, providing bedside care, preparing medication for diverse ailments, maintaining proper hygiene to prevent contagion, supervising hospital personnel, balancing the books, and attending to matters of institutional governance.[26] Despite the importance of these activities to recovery, the low social origins of Italian hospital nurses, coupled with the naturalization of care practices, consigned their work to the bottom of the Renaissance medical hierarchy. Still, their skills proved vital to emerging structures of public health. Towns and cities across Europe capitalized on this pool of cheap female labor to strengthen their hand at plague management and meet increased demands for practical charity.

Anchoring the discussion are the specialized hospitals devoted to the baffling new disease of the "pox"—commonly associated with venereal syphilis—which ravaged Europe after 1494. Blending moral, spiritual, and therapeutic goals, these purpose-built facilities grafted innovative medical thinking onto an older institutional base. All-encompassing nursing routines for pox patients integrated the use of new wonder drugs, stringent management of the non-naturals, and reliance on the senses as diagnostic tools. A fine-grained study of the female staff at the Florentine pox hospital highlights demographic patterns associated with nursing across the *longue durée*. Orphaned teenagers with few prospects made a life in hospital service, where they developed into some of the most proficient pharmacists of their day. Hospital nursing carried great redemptive weight in Counter-Reformation Italy, in part because the work was perceived as repulsive and posed substantial health risks. By the seventeenth century, however, hospital nursing began to coalesce into a recognizable body of knowledge and practice. By showing how Renaissance hospital nurses put new concerns about contagion, cleanliness, and ambient air into practice, my findings demonstrate the penetration of innovative medical thinking into institutional healthcare.

In sum, this study situates female agents of health squarely at the intersection of medical and religious discourse, social welfare initiatives, consumer culture, and the new information economy sweeping late Renaissance Italy. These forgotten healers were vital participants in the broader social matrix linking the early modern medical economy to cultures of experimentation. Their activities within and beyond domestic settings point to the exercise of significant medical agency; examining them enlarges our understanding of the healthcare resources available to Italian urban society. Renaissance courts, convents, and hospitals offered epistemic spaces in which women from different social strata produced and circulated practice-based knowledge about health, healing, and the body. Hence this study not only recuperates a broad palette of medical work but extends the reach of early modern empiricism into unsuspected corners. The result is a fresh view of the provisioning of healthcare in late Renaissance Italy and its relationship to early experimental culture.

CHAPTER ONE

The Politics of Health at the Early Medici Court

NOTHING HAS DONE MORE to transform our understanding of early modern medical practice in recent years than the study of household medicine. Home-based healing remained the "first port of call" for most Europeans until the nineteenth century, despite the proliferation of hospitals and medical licensing throughout the early modern period.[1] In Renaissance Europe, it was expected that women from all walks of life would know how to make remedies and treat family members for common ailments. Some household practitioners served a wider clientele as a way to supplement their livelihood, or simply out of charity. As one of many options available in a pluralistic medical environment, home-based medicine remained firmly entrenched in the hierarchy of resort. Even affluent Europeans who could afford expensive treatment by learned practitioners utilized domestic care as an alternative or adjunct to other therapeutic measures.[2] In contrast to the theoretical expertise claimed by academic physicians, household practitioners amassed a body of knowledge that was largely "orally transmitted, experience-based, concrete and bodily oriented."[3] Moreover, the renewed interest in healthy living in late Renaissance Italy assigned householders a prominent role in health promotion. As domestic interiors became the principal locus for managing the six "non-naturals" affecting health—air, sleep, diet, exercise, evacuation, and the passions—early modern women assumed particular responsibilities as guardians of healthy living.[4]

Domestic care had unusual political significance in Renaissance princely households, where health concerns affected both dynastic interests and the stability of the state. In a world where political power devolved vertically through birth, the health of the prince and his family became inextricably tied to affairs of state.[5] Even the appearance of health—whether a smooth complexion or the exaggerated swagger of virility—could be politicized to suggest reproductive promise and fitness for office. Keeping Renaissance rulers healthy required a consortium of practitioners ranging from court physicians to astrologers, cooks, and wet nurses. Despite the importance of household care to Renaissance health and healing, its place within court settings has only begun to attract critical attention. Recent studies have shown that European noblewomen displayed high levels of health literacy, which often grew out of hands-on practice and experimentation.[6] Yet we know little about how these women participated in everyday care practices and medical decision-making at court. Understanding their involvement as practitioners reveals the sometimes contentious relationship between experience-based medicine and learned professional practice that gave rise to a local politics of health. Without taking the health interventions of these medically informed women into account, we risk misconstruing the practice of Renaissance court medicine and the ways in which courts operated as sites of knowledge exchange.

The early Medici court in mid-sixteenth-century Tuscany offers a rich point of entry into these questions for several reasons. Its very newness as a political structure showcases the evolution of a distinctive medical court culture. In 1532, the centuries-old Florentine republic was replaced by a dynastic principate, led by Duke Alessandro de' Medici. Although Florence quickly became one of the principal courts of Renaissance Europe, the Medici were still newcomers to the dynastic stage when seventeen-year-old Cosimo I de' Medici (1519–1574) assumed the ducal throne following the assassination of his predecessor in 1537. Cosimo's immediate hire of Andrea Pasquali as court physician showed his ability to build a regime that was both separate from the republican past yet continuous with it. The doctor's long service at the Florentine civic hospital of Santa Maria Nuova gave him strong republican credentials, while his more recent attendance on the murdered duke indicated his openness to new political loyalties.[7] Regular medical staff remained limited throughout most of Cosimo's reign, especially by comparison with more established Italian courts.[8]

Facilitating this fraught political transition were two Medici women: Cosimo's mother, Maria Salviati (1499–1543), and his Spanish-born wife, Eleonora of Toledo (1522–1562). Much has been written about them as political figures and cultural patrons, but their medical agency has barely been explored. Both women exercised enormous influence over daily care routines and critical decision-making processes, which brought them into frequent interaction with court physicians and other practitioners. Yet their personalities were a study in contrasts. Granddaughter of Lorenzo the Magnificent, Salviati was renowned for her piety, modesty, and sharp political instincts. In 1516, at the age of seventeen, she married the popular military captain Giovanni de' Medici (later called delle Bande Nere). Widowed ten years later, Maria began grooming her only child, Cosimo, as the rightful heir to Medici political ambitions. Contemporaries described Salviati as an exemplary widow fiercely protective of her son's interests, who nevertheless exercised her political skills behind the scenes. Cosimo revered her as a "mortal goddess" utterly devoted to advancing his personal and dynastic fortunes.[9] Strong bonds of affection between mother and son undoubtedly magnified her influence in health matters, although she held no formal position at court.[10]

Unlike her mother-in-law, Eleonora was born and bred to court life. In 1534, when she was twelve years old, the future duchess moved from her birthplace in León to Naples, where her father, Don Pedro, had recently become viceroy. Steeped in Spanish court ways, Eleonora was raised with the assumption that female political authority could—and perhaps should—be highly visible. This expectation flew in the face of Florentine republican conventions, which for centuries had denied women a public political voice. The young duke supposedly welcomed this choice of bride in the hope that she could instill habits adopted from the fashionable Spanish and Neapolitan courts.[11] Among her many competencies was the preparation of standard medical remedies, as well as luxury products like perfumes and cosmetics; she also quickly became acquainted with new medicinals flowing into Europe from Asia and the Americas. Unlike later Medici consorts, however, the teenage duchess did not bring a personal physician in her entourage when she married Cosimo in 1539—a practice that both facilitated the circulation of medical knowledge and positioned Medici wives as cultural mediators.[12] Still, Eleonora, her father, and the young duke frequently exchanged medical advice in private correspondence.[13] An astute businesswoman, the duchess amassed a considerable fortune from her financial interests in mining, the grain market, and productive estate management.[14]

This medical pluralism—lay and learned, courtly and domestic, Italian and Spanish—made the early Medici court a micro-contact zone for the production and circulation of medical knowledge. Consequently, the pursuit of health in this new court setting became a negotiated, politicized process. Despite inherent frictions, household and learned medical practice generally worked together as adjunctive frameworks whose different approaches to health and healing were mediated by considerations of power. Tracking both these tensions and their resolution are the voluminous letters generated by the Medici court. Routine correspondence normally included health updates about members of the ruling family. When illness struck, however, Medici relatives and staff penned several letters a day, with the hour clearly marked, in order to keep recipients abreast of developments.[15] Recent studies have highlighted the importance of letters and epistolary networks to early modern women, who frequently exchanged healing remedies and advice within "trusted communities of knowers."[16] Extant letters written by and to Medici court women give them exceptional voice as medical agents, while shedding new light on everyday court practice. Whether discussing ordinary ailments or crisis situations, their detailed correspondence reveals the centrality of women's practical knowledge to the establishment of care routines as well as medical decision-making at the early Medici court.

The discussion begins by exploring Maria Salviati's activities as household healer in the 1530s and 1540s, focusing on her familiarity with remedies and health regimens. Salviati left a lasting imprint on medical court culture through her knowledge of pharmacy, pediatrics, and local healing networks. This patrician widow stands out from other household practitioners of her day largely because of the exceptional resources she could access as the duke's mother. I then situate Salviati's expertise in relation to the emergence of pediatrics as a distinct body of knowledge and practice in Renaissance Europe. Although the subject of infant care had deep roots in the Western medical tradition, the fifteenth and sixteenth centuries marked a watershed in vernacular medical thinking about the health needs of children as a group. Then follows an analysis of her complex interactions with court physicians like Andrea Pasquali, both in her capacity as household practitioner and in her role as patient. The closing discussion shows how Duchess Eleonora took the practice of household medicine in new directions when orienting medical court culture along more hierarchical lines in the 1540s and 1550s. Eleonora not only used medical charity as a vehicle for self-fashioning, but also put healthcare to direct political uses when extending favors within and beyond the court. Despite their different styles, both women stood at the center of medical exchanges as the Medici court took shape.

Remedies and Regimens: Maria Salviati as Household Healer

Renaissance women like Maria Salviati (Fig. 1.1) had a wealth of textual information at their disposal that fostered health literacy. Manuscripts circulating widely across Italy included vernacular health regimens like the thirteenth-century *Tesoro dei poveri,* which enjoyed enormous popularity as a handbook for home-based care. Written in Latin by Pietro Ispano (Pope John XXI) and translated into numerous vernaculars, this old standby traversed the body from head to toe, offering remedies for an impressive range of health issues such as fevers, hair loss, and reproductive problems.[17] The staying power of this work is documented by its incorporation into the 1515 recipe book compiled by the Florentine civic hospital of Santa Maria Nuova. Other practical healthcare information circulating in manuscript around 1500 included numerous herbals, plague treatises, astrological materials, and calendars noting appropriate times for bloodletting.[18]

Supplementing manuscript works was a torrent of medical print rolling off Italian presses as early as the 1470s. These materials ranged from cheap pamphlets to the first official civic formulary printed in Europe, issued in Florence in 1499. Especially plentiful were vernacular health regimens that increasingly catered to a lay reading public. Building

Fig. 1.1. Jacopo Pontormo, *Portrait of Maria Salviati de' Medici with Giulia de' Medici*, circa 1539. Courtesy of the Walters Art Museum, Baltimore

on medieval antecedents, printed health regimens exposed readers not only to symptoms and cures but also to principles of healthy living embodied in the non-naturals. Sandra Cavallo and Tessa Storey have argued that the burgeoning popularity of these texts by 1550 promoted a vibrant culture of prevention in late Renaissance Italy. Despite continuities in medical thinking, Renaissance health manuals differed from earlier works like the *Tesoro dei poveri* by taking a more explanatory approach to the body and health maintenance. Many successful health manuals printed in sixteenth-century Italy were authored by physicians, who refashioned themselves as professional health advisers marketing information rather than cures. Consequently, late Renaissance health regimens written in the vernacular both expanded the health literacy of the reading public and amplified the material culture of healthy living.[19]

Among the key texts shaping the practice of household medicine were handwritten recipe books. These compilations were the most common form of women's medical writing in Renaissance Europe, often providing the textual basis for female medical authority.[20] Recipe books offered a way for literate women to navigate everyday problems while perfecting their practical pharmaceutical skills. Often organized in eclectic ways, these valuable guides to household management bundled the culinary arts with instructions for making medicaments and ordinary products like ink, soap, and stain removers. As embodiments of household wisdom and proprietary craft techniques, recipe books were often transmitted as family heirlooms or included in artisan women's dowries throughout the early modern period. As they passed from one generation to the next, these texts acquired new layers of knowledge in the form of additional recipes and practical tips. Some of these volumes were further customized by the addition of para-textual aids like indices or marginal notations. Still others bridged the gap between manuscript and print by incorporating handwritten extracts copied from published works.[21]

One of the most significant Italian recipe books produced around 1500 was compiled by Salviati's mother-in-law, Caterina Sforza (1463–1509), regent of Forlì and Imola. The paternal grandmother of Cosimo I, Sforza was an avid prince-practitioner who sustained keen interests in medicine and alchemy while shuttling among residences in Milan, Rome, Florence, and her own dominion. Like other early modern noblewomen, she used recipes and remedies as a form of currency within a broad epistolary network. Her medical correspondents encompassed noble relations, local apothecaries, political agents, and irregular practitioners. Sforza's massive compendium of 454 recipes, the so-called *Experiments,* integrated the fruits of hands-on experimentation with secrets procured from extensive court contacts across Europe.[22] The vast majority of recipes in her collection were medicinal in nature. They included pills and powders to cure fevers; unguents to treat gout, tumors, sciatica, and wounds; distillates to relieve infection; and elixirs to strengthen the body. Other recipes addressed beauty secrets, veterinary medicine, and the transmutation of metals. A considerable number were rendered either in cipher or in Latin to better preserve their secrecy. Throughout the collection, Sforza attested to the efficacy of various remedies by noting that they had been tried and tested by her own hand, using such phrases as "proven remedy," "truly tested and proven," "proven and certain."

Circumstantial evidence suggests that Maria Salviati had access to this prized collection and probably was instrumental in its transmission. Sforza bequeathed the volume to her youngest son Giovanni (Salviati's husband) when she died in 1509; the volume apparently had come into his hands by 1525, well into the Medici-Salviati marriage. Writing to a family friend that year, Giovanni noted that this precious heirloom had disappeared from his Roman strongboxes. "We must find it," he declared, "because one way or another we want it."[23] Whether Giovanni was referencing the now-lost codex, written in Caterina's own hand, or the copy made by his lieutenant Lucantonio Cuppano is unknown. Cuppano—a colonel in Giovanni's army and his major domo away from home—had transcribed the entire recipe collection, probably on his master's orders. New evidence suggests that the lieutenant divided the copious material into several different volumes. The manuscript volume designated by the initial "C" on its cover forms the basis of the published edition of the *Experiments.* Currently held in a private archive in Ravenna, this codex was long thought to be the only extant copy of Sforza's *ricettario.*

Recently, however, two additional manuscripts copied by Cuppano have been unearthed in the Florentine National Library.[24] One contains a partial index to Sforza's extensive recipe collection; the other, marked with the initial "B" on its cover, adds more than 400 recipes to Sforza's known corpus. Considering all three manuscripts as a whole, Sforza devised, recorded, and experimented with almost one thousand recipes. The newly discovered B codex resembled its published counterpart (the C manuscript) in several ways. Both include remedies to improve human and animal health, as well as recipes for beauty products and incantations for love magic and performing supernatural feats. Both codices confirm Sforza's pluralistic approach to the natural world, utilizing explanatory frameworks that ranged from the humoral to the occult. Both also attest to the extensive use of empirical testing to establish efficacy. Importantly, the B manuscript was censored at an unknown date; as a result, some of the magical recipes noted in the index were excised. A more detailed study of these volumes is currently under way.[25]

Since Salviati acted as executor of her husband's estate, it is likely that some of these materials passed through her hands following Giovanni's death in 1526. As executor, she oversaw property transfers related to her son's inheritance and maintained regular contact with Cuppano about business matters between 1523 and 1527.[26] Whatever its itinerary, Cosimo had acquired this collection by the time he ascended the ducal throne and used it as the basis for his own experiments, adding other recipes over the course of his life. Sforza's recipe book thus became a landmark in the Medici genealogy of experimentation spanning the early modern period. Several of her secrets were enshrined in the Medici court pharmacopeia, including the famed family recipe for scorpion oil, a poison antidote in high demand as a gift.[27]

In their pursuit of health, household practitioners like Salviati gained added legitimacy from Renaissance humanists, who harkened back to ancient Greek notions of household economy. The influential conduct manual written by the Spanish humanist Juan Luis Vives recommended that married women learn how to treat ordinary ailments as an essential part of household management. Since purchasing medicines and services could be costly, home-based care that centered on prevention, pharmacy, and "physic"—the pillars of household medicine—conserved familial resources, making it a thrifty choice for the merchant and artisan classes. "Since the care of the inhabitants of the house falls upon the woman," Vives wrote in *The Education of a Christian Woman* (1524), "she will keep remedies on hand for common and almost daily maladies and will have them ready in a larder." That way, the good wife "will not have to send for the doctor often and buy everything from the apothecary." Practical healing skills could be acquired "from the experience of other prudent matrons," supplemented by reading "some simple handbook on that subject." Vives also tasked female householders with "the regulation of the daily diet," which he considered to be "of the greatest importance" in preventive health maintenance. Yet at the same time he cautioned wives against placing "too much confidence" in their own skills. Married women should "be familiar with the remedies for frequent and everyday illnesses" but should refer more complex conditions to physicians.[28]

Home-based care played a vital role in good household management throughout the early modern period. At the Florentine conservatory of La Quiete, founded circa 1650, training in pharmacy and physic were integrated into the standard curriculum. Young aristocratic girls who boarded there learned how to apply salves, treat wounds, and make distillates. Special attention was paid to mastering the medicinal qualities or "virtues" of various herbs and the art of compounding remedies. Instructed by older women trained in similar institutional settings, the girls mastered proper techniques for cupping and were schooled in the art of purges.[29] Learning these embodied skills was best supplemented by hands-on instruction. Simply reading a recipe was not always enough for understanding the basic properties of ingredients or how they behaved during processing.[30] Guidance by more skilled practitioners helped new learners grasp the fine points of distilling liquids, blending ointments, rolling pills, and making plasters. Such practical training not only enabled young girls to become proficient in pharmacy but also created knowledgeable consumers. Even if the affluent pupils at La Quiete eventually delegated hand-on tasks to their servants, these skills remained foundational to running an orderly household.

Maria Salviati's correspondence brims with exactly this sort of practical knowledge. She clearly was conversant with common distillation techniques for making rosewater and various health-giving tonics, while her willingness to experiment with traditional remedies is suggested by the "secret" remedy for intestinal worms she developed.[31] Worms was a common but serious ailment that could lead to wasting and death, especially in children. Recipes for vermifuges abounded among Florentine merchant families, and ready-made remedies for worms were widely available in apothecary shops.[32] Salviati's cure must have been particularly prized by family members: it was not only used frequently during her lifetime but also enjoyed staying power after her death. When her granddaughter Isabella suffered a bout of worms in 1544, shortly after Salviati died, the matrons tending the infant "had certain stomach remedies prepared according to [her] usual orders."[33] Similar household remedies were developed by Medici relatives. Her cousin Caterina

Cibo—another granddaughter of Lorenzo the Magnificent—created a cure to clear intestinal blockages that later found its way into Medici court formularies.[34]

Despite this familiarity with making medicines, Italian women seem less visible as medical agents in this regard than their German and English counterparts. One scholar has argued that Italian prescriptive texts discouraged elite women from dirtying their hands in the kitchen, where most medicinal remedies were prepared.[35] Although popular texts may have inhibited Italian urban women from making medicines in volume, market structures resulting from greater urbanization probably account for perceived differentials. Regardless, it is important to remember that these are simply differences in degree, not in kind. German noblewomen living on rural estates regularly distributed homemade remedies to tenants and the local poor as part of estate management. These localized hubs allowed aristocratic women with skilled hands and inquiring minds to satisfy their intellectual curiosity while filling important gaps in local healthcare provisioning.[36] By contrast, Italian Renaissance cities boasted highly commercialized, consumer-oriented markets selling remedies at different price points. Residents could range broadly across vendors, procuring wares from favorite shops or buying cures from itinerant healers. Italian female householders still manufactured syrups and salves in their own kitchens, but easy access to ready-made products may have reduced the incentive for them to produce large volumes or great varieties of medicaments. Moreover, Italians city dwellers could capitalize on well-developed systems of credit when making unexpected medical purchases.[37]

Because of the commercialized nature of the Italian medical marketplace, patrician women like Salviati influenced household care as both producers and consumers of medicaments. The social relationships Salviati established with particular purveyors before and after Cosimo ascended the throne left a lasting imprint on subsequent medical provisioning at court. Among her preferred suppliers of ready-made pills, tonics, and distillates were several Florentine convents that had launched commercial pharmacies around 1500—a development discussed in Chapter 3. Most notable was Salviati's decades-long relationship with the Franciscan nuns of Sant'Orsola. The duke's mother had resided in the convent during her husband's extended absence in 1518 and even considered becoming a tertiary there. Salviati regularly purchased remedies from the nuns and ordered some of her favorite rose distillate from them as she lay dying.[38] During her lifetime, the ducal family entrusted the convent apothecary, Sister Benedetta Bettini, with compounding "anything of importance" for their use. At issue were matters of trust as well as expertise: Cosimo was understandably wary when purchasing medicines after being the target of several assassination attempts. This medical partnership between convent and court endured into the next generation. Cosimo's successor, Francesco, and Francesco's first wife, Giovanna of Austria, continued to purchase medicines from the Sant'Orsola pharmacy, including their proprietary "ducal pills."[39] Salviati cultivated similar business relationships with the nuns of Santa Caterina da Siena, who fabricated high-quality medicinal products. Maintaining these commercial attachments after Salviati's death enabled the Medici dukes to take advantage of important medical resources while drawing convents more tightly into court circles. Hence the duke's mother influenced the development of Medici court medicine as both skilled practitioner and discriminating consumer who threw her weight behind preferred suppliers.

Salviati and the Rise of Renaissance Pediatrics

Salviati's medical expertise was most evident in the realm of pediatrics, which emerged as a distinct body of medical knowledge and practice during her lifetime. Although children's health needs had long been recognized as a medical specialty, the fifteenth and sixteenth centuries marked a watershed in thinking about pediatric medicine. This intellectual current had both vernacular and Latinate roots. The explosive interest in the family among merchants and humanists alike was a hallmark of Renaissance Italy, evident in scores of genealogies, family diaries, prescriptive texts, and celebratory images. Court physicians also produced specialized texts dealing with children's health issues in response to concerns about dynastic continuity.[40] Some fifteenth-century medical writings emanating from Italian courts bundled pediatrics with female reproductive health. The influential treatise written circa 1460 by Michele Savonarola, court physician to the Este dukes, for instance, articulated a comprehensive program of medical, dietetic, and pedagogical advice for both pregnancy and child care. In addressing "the women of Ferrara," Savonarola fleshed out medical theory with information derived from his own experience as a practicing physician when discussing problems afflicting expectant mothers, newborns, and children under the age of seven.[41]

Although Renaissance pediatrics retained many connections to this earlier material, physicians began to distinguish the specific health needs of children in the earliest days of print. The Italian physician Paolo Bagellardo da Fiume generally is credited with publishing the first Western treatise devoted solely to pediatrics in 1472. First printed in Padua, this Latin work enjoyed great success, with an Italian translation appearing in 1486. Generally Bagellardo followed the Arab physician al-Razi when discussing common childhood ailments, but he also took a new tack by interpolating advice drawn from his own cases. His treatise is an important example of the growing willingness of Renaissance physicians to assert personal experience as a basis of authority.[42] Some early printed volumes on pediatrics were simply vernacularized excerpts or commentaries on canonical texts, but most were original compositions geared toward the growing demands of a reading public. By the time the humanist court physician Simon de Vallambert published the first French treatise on pediatrics in 1565, similar works had already appeared in German, English, Spanish, and Italian.[43]

The recognition of Salviati's competencies in pediatrics by physicians and family members must be seen against this backdrop. Between the birth of the first Medici princess in April 1540 and her own death in December 1543, the duke's mother supervised the family nurseries at Castello, the Medici villa located a few miles northwest of the city. This caregiving arrangement allowed Eleonora to accompany her husband on travels around the dominion for several weeks at a time. The young couple reportedly enjoyed each other's company and shared a love of gambling, hunting, and other pursuits, but the primary motive for these journeys was political. The first decades of Medici rule remained precarious, filled with anti-Medicean campaigns and anti-foreign sentiment—much of it directed at Eleonora. Entrusting the nurseries to the widowed Salviati also allowed Eleonora to concentrate on matters of state when she served as regent in 1541 and 1543 during her husband's trips to Genoa.

Salviati's growing brood of charges in these years included the three legitimate children born in rapid succession to the ducal couple. Firstborn Princess Maria, her grandmother's namesake, arrived in 1540, followed by the future grand duke Francesco in 1541 and Princess Isabella the following year. A fourth infant, Giovanni, was born just a few months before Maria's death in 1543. Bearing eleven children in all, young Eleonora won praise as "the most fecund Lady Duchess," whose fertility assured dynastic success.[44] The Medici nurseries also included Cosimo's illegitimate daughter Bia (b. 1536 or 1537), for whom Salviati showed great affection, as well as the two illegitimate children fathered by the murdered Duke Alessandro. All of these offspring were fully integrated into evolving court routines, both at Castello and in the family's new quarters in Palazzo Vecchio. Salviati's cousin Caterina Cibo commented that the children lived "in great pomp" at Castello in "rooms hung with gold-stamped leather," although their quarters at the old city hall were more spartan.[45] Maria's assiduous care of Duke Alessandro's illegitimate offspring assumed special significance during this fraught political transition. As their legal guardian, Cosimo promised the Florentine Senate that he would treat his wards well, rather than persecuting them as dynastic competitors. Political attempts to discredit the young duke on these grounds gained little traction thanks to his mother's diligent oversight.[46]

As the nursery expanded, Maria served as the eyes and ears of the ducal couple. Blending the medical and the managerial, she supervised a raft of experienced female caregivers, ranging from matrons and governesses to nurses and chambermaids. The duke's mother was the linchpin in coordinating efforts among caregivers, court physicians, and secretaries. She selected suitable wet nurses for the first four infants born to the Medici couple after canvasing local networks about their moral character; it was the duke's mother who secured their services in advance and reportedly paid them out of pocket when state finances faltered.[47] Her personal involvement with hiring wet nurses diverged from customary Florentine practice, which normally delegated financial arrangements to fathers. Although this situation reflected immediate circumstances, it nevertheless anticipated developments in seventeenth-century Roman aristocratic households in which mothers took a more hands-on role in child-rearing than they had in previous centuries.[48]

In selecting wet nurses for the Medici infants, Salviati could draw on a wealth of medical thinking. Renaissance physicians published entire Latin booklets on the topic of suckling, although they disagreed strenuously about whether maternal breastfeeding was superior to wet nursing.[49] Similar disputes about suckling rippled through vernacular texts. To Renaissance merchants and humanists, milk not only nourished newborns but also imprinted moral and social characteristics linked to class. Hence wet nurses—generally poor women from the countryside—had to be chosen with care.[50] Reflecting this ambivalence, the Florentine humanist Leon Battista Alberti advocated maternal breastfeeding in his treatise on the family (1432), yet at the same time offered lengthy advice about choosing a wet nurse.[51] The practice of wet nursing brought other dangers too. Preachers and moralists feared that it freed mothers to pursue carnal pleasures, given that sexual intercourse during lactation was thought to spoil the milk.[52]

These contradictions made the wet nurse a culturally ambivalent figure in Renaissance Italy. On the one hand, she was "a powerful symbol of plenty" whose abundant, youthful breast milk helped keep infants "fat and fresh as a rose," as one ducal secretary put it.[53] On the other, these women were commonly associated with fraud, sexual licentiousness, and even infanticide. To alleviate these anxieties, Salviati brought wet nurses to live at Castello, rather than placing the infant with them in the countryside—an arrangement that only the wealthiest families could afford. Physical proximity gave her oversight of the nurse's daily routines and sexual activity, which reinforced social hierarchies of dependence and subordination. Live-in arrangements also allowed the duke's mother to correct any irregularities that affected suckling. In early November 1542, for instance, she reported to Eleonora that eighteen-month-old Princess Maria had grown irritable because she was teething. Salviati thought that the wet nurse was partly to blame. Displaying her grasp of humoral principles, Maria wrote to the absent mother: "I don't think [little Maria] is suffering terribly, but she wouldn't suffer any more if she no longer took the wet nurse's milk, which is judged to be fiery." Unseasonable weather apparently had disturbed the nurse's humoral balance, which was remedied by light purges and dietary alterations.[54]

The duration of Salviati's oversight in the nurseries was extended by the Medici practice of breastfeeding their children longer than other patrician families. Both textual evidence and bioarchaeological findings reveal that the Medici weaned their children at roughly thirty-two months of age, in contrast to the more usual age of eighteen to twenty-four months.[55] The rationale for this practice remains unclear, but it apparently created prolonged attachments between infant and nurse that complicated their separation. Once Princess Maria was finally weaned, after thirty months at the breast, Salviati reported to the duchess that "sometimes during the day when the memory and touch of her wet nurse spring to mind, she starts crying immediately." Speaking from experience, she reassured Eleonora that the problem "will stop little by little."[56] As the Medici court developed, former wet nurses were kept on the payroll even after infants were weaned, resulting in elaborate ties of clientage.[57]

Despite these supervisory duties, it appears that Salviati did not play a direct role in the birthing process itself. Most births in fifteenth-century Florence were managed by a midwife, with other female relatives and friends in attendance. Only rarely did male physicians participate in this female-centered event.[58] The Medici court did not retain a salaried midwife but instead called one as needed—a practice that was commonplace in Italian Renaissance courts, in contrast to the royal midwives employed in northern Europe. In 1586, for instance, Grand Duke Francesco praised the "skillful hands" *(buona mano)* of the same itinerant midwife who successfully delivered heirs to the Gonzaga and house of Savoy in short order, despite their physical distance. Duchess Eleonora herself angrily dismissed the midwife secretly called by her staff when she was pregnant in July 1549, because she considered her premature presence a bad omen.[59] Later Medici consorts relied on broader geopolitical connections to assure a safe delivery. Giovanna of Austria experienced numerous difficult pregnancies resulting in the birth of only one sickly heir, Filippo (1577). Given her medical history, a midwife was brought from her native Austria when she was pregnant again in 1578. After a long, complicated labor, the German-speaking midwife partially delivered a baby that presented in transverse position. She was unable to turn the infant in the womb, and a group of male doctors intervened but could not save either mother or child.[60] The absence of a court midwife meant that Medici consorts customarily depended on their female networks—friends, relatives, court ladies, nun pharmacists—for practical advice about pregnancy and childbirth, as discussed in Chapter 2.

As the Medici children matured, Salviati groomed them in habits of healthy living. Part of her role as household guardian centered on instilling health precepts in her young charges that could be sustained throughout their lives. She encouraged the Medici children to engage in outdoor pursuits such as riding and walking, which helped maintain their humoral equilibrium.[61] At the same time, her interest in preventive health measures sparked frictions with her Spanish daughter-in-law. Although the two women reportedly got along well, they occasionally locked horns over how to implement health advice, especially regarding the vexed subject of "good air." This non-natural attracted great attention in Renaissance health manuals and plague treatises alike. Noxious air, in the form of miasma, had long been recognized as a disease-causing agent. Indeed, Eleonora expressed displeasure that her mother-in-law kept the children with her in Florence, "where the air is known to be exceedingly bad for them," during a suspected plague outbreak in 1542. The duchess moved them to a suite of rooms at the Badia Fiesolana in the surrounding hills, where the air reportedly was better for their health.[62] Concerns about ambient air, which surfaced with growing frequency in sixteenth-century printed regimens, are apparent in Salviati's thinking as well. After the ducal family moved into the Palazzo Vecchio in late 1540, she voiced concern that the recently refurbished rooms were too noisy, cramped, and dark to be healthy for the children and their wet nurses. Enclosed spaces supposedly trapped the exhalations of inhabitants, producing an unhealthy environment.[63]

Salviati continued to monitor the health status of the Medici offspring throughout their early childhood. Her frequent health bulletins to the traveling duke and duchess offered her own diagnostic assessments, rather than simply reporting what physicians had said. In this regard, she continued a tradition of independent medical evaluation begun some two decades earlier in her letters to husband Giovanni regarding the health status of their infant son.[64] In July 1540, for instance, Salviati assured the duke that Princess Maria was recovering well from an unnamed illness. A few weeks later, she notified him that the boils on the chest of his young ward Giulia had healed completely.[65] Although health information routinely circulated in Medici court correspondence, letter writers controlled the narrative representation of illness or recovery. Distant readers could gauge the gravity of a situation by scrutinizing the organization, tone, word choice, and other details offered in a letter. Part of Salviati's duties as household practitioner thus centered on relaying vital information in a consistent authorial voice that the duke and duchess could interpret with confidence. Her letters aimed to strike a balance between reporting symptoms accurately and avoiding undue anxiety for absent parents. By contrast, trusted court secretaries rarely used their own voice when describing sensitive health matters, instead attributing viewpoints to particular physicians in keeping with their professional role.

The detailed nature of Salviati's letters permits a closer analysis of her medical thinking. Gianna Pomata has argued that lay and learned practitioners used distinct but overlapping theoretical language and diagnostic terminology when describing the early modern body.[66] Physicians explained sickness in terms of humoral imbalances, whereas laypersons focused more on subjective sensations within the body, describing illness in terms of flows and fluxes.[67] As Rankin has observed, however, these discourses did not always map neatly onto a binary divide between lay and learned practitioners.[68] Medici court correspondence confirms that both patients and practitioners often spoke in mixed registers when discussing health and illness. The court physician Andrea Pasquali referred to fluxes far more frequently than did Salviati in his letters to the duke and duchess; conversely, Salviati frequently noted humoral imbalances in her health bulletins.[69] Therapeutic choices made by lay and learned practitioners also overlapped. Pasquali was known to use forms of sympathetic magic to supplement naturalistic healing, ordering that a fragrant, heart-shaped cloth poultice be applied to Cosimo's heart to alleviate "windiness." When writing the prescription, he sketched a pattern for the poultice, instructing the pharmacist to "make this right away, a half-finger larger rather than smaller."[70]

One of the terms Salviati used most frequently in her letters to denote ill health was "indisposition." This capacious, often purposely vague concept also was deployed by Florentine ducal secretaries and by court physicians in Renaissance Mantua when sending health bulletins.[71] The duke's mother utilized this term in several different ways: as a euphemism to describe culturally sensitive conditions like syphilis or rectal bleeding; as a way to mask the precise nature of an illness from prying eyes; or simply as a placeholder to mark a condition whose cause was unknown. For Salviati, being indisposed signaled above all a loss of functionality—an inability to perform one's everyday duties—rather than expressing the subjective sensations of corporeal experiences. Being indisposed was a practical measure of debility rather than an introspective one.

In explaining illness to the absent ducal couple, Salviati integrated widely held beliefs about the influence of celestial bodies into a naturalistic framework. Renaissance Europeans took a holistic view of health and disease causation that situated the human body within the macrocosm of the cosmos. Consequently, it was commonplace to attribute large-scale epidemics, as well as individual ailments, to the stars or other astral powers. However, the duke's mother apparently grounded these principles in more localized systems of knowledge, since Florentines ascribed particular agency to the powers of the moon as a determinant of children's health.[72] In October 1542, when Princess Isabella was only forty days old, Salviati notified Cosimo that the newborn "has been coughing up black phlegm *(apparir negra)*. I blame the passage of the moon, but last night we watched her carefully and I have not lost hope. She coughed up a great deal, but this morning seems much improved." The following month, Salviati continued in the same vein, this time more certain than ever "that the moon ha[d] caused some alteration" in the baby's health. In the intervening weeks, however, she cited humoral principles as well as human agency in her search for explanations. This time the duke's mother blamed "excessive feeding" by the wet nurse to explain the harrowing array of symptoms that continued to afflict the infant.[73] Princess Isabella's condition finally improved a few weeks later following some light purges.

This frightening episode, coming only six months after the death of Cosimo's daughter Bia, prompted Salviati to reflect on the difficulty of diagnosing internal ailments with any certainty. An internal problem "is always something to be more feared," she mused to her beloved son, "because one is able to know exterior conditions and concoct remedies for them more easily than for an internal ailment."[74] This rare moment of reflection on the healing arts echoed the customary division of medical labor, which reserved internal medicine for physicians and consigned external treatments to surgeons and empirics. Yet in this passage, Salviati seems less concerned with policing boundaries between practitioners than with recognizing the problem in knowing what lay hidden beneath the skin.

Medical Decision-Making at the Early Medici Court

Medical decision-making at court was complicated by the overlapping remit of Renaissance court physicians and the conventional duties of household practitioners. The small size of most Renaissance courts meant that medical professionals were put to work in other capacities as pedagogues or diplomatic agents.[75] In their role as mentors and companions to young princes, court physicians constructed daily regimens for growing children and familiarized future rulers with core principles of health and hygiene. These routines not only instilled habits of healthy living in children but also contributed to their moral formation.[76] Physicians at the Sforza court in fifteenth-century Milan, for example, governed every waking action of young duke Lodovico in hopes of laying the foundation for an orderly life. Diet was a particularly crucial area of oversight, since court life imposed social obligations to partake in impressive banquets that could undermine health.[77] Drinking wine was another area requiring close attention. Contemporary dietary theory held that children who were still growing should not drink undiluted wine because its "hot" humoral quality exacerbated their naturally hot constitution.[78] When young princes made their first forays into diplomatic life, court physicians proved to be ideal traveling companions who acted in loco parentis. When the physician Dionisio Reguardati chaperoned Lodovico Sforza's visit to Cremona in 1467, he reported the boy's health status to his mother Bianca Maria Visconti several times a day.[79]

One of the unwritten duties of court physicians was to promote a budding sense of manhood in future heads of state. As mentors and companions, court physicians helped prepubescent boys make the transition from a predominantly female world of caregivers to the oversight of men around the age of six or seven—the canonical age of reason.[80] This tacit remit is neatly illustrated by the diplomatic mission to Genoa undertaken by Duke Cosimo's seven-year-old son Francesco in 1548. Political circumstances made it expedient to send the boy as the duke's proxy to meet the future king Phillip II of Spain. The mission had the added bonus of exposing Francesco to the highest levels of diplomatic life for the first time. Included in the entourage was Medici court physician Andrea Pasquali, who took responsibility for the prince's well-being. The ducal secretary Lorenzo Pagni reported that Pasquali was an exacting but affectionate mentor who helped Francesco navigate these new experiences. Sumptuous banquets—some of which included more than ninety different dishes—were a particular challenge. According to Pagni, the doctor skillfully steered the prince between sound medical principles and aristocratic pleasures, showing him how to restrain his appetite by counseling that "this dish is too tough, this one undercooked, or that one hard to digest."[81] Under Pasquali's tutelage, young Francesco relished the trip immensely. In his final dispatch to the duke, Pagni remarked with some amusement that "upon his return, Madonna Giulia"—head of the Medici nurseries after Salviati's death—"will have great difficulty and effort in governing [Francesco] and keeping him under her usual care, because he has begun to enjoy the grandeur, freedom and latitude that he has at present."[82] Pasquali apparently had done his job well.

We know less about the role that court physicians played in the health education of princesses, but it appears that they shared responsibilities with medically informed women to a much greater degree. Familiarizing noble girls with health, hygiene, and the functioning of their bodies was a collaborative effort by both learned and lay practitioners. This arrangement was driven not only by gendered considerations of propriety but also by the pressing need to ensure that noble girls fully internalized core health principles at a young age, since marriage invariably uprooted them from home. Of particular importance was understanding how ordinary behaviors affected their reproductive health and moral reputation. Because the body of a princess was a public, social instrument, medical and moral discourses were interwoven much more tightly for elite girls than for their brothers. By following an orderly health regimen, future consorts could ensure their fecundity while setting a high moral standard for the court.[83]

The medical education of young Suzanne of Bourbon by the Lyonnais physician Symphorien Champier and her mother Anne of France provides insight into what was probably a common arrangement at European courts. In 1503, when twelve-year-old Suzanne was approaching the age of marriage, Champier addressed an extended chapter of a vernacular health manual titled *The Ship of Virtuous Ladies* to her. This section prescribed rules for healthy living within the context of marriage, emphasizing behaviors that affected Suzanne's reproductive potential.[84] Interweaving moral and medical precepts, this humanist-educated physician warned that excessive drink and immoderate sexual activity could result in illness or even sterility. He also alerted Suzanne to the physical changes associated with puberty, such as changes in breast size, and even advised her about the color of healthy menstrual blood, using Aristotle as his guide. Didactic literature normally was silent on these matters, assuming instead that princesses would learn about the workings of their body from women in their entourage.[85] Shortly thereafter (1505), Suzanne's mother issued separate instructions to prepare her daughter for marriage later that year. Drawing on a wealth of vernacular advice literature, Anne extended her counsel to clothing and other areas of comportment that affected health and fertility. She cautioned her daughter against wearing low-cut gowns, for instance, because they invited chills that might impede conception.[86] Anne also reinforced the importance of adhering to a gendered gestural code, such as lowered eyes and restrained physical movements that signaled bodily control. Together these treatises created a comprehensive regimen that interwove medical principles and behavioral norms.

The intersecting domains of responsibility between court physicians and household practitioners framed the medical interactions between Maria Salviati and Andrea Pasquali in the early 1540s. Court letters paint a picture of mutual respect for each other's position and expertise. Pasquali himself was something of a hybrid practitioner who served in multiple capacities.[87] Following various academic appointments at the universities of Bologna, Ferrara, and Padua from 1510 to 1516, he began work at the Florentine civic hospital of Santa Maria Nuova. His experiences with the hundred or so skilled female practitioners on house staff may have colored his interactions with the duke's mother.[88] Friendships with noted humanists like Paolo Giovio and Benedetto Varchi helped him bridge academic, institutional, and court circles. Pasquali enjoyed the full confidence of the duke, who named him to the prestigious Florentine Academy as well as several high civic offices. One measure of esteem was the monthly salary of forty-two ducats he drew as court physician, far more than the seven to ten ducats earned by ducal secretaries.

Nevertheless, the lack of published medical writings has cost him lasting fame.[89]

Medical interactions between Salviati and Pasquali can be characterized as a partnership of unequals, whose overlapping remits required skillful negotiation by both parties. When one of the Medici children fell ill, Salviati was the first to be consulted as informal head of household in the absence of the ducal couple. It was the duke's mother who decided whether a child's health problems warranted professional intervention, summoning Pasquali only after her own remedies had proved ineffective or the situation worsened. After the physician arrived, Maria negotiated treatment options with him and ensured that agreed-upon therapies were followed. Then she tracked the young patient's recovery, offering selective updates and reassurances to the ducal couple as already noted. In other words, Salviati simultaneously acted as a family health sentinel and medical broker responsible for the full range of caregiving within the nascent court. As a household practitioner, she distinguished transient conditions from more serious ones, constructed an intelligible diagnosis, and developed an appropriate treatment plan. Above all, her experience told her when to act. Timely interventions were crucial in infant care, since newborns still at the breast generally lacked the resilience of older children and adults. Household medicine thus involved the art of right timing—a widespread cultural preoccupation in the early modern period, evident everywhere from fencing manuals to theatrical productions.

Two episodes clarify both the process of medical decision-making and the integration of household medicine into the hierarchy of resort at court. In neither case were professional services prioritized over household care in treating the Medici heirs; instead these competencies worked in tandem. The first case involved Princess Maria, who proved to be a fussy baby throughout her infancy. In February 1542, Salviati herself dismissed the feverish wet nurse who was suckling the child—then almost two years old—to avoid harming the infant. When a substitute nurse was brought in, however, little Maria "wouldn't even consider taking the new breast." It was only at this point that Salviati summoned the physician il Ripa, who gave the new nurse a light purgative to rebalance her humors in hopes of inducing the baby to suckle. Still, the tiny princess refused the new breast, reportedly saying that it tasted "like poo"—a claim that seemed medically intelligible to her caregivers. Frustrated by this outcome, the physician relented and recalled the original wet nurse.[90]

The second, more extended, case concerned the ailing Princess Isabella, which generated a detailed report from the trusted ducal secretary Pierfrancesco Riccio to the duke. Here the main protagonist was not the duke's mother but the nursing network she had established at court. Although this episode does not involve Salviati directly, it nevertheless offers an important perspective on the dynamics between household expertise and professional services at court. In January 1544, a month after Salviati died, seventeen-month-old Isabella spent a restless night in the arms of her nurse, Domenica di Bonichi. After the baby woke up coughing several times, Domenica and the infant's wet nurse summoned Isabella Rainosa, Eleonora's chief lady-in-waiting. Together they decided to administer Salviati's trusted vermifuge before calling the doctor because "it seemed to them that her anus smelled like worms."[91] Apparently the nurses had prudently stockpiled this medicament in the household medicine chest for future use. The three women mobilized their collective experience to diagnose the ailment, using their olfactory sensitivity and habituation to the perceived medical meanings of particular odors.[92] To some extent, this use of sensory data paralleled the diagnostic process followed by learned physicians when they took a pulse or visually evaluated urine. Instead of comparing their observations to authoritative texts, however, these caregivers situated their perceptions within culturally constructed sensory regimes.

Only after the nurses had evaluated the situation and tried Salviati's remedy did the physician Francesco da Gamberaia appear on the scene. Riccio's letter does not specify the exact interval between the initial summons and the doctor's arrival, but at least several hours must have elapsed. Once at Castello, the physician "interrogated the nurse about all that had happened during the night." In the presence of Isabella Rainosa, he inquired about "the [baby's] coughing and disturbance, and every other thing that he felt it necessary to ask." Building on this information, Da Gamberaia then proceeded with his examination. He concluded "that he could not find a fever, but that her stools and her teeth showed the effects of the pain of worms, and recognized that her restlessness and fractiousness were a result of the condition of her anus and recommended giving her a draught of couch grass and unicorn horn."[93] Thus, for the most part, his diagnosis confirmed the women's findings. Still, the ailing infant refused to take the prescribed remedy until Rainosa "used her ingenuity" to successfully administer "four or five drops." After the baby's stomach was anointed with wormwood oil, she finally settled down, started suckling again, and dropped off to sleep.

Although Riccio ascribed a greater curative role to the consulting physician, both the nurse Domenica di Bonichi and Isabella Rainosa acted as important narrative agents and medical decision-makers in this scenario. These caregivers not only diagnosed the baby's condition correctly but also constructed it as an intelligible illness through the narrative process. Close reading of bodily signs represented a significant form of information gathering, while organizing illness narratives provided yet another generative process. Pooling their accumulated expertise, these women filtered their reading of symptoms through a grid of meaning constructed by experiential knowers over time. Nurse Domenica mediated the flow of medical communication in the way she selected, assembled, and relayed specific details about baby Isabella's condition to the physician. Since the child could not speak for herself, Domenica served as her proxy, much like the early modern German laywomen who acted as "voices of the sick" in conveying critical information to physicians about housebound relatives or neighbors.[94]

Similar perspectives on medical decision-making at court emerge from encounters between Pasquali and the duke's mother during her own bouts of illness. Patients in early modern Europe exercised considerable autonomy when choosing healers and negotiating therapies; self-care was commonplace, even at the highest social levels.[95] Duke Cosimo was a typical early modern patient in this regard. On more than one occasion he modified or ignored Pasquali's advice outright, especially when an illness persisted. In September 1543, Eleonora noted that her husband had agreed to move to the Medici villa at Poggio a Caiano, where the good air might hasten his recovery from a recent illness.[96] Yet the following month, Cosimo stubbornly rejected Pasquali's advice to take clysters, which the doctor considered to be the "true medicine" to break his fever. The ducal secretary reported that Cosimo "wants to treat everything in his own way, so that [Pasquali] is desperate and the duke's illustrious mother even more so." A few weeks later, the duke willingly drank "three or four glasses" of therapeutic spring water on doctor's orders, but refused to take more. He then informed Pasquali that under no circumstances would he submit to more clysters, since he already felt sufficiently "flayed" by these measures.[97]

Salviati exercised similar autonomy in medical encounters with court physicians. In her final years, she suffered from numerous complaints such as bleeding hemorrhoids and recurrent fevers that often left her debilitated for weeks at a time. From 1540 until her death three years later, she was attended by Pasquali and a handful of other physicians at Castello. Court correspondence captures some of the negotiations between patient and healer, which echo the more formal cure agreements studied by Pomata.[98] Maria engaged in a verbal give-and-take with physicians to develop therapies that suited her iron will and exceptional sense of modesty. She began mapping her own treatment plan almost as soon as her health problems began in earnest. In September 1540, Pasquali prescribed a liquid remedy that would "comfort" her stomach and staunch her rectal bleeding. The following day, he reported to Cosimo that her kidneys were still inflamed but that there was no additional flux of blood. The court physician remained optimistic about a cure, "even though she doesn't want to take anything by mouth but only wants to use external remedies, in which I have little confidence." Salviati's preference for such therapies forced him to adjust his protocols. After consulting with other doctors, Pasquali devised "a bath that [would] refresh and constrict that area."[99] At the same time, however, her preference for external remedies clashed head-on with her sense of modesty. Despite her chronic pain, this proper widow refused to let physicians examine her rectal area directly. Consequently, Salviati's chambermaid Antonia administered the soothing clysters Pasquali prescribed, using a slender silver tube. In this instance, the allocation of medical labor was driven less by issues of expertise than by gendered sensibilities.

As she lay dying in early December 1543, Salviati modelled Christian resignation in the face of suffering while continuing to make decisions about her own care. Recent scholars have remarked on the performative nature of suffering in the early modern period. The cultural categories through which bodily affliction were described and understood index larger belief systems as well as individual subjectivity. By stressing pious forbearance and self-effacement, Salviati inflected

Christian suffering in particularly feminine ways. Her autograph letters displayed a sense of selflessness tinged with Christian optimism, thereby creating an exemplary "textual display of suffering for posterity."[100] Good remedies might facilitate healing, but a return to "pristine health" could be credited only to God's grace. Because Salviati tended to deflect concerns about her health problems, there are no lengthy, introspective patient narratives that speak to her subjective experience. Much of what we know about her ailments comes from her son and doctors, whose illness narratives bolstered her representation as a pious widow. When Salviati did offer firsthand reports in her letters, she often adopted an indirect mode of reporting that stood in stark contrast to her meticulous health bulletins about the Medici children. Writing to Cosimo in May 1542, for instance, Maria reported the progress of her current "indisposition" through Pasquali's eyes, telling her son that the physician was pleased with her progress and was especially gratified that her "aches and pains" had ceased.[101]

Salviati demonstrated similar behavioral patterns even on her deathbed—one of the most important sites for self-representation in the early modern world. A week before she died, she wrote to Eleonora, who had recently fallen ill. Maria voiced regret that her own illness prevented Pasquali from attending the duchess in person. At the same time, she tried to allay Eleonora's fears by reminding her that the doctor was exceedingly familiar with her ailments. As news of her decline reached the ducal couple in Pisa, they lamented her desire "to be treated according to her own advice" instead of giving Pasquali free rein.[102] No doubt this letter was read aloud in the sickroom, since expressions of affection and encouragement conveyed in writing had important therapeutic value that could affect recovery.[103] In the early modern period, illness was not a solitary experience but a complex social event mediated by social norms and expectations. Attending Salviati in her final weeks were her sister Francesca, her cousin Caterina Cibo, a certain Mona Cassandra, and other chambermaids.[104] Besides tending the dying woman's needs, these women performed the essential act of witnessing the "good death" she made. Three hours before she died, Salviati heard mass, sat up in bed and ate some gruel. In a last-ditch effort to save her, Pasquali ordered some of her favorite distillate from a Florentine convent pharmacy, but to no avail.[105] Her body was transported from Castello to her cherished convent of Santa Caterina da Siena, before being buried in the Medici crypt at San Lorenzo. Deprived of his closest adviser, Cosimo made Maria's commemoration part of his cultural politics in later years.[106]

Eleonora of Toledo: Constructing a Medical Court Culture

After Salviati's death on December 12, 1543, the nascent court entered a new phase. Pasquali stayed on as court physician, but the day-to-day care of the nurseries fell to Spanish court ladies working with local female practitioners. Among the most important were Isabella Rainosa (or de Reinoso), Eleonora's principal lady-in-waiting, and the Tuscan caregiver Giulia degli Amadori da Prato, who served as head matron and governess. Their purview continued to expand with the birth of new Medici offspring almost annually until 1554. Although household and professional care continued to work in productive partnership, the nature of their relationship gradually shifted under Eleonora's watchful eye. Some of these changes were linked to her imperious personality; others reflected the maturation of the Medici court itself. The duchess played a key role in instituting a hierarchical medical culture at court. One scholar has commented on her tendency to regard physicians as little more than court functionaries who could instrumentalize her choice of treatments.[107]

Like her mother-in-law, Eleonora was schooled in pharmacy. Her familiarity with remedies and recipes pervades court correspondence, whether linked to household care or embedded in a transnational gift economy. The duchess was frequently on the receiving end of valuable medicinals and proprietary recipes sent by members of the Spanish Neapolitan court and far-flung correspondents. This exposure kept her apprised of new medical trends as well as alterations to time-honored remedies. In October 1543, Eleonora instructed the court secretary to "have the apothecary summoned, and commanded that he distill eighteen flasks of watermelon distillate in the manner written on the attached sheet that should be given to him." The recipe has not survived, but she must have been familiar with this plant, which had been cultivated in southern Europe for centuries before being exported to the New World. Recognizing that the remedy entailed an exacting manufacturing process, Eleonora closed by saying, "Tell the apothecary that everything should be executed with great care." The following year, she insisted that

Pasquali spell out the composition, usage, and effects of the unguents prescribed for her chest affliction, possibly because they were unfamiliar to her as a Spaniard.[108]

A domestic accident in February 1544 underscores Eleonora's mastery of household remedies. On this occasion, four-year-old Francesco had hit his older sister Maria with a stick while playing, opening up a nasty gash on her forehead. Eleonora immediately directed Pasquali to make a salve containing distilled human fat to treat it. In specifying this remedy, she stressed that she wanted to prevent any scarring that would mar her daughter's looks—a candid recognition of the social importance that female beauty carried at court.[109] Remedies containing human substances—skin, bone, flesh, fat—could be found in apothecary shops and personal medicine cabinets across Renaissance Europe.[110] Despite the gruesome nature of the cure, the duchess had made an informed therapeutic choice that partly reflected her Iberian heritage. Human fat procured from the bodies of executed criminals was the basic ingredient in high-quality ointments used to dress wounds and diminish scarring from smallpox and various skin conditions. It was a standard method of wound care in Iberian medical practice, frequently used by Spanish colonizers in the New World.[111] Undoubtedly the duchess did not get her hands dirty in making this remedy, but the immediacy of her response testifies to her impressive health literacy.

Eleonora's awareness of new materia medica was heightened by her exposure to the transnational economy of sixteenth-century Italy. Scholars disagree about the extent to which the influx of new drugs from Asia and the Americas transformed the traditional European medicine chest in the first half of the century, since factors such as access, supply, and price variations affected the adoption of specific medicaments.[112] By 1550, however, traditional guilds had adjusted their formularies to reflect the introduction of new pharmaceuticals. That year, the Florentine *ricettario* was reissued because "time has revealed new sorts of medicines." When it was revised again in 1567, the number of approved recipes had almost doubled, with fifty-eight new recipes added to the original sixty-nine.[113] Certainly Eleonora's extensive diplomatic connections facilitated the circulation of precious medicinals between courts, transforming these goods into a kind of political currency. In 1553, the duchess received a gift of luxury articles from Istanbul, including a box containing terra sigillata, a prized medicinal clay used as a poison antidote and general cure-all. This precious commodity had been jointly certified as "perfect" by a physician to the Turkish sultan and Bernardino Porcellini, member of a trusted Florentine apothecary firm then residing in Istanbul.[114] Terra sigillata had long been a staple in the Mediterranean medicine chest, but access to its sources in the eastern Mediterranean, which produced the finest grade of clay, had been hampered by conflict with the Ottoman Turks. As the recipient of exciting new rarities or precious old standbys, the duchess anchored a wider hub of transnational exchanges that affected the practice of medicine at the Medici court.

Eleonora's familiarity with pharmacy helps contextualize the secret recipe for a face tonic commonly associated with her name. Although products for the routine care and beautification of bodily surfaces occupied an honored place in medieval healthcare, cosmetics gained enormous popularity in Renaissance Europe.[115] In fact, the use of beauty products and perfumes frequently acted as a stimulus for sixteenth-century Italian court women to experiment with new recipes or ingredients themselves. Isabella d'Este, Marie de' Medici, and Bianca Cappello were all known to have created face lotions and signature perfumes that they integrated into larger circuits of gift exchange.[116] Despite the hygienic value of these items, they remained freighted with moral objections because of their associations with vanity and artifice. The use of rouge in particular, which Eleonora was known to apply, provoked unwelcome parallels with prostitutes and courtesans.[117] Nevertheless, the desire to improve surface appearances found new backing because the skin itself took on new social significations in the sixteenth century. Ideal skin for both women and men was smooth and unblemished, indicating a well-balanced inner complexion. In court settings, perfect skin increasingly became a marker of noble status, requiring the use of cosmetic practices by both sexes. The advent of syphilis also played an important role in assigning new cultural meanings to skin, since the pox was a highly visible condition marked by disfiguring lesions and the erosion of soft facial tissues. European encounters with indigenous populations focused keen attention on skin color, which began to define racial differences by the seventeenth century.[118] Traditional objections to beauty products gradually gave way under these combined pressures.

The recipe for the signature face tonic "used by Duchess Eleonora" was handed down in the collection started in 1569 by Stefano Rosselli, the Medici court provisioner who opened a bustling apothecary shop at the Canto del Giglio that year.[119] Made from ordinary ingredients such as lemon, eggs, and goat milk, this product formed part of a broader health regimen. It is unclear, however, whether the duchess concocted this recipe herself or commissioned its manufacture from Rosselli. Since Eleonora was renowned for her luminous complexion that more than satisfied Petrarchan standards of beauty, attributing this product to her may have sparked the interest of other consumers hoping to achieve similar results. In recording several hundred recipes over a twenty-year period, Rosselli noted the provenance of only a dozen remedies, giving the duchess's beauty secret some distinction. Certainly by the late sixteenth century, associations with noble makers or users had become a standard way to enhance a recipe's value.[120]

Although Eleonora's reign saw the growing use of cosmetics and new medicinals, the primary feature that distinguished household medicine under her tutelage was the political uses to which it was put. Both Cosimo and Eleonora used healthcare strategically to distribute favors within court circles and to create patronage networks outside them. In this sense, medicine accomplished important political work in consolidating Medici rule and cementing a complex ordering of rank within the court. Eleonora's deep understanding of court protocols allowed the new regime to simultaneously mark fine gradations of status and extend its diplomatic networks by allocating medical resources in particular ways. Still, the duke and duchess oversaw different domains, in keeping with their respective political positions and contemporary gender norms. The late Renaissance court was not only a center of knowledge and power but also a household that replicated gendered expectations about appropriate duties and activities.

As sovereign, Cosimo frequently used medical resources as a diplomatic tool, dispatching Pasquali and other court physicians to political allies near and far. The duke readily satisfied the 1551 request of Diego Hurtado de Mendoza, the Spanish military governor of Siena, for help from the Medici court physician and a good apothecary "because he doesn't trust the ones in Siena." Granting this favor was a convenient way to strengthen political alliances and keep lines of communication open, since medical personnel often served as go-betweens. The duke cultivated relationships with former political opponents in similar fashion. Cardinal Innocenzio Cibo initially had opposed Cosimo's appointment as duke; yet when he fell ill in Carrara, the duke sent the time-honored medicinals rhubarb and manna—"the best and most perfect that there is on hand"—before dispatching Pasquali himself to the sickbed.[121]

By contrast, Eleonora practiced a kind of household medicine writ large. She simultaneously fulfilled her wifely duties and showcased her magnanimity by extending care to relatives, ladies-in-waiting, the children's nurses, court visitors, and local clients. It was customary for Renaissance princely courts and great ecclesiastical households to subsidize some of the medical needs of their entourage.[122] Princes and cardinals normally sent court physicians to care for relatives and friends when they fell ill and often paid for expensive medicines as well. Yet medical care for court staff was never a given. In the 1460s and 1470s, the Milanese duke Galeazzo Maria Sforza confined his support to family and favorites, dismissing requests from staff beyond his inner circle. Some *signori* lent their physicians to dependents but did not pay for their medicines. The early Medici court was unusually generous in underwriting the cost of medicines for courtiers, which helped assure their political support. The Medici duke kept a running account on their behalf at the Giglio apothecary shop from 1537 to 1568.[123]

One of the main recipients of Eleonora's largesse was her older brother Don Luis, who suffered from syphilis and other maladies. Don Luis joined his sister at court soon after her marriage, often traveling with the ducal couple and acquiring property for a pleasure garden in Florence through ducal intervention in 1545.[124] In October of that year, the duchess instructed Pasquali to send "a poultice or some kind of ointment" right away to treat her brother's stomach ailment.[125] A few months earlier, she had instructed the physician to rejoin the court immediately after treating Cardinal Cibo in Carrara. There was good reason for this urgency: Eleonora wanted her brother to undergo guaiac treatment under the doctor's supervision. This important New World drug became a standard treatment for syphilis, although it could be prescribed for other ailments including joint pain and epilepsy. Eleonora herself took guaiac on Pasquali's orders to calm the nausea associated with pregnancy, although it did not have the desired effect.[126]

As syphilis ravaged the continent after 1500, heated controversy arose over its origins and treatment. The two most popular remedies for pox were mercury and guaiac, each of which had staunch advocates. Physicians who emphasized the resemblance between syphilis and the more familiar disease of leprosy generally supported the use of mercury, which had a long history in treating skin conditions.[127] Applied externally in the form of ointments or rubs, mercury had brutal side effects such as excessive salivation and loosened teeth, which were considered positive signs that peccant matter was being excreted. By contrast, advocates of guaiac stressed both its efficacy and benign nature. Among its ardent supporters was the German knight Ulrich von Hutten, whose 1519 treatise on the pox helped popularize the drug. Part of the success of "holy wood," as guaiac was called, rested on the belief that pox was a new disease originating in the Americas. It was commonly argued that a beneficent God had placed a cure in the same areas where the disease flourished.[128] Taking guaiac involved an arduous regimen (discussed further in Chapter 5). Sufferers undergoing treatment ingested decoctions daily for thirty to forty days, followed a severely restricted diet, and subjected themselves to copious sweating—a grueling therapy that left them debilitated for weeks. Moreover, guaiac was exorbitantly expensive in its early years.[129]

Apparently Eleonora held firm opinions about these controversies. The ducal secretary remarked that the duchess considered guaiac the best way to cure her brother's "indisposition," adding emphatically that "one doesn't dare differ with her."[130] Eleonora clearly recognized the public health implications of syphilis. After 1545 she contributed 130 scudi annually from personal funds to subsidize the Florentine pox hospital, including its purchase of guaiac.[131] Her belief in the curative powers of this medicinal reflected her close ties to the Spanish court and its commercial interests. Numerous Spanish authors identified the pox as a New World disease in need of a New World cure. The Spanish priest Francisco Delicado stressed this link in his popular "how-to" manual, *El modo de adoperare el legno de India occidentale* (1525), in which he instructed readers how to prepare and administer guaiac.[132] Strengthening the case was Gonzalez Fernandez de Oviedo's history of the Indies, in which he claimed that indigenous populations had long used guaiac to cure syphilis. Significant revenues were at stake, since the king of Spain held rights to this lucrative commodity. Contrary to popular belief, however, the Fugger banking family did not hold a monopoly on its trade.[133]

Two years after she first asked Pasquali to administer the wood cure to her brother, the duchess locked horns with him again over the same matter. It was not uncommon to take the cure more than once if it failed to produce the desired results. In late March 1547, Eleonora told Pasquali to send the full protocol for guaiac treatment so that her brother could take the cure on his own. Don Luis needed to complete treatment and recover his strength before beginning the taxing journey to Naples, where he was being groomed as regent.[134] When the materials still had not arrived two weeks later, Eleonora exploded angrily. She "marveled" that Pasquali had failed to fulfill her request and ordered him to appear in person immediately. This artful medical courtier sidestepped the summons by replying that he would appear at court right after Easter; in the meantime he was dispatching an apothecary with the medicine.[135] The resolution of this contretemps is not recorded in court correspondence, but the episode reveals the new footing on which learned medicine operated at the early Medici court under Eleonora's sway—one that was increasingly subject to the demands of power.

The duchess also oversaw the health needs of all her female attendants but lavished special care on Isabella Figueroa, her head lady of chamber.[136] Figueroa's privileged position made her privy to the duchess's body as well as her most prized possessions; their shared Spanish origins probably enhanced their bond. When Isabella fell ill at Castello in August 1544, Eleonora instructed the ducal secretary to send two physicians to attend her.[137] The duchess expressed her "infinite displeasure" upon hearing that Figueroa had relapsed in October 1545, and ordered Pasquali to remain by her side "to make sure that she lacks for nothing." The same day, the ducal couple canvased monastic pharmacies for special compotes and other health-giving foodstuffs to sustain Isabella through the illness. Cosimo displayed his largesse by paying for these remedies; Eleonora added two pair of partridge as a personal gift. When Isabella still had not improved three days later, the couple repeated their instructions that Pasquali should provide every possible treatment. After two anxious weeks, Figueroa started to recover and resolved to "take the water," much to Eleonora's delight.[138]

Thanks to her private fortune, the duchess also had the means to dispense medical charity beyond court circles. Distributing remedies and therapeutic advice was a permissible activity for noblewomen and consorts because of its perceived charitable intent. Yet Eleonora used medical charity in politically strategic ways to overcome residual opposition to a princely regime and foreign bride. By dispensing medical resources from private funds, the duchess amplified a public persona centered on abundance and generosity, paralleling iconographic programs that linked her personal wealth to the prosperity of the Medici state.[139] Satisfying requests for medical assistance both expanded Eleonora's role in day-to-day governance and opened another avenue for political patronage and self-fashioning. Since many of these requests reached her via private correspondence or intermediaries, they also strengthened Eleonora's grasp of local information networks. For example, the duchess activated a chain of command to ensure that a local nun would receive much-needed care in August 1545. Working through the matron Giulia degli Amadori, Eleonora instructed Pasquali to compound a prescription for her and "to make sure she is given whatever she needs." The following year the duchess sent a jar of medicinal rose sugar to an ailing Pisan notary, whose illness left him housebound for three days. When he recovered, this grateful client thanked "our most illustrious patron" for her generosity.[140] Given the frequent residence of the Medici court in Pisa, it was politically expedient to cultivate good relations with local inhabitants. Under normal circumstances, however, neither Eleonora nor Cosimo distributed medicines gratis to their subjects.[141]

The distribution of medical charity both inside and outside the court multiplied Eleonora's interactions with Pasquali and other court physicians, who remained few in number. Despite its later splendor, the early Medici court was simply too new, too small, and too provincial to maintain a small army of healers like the ones found at the great royal houses of Europe. The notoriously demanding Eleonora was every inch the duchess in dealing with these experts. In April 1543, for instance, she ordered Pasquali to attend the imperial diplomat Francesco Alvarez de Toledo, a frequent visitor at court who had fallen ill after a meal. The duchess was none too pleased with this turn of events, since she had recently supplied him with medicines. Pasquali was instructed to care for this important ally "with the same diligence" that he would show toward the ducal couple, accompanying him on the next leg of his journey to Bologna if necessary. The communiqué ended by saying, "These are official orders."[142] Eleonora certainly fulfilled her wifely duties in caring for the health needs of her guests, but she did so with a keen grasp of their political implications. Under the duchess' direction, the politics of health at the early Medici court became more overtly politicized.

This assessment has shown that the practice of household medicine played a vital role in health and healing at the early Medici court. The competencies displayed by Maria Salviati and Eleonora of Toledo in pharmacy, "physic," and preventive health gave them a strong platform for informed decision-making when interacting with court physicians. By and large, the practice-based knowledge in which these Medici women were steeped worked in tandem with professional medical services. The intense concern with family life evident throughout the Renaissance period gave medically informed women like Salviati a recognized niche within pediatric care, which emerged as a distinct body of knowledge and practice in her lifetime. Nevertheless, the medical purview of these Medici court women was not limited solely to infants and children but encompassed the household writ large. Working with a consortium of skilled practitioners, their medical agency extended to narrative representations and medical provisioning, as well as hands-on care.

Sharp contrasts in the ways these two women interacted with medical professionals not only reflect their different personalities, but also point to fundamental differences in the uses to which healthcare was put by the court after 1540. Wealthy, aristocratic Eleonora dispensed medical favors as a means of calibrating rank within the court and building client networks outside it. Her liberal use of medical charity buttressed other dimensions of her public persona as a model of generosity who helped guarantee the prosperity of the Medici state. Unlike her mother-in-law, Eleonora instrumentalized physicians' services as part of an emerging chain of command. Examining the politics of health at the early Medici court not only demonstrates the impressive health literacy and medical agency of these two noblewomen; it also showcases how the organization of healthcare helped constitute the Renaissance court itself.

CHAPTER TWO

Gifts of Health

Medical Exchanges between Court and Convent

GIFT-GIVING HAS BEEN RECOGNIZED as a central element of Renaissance culture since the anthropological turn some thirty years ago. Gifts helped constitute a vast economy of exchange by building social capital, stabilizing hierarchies, and affirming ties between peoples and states. Numerous studies have shown that Renaissance gift-giving generally accentuated status differences and heightened competition between equals. Diplomatic gifts of high artistic merit—paintings, sculpture, jewelry, metalwork—have captured the lion's share of attention, in part because they slot comfortably into established artistic categories and familiar political narratives. These exquisite objects contributed to the calculated liberality at work in Renaissance courts. Grand Duke Ferdinando I de' Medici voiced the instrumentality of gift practices when he informed the Tuscan ambassador to Spain that "we don't give gifts for the sake of magnanimity, but out of self-interest and to those who render us services."[1] In contrast to magnificent objects, the token nature of small, personal gifts communicated affection or rewarded dependents for services well-rendered.[2] Regardless, gift-giving sparked reciprocity in the form of continued loyalty, material counter-gifts, or future favors.

To date, gifts associated with health and healing have received scant attention, partly because their sheer number and variety resists a clear conceptual identity. In Renaissance Europe, gifts intended to heal, promote recovery, or foster good health encompassed an enormous range. Gifts of health ran the gamut from material goods such as remedies, recipes, and tools for self-care to ephemeral offerings of consolation and therapeutic advice. It is tempting to focus on material objects, since they left a large footprint in the historical record. Certainly Renaissance elites trafficked heavily in such items as a means of solidifying political networks, circulating innovative products, and redistributing the fruits of empire. The French queen Marie de' Medici, for instance, frequently received medicinal goods from Italian courts, which she then regifted to local allies and political favorites to create larger networks of favor.[3] In fact, Italian aristocratic women experimented with pharmacy partly as a way to expand their gifting potential.[4]

Yet we should not overlook the value attached to intangible gifts of health such as consolatory letters or soothing words of solace. Isabella d'Este—one of the most lavish consumers in Renaissance Italy—viewed her personal letters as efficacious medicine capable of refreshing body and spirit. By conveying her concern and depth of affection to an ailing recipient, these "medicinal missives" worked as antidotes to pain and suffering.[5] Although expressions of friendship and social support often left sparse textual traces, they nevertheless offered powerful hope that strengthened resilience or enacted physical changes in the bodies of sufferers. This type of affective healing is associated most commonly with medieval saints' cults and miraculous cures, but intangible gifts of support and encouragement formed an essential part of Renaissance care practices as well.[6] The value ascribed to ephemeral gifts highlights the complex nature of healing that extended beyond curative remedies or therapeutic interventions.

Whatever form they took, gifts of health assumed particular importance for Renaissance noblewomen for several reasons. First, they opened significant opportunities for female agency and self-fashioning, since assisting the sick was both a Christian obligation and a household duty. These charitable activities gained added breadth and legitimacy among Catholics and Protestants alike as social welfare assumed greater prominence in religious reform programs. As discussed earlier, female consorts like Eleonora of Toledo used the performative gestures of medical gifting to promote a beneficent image of self and state. Even mediating such exchanges enabled noblewomen to fashion a public persona. When the Spanish countess Leonor Pimentel navigated diplomatic channels to obtain "a small vial of stomach oil for her husband" from a local Medici agent, she secured her reputation as a good wife within court circles while solidifying international alliances.[7]

Second, gifts of health positioned Renaissance noblewomen as cultural brokers between courts. Brides moving in to or out of the Medici court customarily brought family recipe books, prized remedies, medical paraphernalia, and personal physicians in their entourage.[8] The 1617 marriage of Caterina de' Medici, daughter of Christine of Lorraine and Ferdinando I, to the Mantuan duke Ferdinando I Gonzaga, exemplified how elite women facilitated the circulation of goods and knowledge. Caterina packed various Medici court remedies as well as exquisite articles for self-care in her trousseau. Among the health-related items she took to her new home were a medicine chest "filled with small silver pots and other small items pertaining to pharmacy and surgery," which was valued at 200 scudi. She also brought a silver perfume burner to ensure healthful ambient air in her chambers; a silver spittoon for voiding phlegm; a syringe, cauldron, and portable stove for making and administering clysters according to personal preferences; and various tools for cleaning the ears, tongue, and teeth.[9] All these items helped constitute the rich material culture of prevention in evidence by 1600. Her mother later asked Caterina to obtain the Gonzaga court recipe for gout ointment, despite having access to a similar remedy developed by "the House of Austria" through her Hapsburg daughter-in-law, Maria Maddalena. Apparently Christine disliked its odor, so decided to look elsewhere.[10]

Keeping these considerations in mind, in this chapter I examine the medical agency of Renaissance noblewomen through the lens of their gift exchange with another subset of women: cloistered nuns. At first glance, these two groups stood at opposite ends of the spectrum in terms of wealth, political power, and social visibility. However, scholars have noted the many circuits of power linking Italian courts and convents that made these sites increasingly interdependent by the sixteenth century.[11] In fact, convents were essential to the social formation of Italian elites by virtue of their role in family marriage strategies, which consigned roughly half of all aristocratic girls to celibacy after 1500.[12] Moreover, sixteenth-century Italian noblewomen often colonized religious communities as extensions of aristocratic living spaces, taking shelter there from unhappy or abusive marriages and transforming these institutions into focal points of artistic and political patronage.[13] Deepening this interdependence still further were the political investments Renaissance nuns and noblewomen made in medical charity. Although Italian aristocratic women often distributed recipes and remedies to friends and clients, these gifts were only tenuously tied to broader social welfare schemes. In order to achieve greater impact, noblewomen often relied on nuns to mediate large-scale medical charity, such as running hospitals, distributing medicines to the poor, and nursing the sick. In other words, nuns could be instrumentalized as medical proxies by female benefactors within a Catholic charitable economy; each group contributed different skills and services to the pursuit of health.

Here I develop two detailed case studies as a way to probe the nature, motivations, and workings of medical gift exchange between nuns and noblewomen in late Renaissance Italy. Bookending the period under consideration, these cases shed light on macro-historical concerns such as the circulation of new materia medica resulting from early globalization, as well as microscale questions regarding the therapeutic value of consolation within intimate relationships. Both cases originate in singular, cross-class female friendships between Catholic women that blended practical interests, spiritual identities, and emotional bonds; in both instances, personal alliances ultimately generated wider networks that affected the workings of state and society.[14] Both cases expand our understanding of women's medical competencies, especially in the realm of pharmacy, and add new layers of evidence to the practice of household medicine at court discussed earlier. Both draw extensively on personal letters and women's epistolary networks to map the flow of medical information and its translation into broader communities of practice throughout the early modern period. Even in an age of print, handwritten letters enlarged the circulation of medical knowledge while offering important possibilities for writing the self, especially among cloistered nuns.[15]

At the same time, each case provides a distinct angle of vision on medical gift exchange and the extension of medical charity within female circles. The first case details the transnational commerce conducted between Queen Leonor of Portugal and the Florentine nuns of Le Murate, which spanned more than a half century from the 1490s to 1565. Because long-distance trade required considerable capital as well as extensive business networks, scholars have long argued that the growth of early modern capitalism confined women to local and regional market production. Queen Leonor's impressive reach points the analysis in a different direction. Her core resources in medical exchanges were commodities drawn from Portuguese royal coffers, especially sugar and spices, which the Murate nuns reciprocated with small handmade gifts meant to benefit the queen's spiritual health and communicate a sense of shared religious identity. Focusing on material goods flowing through maritime trade networks, this case shows that the Portuguese queen integrated nuns across the Mediterranean basin into her charitable strategies as both agents and recipients. At the same time, it problematizes tensions inherent in medical gift exchange. There was tremendous potential for conflict between the selfless ideals of Christian charity and cultural notions of noble liberality, especially when thinking about the expected return on the gift.[16]

The second case turns from the impact of new commodities traded over long distances to the small tokens of affection exchanged locally between the Tuscan archduchess Maria Maddalena of Austria and the impoverished, illiterate healer Sister Orsola Fontebuoni. Status differentials between the two women could not have been greater. Nevertheless, Fontebuoni served as both respected health consultant and valued spiritual adviser to the archduchess and other Medici court women in the 1610s and 1620s. She was a leading participant in a multinodal network consolidated by the exchange of precious religious objects and potent fertility aids. One of the major threads linking these women was the pursuit of health through a mixture of naturalistic and spiritual healing. Juxtaposing new and old remedies, material and immaterial gifts, local and transnational exchanges, these two cases show how diverse healing modalities created spaces for female medical agency while giving Renaissance medicine its unique dynamism.

Transnational Exchanges: Queen Leonor of Portugal and the Nuns of Le Murate

Medical exchanges between the Murate nuns and the Portuguese queen had their roots in the emergence of patronage as a governing social practice. By the late fifteenth century, patronage had intensified the traffic in gifts as both patrons and clients sought to manage their networks through gift-giving. Thanks to support from affluent donors—Florentine patricians, ecclesiastical dignitaries, Italian noblewomen—the Murate had blossomed into the largest convent in Renaissance Florence by the end of the fifteenth century.[17] In the 1490s, the Murate nuns widened their social reach by sending small gifts such as homemade confections, small devotional objects, and fruit grown in their orchards to noblewomen within and outside their immediate circle. This strategy was sparked by the perpetual quest for funds, but it also reflected the privatization of ecclesiastical institutions by local and regional elites.[18] In particular, the nuns targeted pious noblewomen already connected to the convent through extended kinship ties or political affiliations. This strategy of proactively giving gifts to wealthy women proved extremely productive. Caterina Sforza, a staunch convent ally discussed in Chapter 1, thanked Abbess Elena Bini for "the pomegranates and other fruit from your orchard," which she may have put to medicinal uses.[19] Eleonora of Aragon, Duchess of Ferrara, pledged her financial assistance after receiving two small gifts from the abbess in these same years.[20] As foreign invasions rocked the peninsula between 1494 and 1530, the Murate nuns increasingly looked to noble patrons for political support and protection.

Perhaps the most unlikely convent patron to emerge in these years was Queen Leonor of Portugal (1458–1525), the widow of King João II. Leonor was the wealthiest woman in the Portuguese kingdom around 1500, commanding a fortune second only to that of her brother, the future king Manuel I (r. 1495–1521). Aside from her vast landholdings, Leonor profited from the early fruits of empire as a principal shareholder in the Casa da India, the royal warehouse where products of the global spice trade were stored. A devout, sickly woman, the queen showed a particular devotion to mendicant piety emphasizing contemplation and social welfare; indeed, two painted portraits depict her wearing the habit of a Franciscan tertiary.[21] Leonor made the extension of medical charity a cornerstone of her public persona, similar to several of her peers. Widowed queens and regents like Catherine de' Medici of France often allayed fears about the independent exercise of female political power by representing themselves as pious exemplars in their cultural and charitable programs.[22]

Like the Medici court women discussed earlier, Leonor was conversant with the medical thinking of her day. In addition to her religious interest in aiding the sick, health issues figured prominently in her own life. Chronic ailments left the queen almost continually bedridden in her later years, prompting her to construct a private oratory next to her bedchamber.[23] Her interests in health, healing, and the natural world are reflected in the medical reading included in her personal library—the largest collection owned by a Portuguese woman and one of the most important in the realm. Among the roughly eighty books inventoried in 1537, twelve years after her death, was a short treatise on plague *(Regimento proveytoso contra ha pestenença)*, written in Portuguese by the Franciscan friar Luis de Ras. Aimed at prevention as well as treatment, this work was one of the first such manuals printed in Lisbon in 1495–1496. Other medical works in her library included the *Compendio de saude humana* (Seville, 1515), a vernacular translation of the *Fasciculus de medicina,* the first illustrated anatomy book, attributed to the fifteenth-century German physician Johannes de Ketham. Leonora also owned two major reference works: Gabriel Alonso de Herrera's *Obra de agricoltura* (1513), which contained information on veterinary medicine, meteorology, and the influence of food on health; and the thirteenth-century encyclopedia *De proprietatibus rerum (On the properties of things),* authored by the Franciscan friar Bartolomeus Anglicus. Taken together, these works provided a basic textual introduction to health, disease, and natural history.[24]

Despite sharing common medical interests with other noblewomen, Leonor's charitable strategies were bound up in unique ways with Portuguese overseas expansion and the global commodities trade. Portuguese royal women enjoyed unmatched resources in constructing charitable programs owing to the Crown's monopoly on maritime trade goods, as well as the independent financial structure of royal households. Leonor's mother, Doña Beatriz (c. 1429–1506), was one of the first Portuguese noblewomen to use colonial commerce as a platform for medical charity. Between her husband's death in 1470 and her son's accession to the throne in 1495, Beatriz held title to the profitable sugar production taking root on the island of Madeira. Following the discovery of a maritime route to India in 1498, Beatriz established a massive new palace pharmacy fully stocked with sugar, spices, and other goods acquired through Portuguese commerce and colonization. Besides employing a physician and apothecary, the pharmacy also incorporated several women into its service, including at least one female slave. At her death in 1506, Doña Beatriz bequeathed the entire pharmacy—staff included—to the convent of Nossa Senhora da Conceiçaio, which she and her husband, Fernando Duke of Beja, had founded some years earlier. Run jointly by nuns and the duchess's female staff, the pharmacy helped supply the entire town of Beja with remedies.[25] Her daughter Leonor followed in her footsteps when founding the convent of Madre de Deus just outside Lisbon, to which she donated part of her well-stocked dispensary, private library, and vast relic collection. These two royal convents represented important administrative nodes in the crown's state-building project.[26]

Between 1480 and 1520, Leonor was instrumental in establishing new charitable institutions supported by the global commodities trade. Of particular importance to their financing were annual allotments of sugar dispensed by the Crown to female members of the royal family. King Manuel's wife Maria received 200 *arrobas* (5,000 pounds) of sugar annually; Manuel's sister, the dowager queen Leonor, was granted half that amount, while his other sister, the widowed duchess of Braganza, got thirty *arrobas*.[27] Each woman was free to dispense her allowance as she saw fit. Using these exceptional resources, Leonor founded a hospital (Nossa Senhora do Pópolo) in 1485 on the site of a therapeutic hot springs, whose statutes were partially modeled on the Florentine civic hospital of Santa Maria Nuova, known for high standards of care.[28] In 1498 she also founded the first confraternity of the Misericórdia, tasked with performing the seven acts of mercy among the needy. These confraternities spread rapidly throughout the Portuguese maritime empire in the sixteenth century until they controlled most charitable activities in the realm. Her husband, King João, created the central royal hospital of Todos os Santos in Lisbon by uniting seventy-four older foundations;

this new unit acted as a clearinghouse for the redistribution of sugar, spices, and medicinals to other charitable facilities. Foundations in imperial outposts from Morocco to India received subsidies of similar goods as a means of expanding Portuguese royal presence and authority. This centralized scheme distinguished Portuguese charitable activity from its Italian counterpart, which remained highly localized in its organization and administration.[29]

Although Leonor's dedication to the sick poor is unmistakable, her relationship with the Murate resulted from a strange twist of fate.[30] The convent chronicle recounted how a mature woman named Eugenia arrived there in 1450 but left before taking vows, in order to minister to the sick in the Holy Land. After caring for pilgrims there for over three decades, Eugenia decided to return to Italy, this time intending to establish a similar hospital in Rome. Blown off course, this religious woman arrived at the Portuguese court instead, where she developed a close friendship with Leonor. The queen promised to support the new project using part of her sugar allotment; when Eugenia returned to Florence instead of Rome, however, this pledge was transferred to the Murate. The two women continued to exchange letters rich in the language of friendship. Addressing the nun as "Eugenia amiga," Leonor turned to her friend for spiritual support when both her husband and son died in 1497.[31] Throughout her letters (Fig. 2.1), the queen expressed admiration for the nuns' self-imposed austerity and involvement with local social welfare initiatives.[32] In the same years, the Murate nuns cultivated good relations with Leonor's brother Manuel through artful expressions of gratitude and prayer.[33]

At the heart of these exchanges was sugar—a commodity that transformed the early modern globe. The importance of sugar to the development of Renaissance medicine can hardly be overstated. James Shaw and Evelyn Welch have observed that the increased use of sugar-based remedies after 1500 constructed taste and influenced broader consumption patterns in the early modern period.[34] Considered by physicians to be a "perfect" food, this commodity testifies to the importance of dietetics in Galenic medicine. When used as a base in syrups, electuaries, and juleps, sugar was thought to conserve or enhance the therapeutic properties of medicinal herbs. Medieval physicians were conversant with sugar-based medicines in Arabic sources, which had both curative and

Fig. 2.1. Letter from Queen Leonor of Portugal to the Abbess of Le Murate, 1500. ASF. CRSGF. 81. Vol. 100, #226. By permission of the Ministero per i beni e le attività culturali

preventive uses. Arnaud of Villanova, for instance, often prescribed sugar for various maladies and recommended the consumption of sugar syrups to invigorate the elderly. Candied fruits consumed in wealthy German households were thought to build up the strength of a convalescent recovering from fever.[35] Because only small quantities of sugar were imported from Sicily and the eastern Mediterranean throughout the Middle Ages, its medical impact remained limited. One scholar has estimated that fifteenth-century Europeans consumed an average of only one teaspoon of sugar annually.[36]

After 1450, Portuguese colonial expansion in Madeira and the Azores increased the supply of cane sugar available throughout the Mediterranean basin. By the 1490s, growing production volumes in those islands dramatically dropped the price of sugar, leading to increased consumption throughout Europe as well as the rise of slave labor in Portuguese Atlantic colonies.[37] The impact of this growth could be felt in local Italian apothecary shops by 1500. At the turn of the century, sugar was the second most common commodity stocked at the bustling Florentine pharmacy of the Giglio, surpassed only by wax. Various grades of sugar were retailed there at different price points. The most expensive was refined white sugar; the 1499 Florentine pharmacopeia required that all sugar-based medicinal syrups be prepared using this grade. Lower-quality sugars could be refined by cooking them three times, a process that probably was handled in-house by apothecaries, leading to significant profit margins of 100 percent or more.[38] This escalating consumption had numerous critics in sixteenth-century Italy. Some physicians decried sugar's association with Arabic medicine, which had "corrupted" pure Galenic practice. Others criticized it as an extravagance for the rich, since a pound of white sugar in the 1490s cost more than a skilled Florentine craftsman could earn in a day. Certainly the vast majority of European poor consumed sugar only while being treated in hospital. Despite these criticisms, sugar had become such a staple of medical practice by the eighteenth century that the saying "like a pharmacy without sugar" became synonymous with extreme poverty.[39]

Leonor's pledge of sugar to the Murate must be understood within several overlapping frameworks—religious, political, economic. Subsidizing a distant convent enabled the queen to enlarge her charitable reputation in ways that were both permissible and familiar for a Catholic noblewoman. The Murate abbess not only thanked Leonor for her gift on many occasions, but also assured her that, in "alleviating our extreme necessity," her reputation for beneficence would spread far and wide.[40] The allotment itself was a source of political legitimacy for Leonor as a member of the royal family; her ability to gift it beyond the boundaries of empire highlighted her royal prerogatives. Even Leonor's brother gifted sugar solely to foundations in new Portuguese territories, not to those in foreign lands. In extending her charitable reach to other parts of the Mediterranean, the queen solidified local relationships with Florentine merchant-brokers resident in Lisbon, several of whom helped manage the Crown's international commercial networks.[41] More importantly, however, her gift advanced Portuguese commercial policies by increasing the visibility of this commodity in new markets. Viewed from a purely economic standpoint, Leonor's gift of health to the Murate facilitated the circulation of medicinal commodities over which the Crown enjoyed monopoly control. Whatever its charitable motivations, the queen's pledge of sugar successfully widened market channels for this profitable imperial foodstuff.

To be sure, the queen's gift remained modest in terms of volume. Originally Leonor had pledged a shipment of forty-eight *rove* of Madeira sugar every three years, the rough equivalent of 560 pounds annually, or 18 percent of her allowance.[42] The Murate never realized the full amount. Utilizing now-lost account books, the convent chronicler Sister Giustina Niccolini calculated that the nuns averaged just a little more than 300 pounds of Madeira sugar annually between 1497 and 1559.[43] Other official sources put the Crown's annual gift of various grades of sugar at roughly 400 to 560 pounds, excluding cash, incense, expensive spices, and comfits.[44] Volumes handled by Florentine commercial apothecaries were substantially larger. The Speziale al Giglio, for instance, purchased six to twelve times that quantity annually (more than 3,700 pounds) in the 1490s.[45] Still, the queen's consignment was enough to supply a midsize hospital maintaining one hundred beds, much like the one Leonor founded at the hot springs. That facility consumed roughly 330 pounds of sugar a year in the early sixteenth century.[46] Similarities between these two subsidies suggest that the queen conceived of her Portuguese hospital and the Florentine convent dispensary as parallel projects meriting comparable support.

When the first shipment of Portuguese royal goods arrived at the Murate in January 1497, the nuns were already equipped to process large volumes of sugar. The convent had a dedicated still room or "kitchen" in which they had been making medicines, perfumes, and comfits commercially for several decades.[47] The initial cargo consisted of 1,000 pounds of fine Madeira sugar, 200 scudi in cash, and other confections like pennets, candied almonds, and quince paste. Later shipments included valuable spices and incense from Asia, as well as low-grade "red sugar" used for making clysters.[48] These exotic gifts not only gave the Murate pharmacy a huge commercial boost but also exposed its customers to a globalizing marketplace at an early date. Thanks to the queen's gift, the Murate became a significant outlet for the redistribution of Portuguese colonial goods in a booming secondary market. By mapping the often-circuitous routes taken by early modern goods, this exchange documents women's activities as distributors within long-distance commodity chains, in which their participation is rarely noted. The convent used this extraordinary windfall to grow its business, making it one of the most important monastic purveyors of fine medicinal items in the city. It is not surprising that Abbess Speranza Signorini pressured their go-between in Lisbon to expedite deliveries of these precious gifts.[49]

Among the standard medicinal items Leonor included in these shipments was quince paste *(cotognato),* known for its tonic properties throughout the Mediterranean since the days of Pliny. Made from cooked quince sweetened with sugar and traditionally eaten at the end of a meal, its dietary function was to seal off the stomach to prevent harmful vapors from rising to the brain.[50] Often molded into fantastical shapes, *cotognato* was a true medicinal listed in the Florentine pharmacopeia. Its diffusion tracks the growth of a more differentiated medical market across Renaissance Europe. Particular varieties of quince paste were produced in Italy, Spain, and Portugal, using slightly different recipes and cooking methods. By 1550, this product had penetrated the English market, where it became known as "marmalade" (from the Portuguese *marmelo,* or quince).[51] A recipe for a pale, refined Portuguese variety was included in the 1593 book of secrets kept by the Medici court confectioner Stefano Rosselli. This specialty product, "which few here know how to fabricate," had been introduced into local court circles by the wife of a prominent Portuguese merchant living in Florence.[52] These fashionable variants of a standard foodstuff represented only a fraction of the health-related products flooding local markets by the end of the century. Grand Duke Ferdinand tried to protect Tuscan producers from foreign competition by setting official prices for every conceivable medicinal, ranging from oils, herbs, and distillates to spring waters, sugars, and comfits. Issued in 1593, this legislation was one of the first comprehensive, state-sponsored attempts to regulate the medical marketplace by means of price fixing.[53] The subject warrants further analysis, but given accelerating commercial competition, it is not unreasonable to see Leonor's gift of Portuguese medicinal goods as an early attempt at product placement.

Part of the queen's stated rationale in donating these items to the Murate was "to assist your infirmary."[54] Food has often been called the forgotten medicine, used for both preventive and curative purposes within a Galenic framework. Consumed as part of a comprehensive dietary regime, delicacies like rose-scented sugar, pennets, and quince paste helped restore the physical and psychological well-being of nuns whose health had been undermined by penitential practices. Indeed, in 1505 the abbess assured the queen that these items "greatly comfort and restore the sick nuns who are numerous and in continuous need."[55] Pennets settled the stomach, while the medicinal sweets known as manus Christi, made from sugar and rosewater, were considered to be good for the heart and even were listed in the Florentine *ricettario*.[56] These gifts must have been especially welcome in light of the food shortages plaguing early sixteenth-century Florence. Because of their therapeutic value, medicinal sweets were exempt from customary dietary strictures governing monastic life. Thomas Aquinas considered sugared items to be more medicine than food; hence, from a theological standpoint, they did not disrupt monastic fasts.

Considering the volume of the queen's consignment relative to the convent population, however, it seems certain that the Murate nuns regifted or retailed some of these goods rather than consuming them entirely in-house. Distributing or selling surplus items was commonplace, fueling a brisk traffic in secondhand goods in most Italian cities. This kind of repurposing was especially frequent among charitable institutions. Leonor took this practice a step further by coordinating medical commerce between convents and hospitals to develop a broader network of care. The convent Leonor founded outside Lisbon fabricated medicaments using royal allotments of sugar and other medicinal substances; in turn these remedies were distributed gratis to her new hospital.[57] In other words, she used the nuns' labor and pharmaceutical skills as well as the productive capacities of their institutions to mediate her medical largesse. In so doing, the queen created a networked enterprise anchored by the royal gift, in which Portuguese nuns acted as both recipients and agents of medical charity. Given the queen's strategic use of resources, it seems likely that she had a similar labor arrangement in mind for the Murate. What appears as a straightforward pious gesture involving a single institution may have been part of royal aspirations to achieve a broader charitable impact.

Brokering the majority of consignments in the 1510s and 1520s was the Florentine slave trader Bartolomeo Marchionni, who had taken Portuguese citizenship in 1482. His role in the exchange underscores how shared religious values within Catholicism itself fostered long-distance business relationships by promoting trust and reducing inherent risks. Marchionni hailed from a line of apothecaries; his father and grandfather ran a pharmacy on Piazza S. Lorenzo, where he had apprenticed before securing employment with the Cambini banking firm in Lisbon. Taking Portuguese citizenship was a deliberate choice that facilitated his integration into the Crown's highest commercial circles.[58] In light of his extensive business contacts and pharmaceutical experience, Marchionni was ideally positioned to broker exchanges of medicinal goods with a favored convent in his native city. This merchant-banker donated his services gratis, underwrote ancillary transportation costs, and paid customs duties on the royal gifts out of pocket. Subsequent Florentine firms, some of which were staffed by Marchionni's relatives, continued these practices after his death circa 1530.[59] Advancing the queen's charitable ventures, Marchionni solicited the nuns' prayers as part of the expected return on the gift.[60]

From the beginning of the exchange, the Murate nuns reciprocated with small tokens offered "as a sign of our affection."[61] Renaissance codes of gift-giving were firm about repaying moral indebtedness; lesser gifts could appease the creditor even if they could not fully erase the debt. Hence it was not unusual for transport costs to exceed the market value of a diplomatic counter-gift.[62] To reciprocate the queen's generosity, the Murate nuns turned to handmade gifts of a spiritual nature, whose exchange value underscores the holistic conception of Renaissance health encompassing body and soul. Prayer books were an obvious choice as counter-gift, since the convent was renowned for producing high-quality manuscripts given to dignitaries.[63] In fact, in 1501 the nuns created a book of hours—a personalized prayer vehicle much favored by noblewomen—for the queen's private oratory. The fact that this manuscript was still housed in Leonor's private library when she died attests to its personal significance.[64]

Other gifts included numerous talismans of varying sizes worked in gold and silver. These objects, called Agnus Dei, held strong eucharistic associations; they were often worn around the neck to protect wearers from harm. Their diminutive size made them even more precious as tokens of affection.[65] In 1505 the Murate nuns gave the Duchess of Ferrara a similar object, which she called "a beautiful and devout thing" and "immediately had placed around [her] neck with devotion."[66] The nuns also sent exquisite garlands of silk flowers—one of their commercial specialties—to embellish the queen's private oratory. These decorative objects track the surging interest in preventive health measures based on the non-naturals. Simply gazing on such *naturalia* was thought to refresh the senses and restore emotional equilibrium.[67] Crafting natural-looking garlands of silk flowers had become a staple of convent economies by the mid-sixteenth century. In 1548 the Tuscan convent playwright Sister Beatrice del Sera boasted about her ability to form different varieties of silk flowers "in their true likeness."[68]

Each of these handmade gifts tapped a different type of healing power within a complex therapeutic system, highlighting the multiplicity of sources on which Renaissance practitioners could draw. The book of hours channeled the power and authority of sacred words; the little talismans invoked magico-religious properties in the pursuit of health; the floral garlands incorporated naturalistic principles prominent in prevention. Rather than trying to erase disparities in wealth and power, these "little things," as the nuns called them, strategically confirmed social distinctions by acting as the material agents in the reproduction of hierarchical social relations.[69] These offerings nevertheless carried high exchange value precisely because they were meant to enhance the queen's overall well-being within holistic conceptions of health. These little things took on additional layers of significance because they were produced by the nuns' own hands in a group setting and communicated a sense of gratitude, affection, and shared religious interests. In other words, the nuns did not simply consign a random group of objects to Leonor, but instead assembled a cohesive bundle of health-related gifts that created a value-to-value exchange. The key to these perceived equivalencies was the tight connection between spiritual and physical health in Renaissance medical thinking. The high regard in which Leonor held these items is evident in the fact that she later bequeathed them to the convent she founded.[70]

By the 1520s, it had become increasingly difficult to collect on the queen's pledge. Ongoing problems with receipt prompted the nuns to importune their brokers on more than one occasion and to stimulate reciprocity by sending gifts in advance of anticipated shipments. Manuel's death in 1521, followed by Leonor's demise four years later, presented stiff challenges. Other Florentine firms in Lisbon tried to sustain the exchange after Marchionni's death, circa 1530, but failed to overcome mounting obstacles.[71] In 1548, the Florentine merchant Luca Giraldi reported to the abbess that he was unable to procure the full pledge of sugar but had managed to obtain three casks through his friends' good graces. He wrangled one last consignment from the Crown in 1559, representing five years of unfulfilled pledges from 1551 through 1555.[72] Despite continued pleas from the nuns and intermediaries, the Portuguese Crown permanently defaulted on its charitable obligation in 1564. As a court functionary explained, the king had been "impoverished by the wars in Africa waged by the Moors," in addition to other expenses.[73]

The cessation of these valuable gifts could not have come at a worse time for the nuns. The epic 1557 flood of the Arno River severely damaged the ground-floor pharmacy, destroyed most of its medicinal stock, and rendered its distilling equipment unusable. In a single blow, the convent lost key business revenues while facing heavy reconstruction costs. Seeking a way forward, the nuns turned to noble patrons for support, particularly members of the Vitelli and Cibo families, who maintained close ties with the Medici dukes. Other donations to rebuild the pharmacy came from Eleonora of Toledo, Pope Paul IV, and assorted cardinals.[74]

The gift exchanges between Queen Leonor and the Murate nuns over a crucial half century highlight the interdependence of nuns and noblewomen in exercising medical charity in a rapidly expanding world. Offering gifts to this far-flung convent extended the queen's reach beyond her immediate orbit, enabling her to display beneficence across the Mediterranean. For Leonor, medical charity functioned as both an expression of piety and an instrument of diplomacy. These charitable investments also advanced Portuguese commercial interests by widening market channels for both local products and global commodities. As an outlet for Portuguese colonial goods, the Murate exposed convent clients to the fruits of empire at an early date, but the nuns also exploited these gifts to expand their pharmacy business and local patronage networks. Their diminutive counter-gifts sustained a long-lasting medical exchange premised on an integrative concept of physical and spiritual health.

Exchanging Intimacies: Maria Maddalena of Austria and Sister Orsola Fontebuoni

The second case examined here contrasts with the previous one in several ways. Rather than entailing long distances across political borders, the gift exchange between the Tuscan grand duchess Maria Maddalena of Austria (1589–1631) and Sister Orsola Fontebuoni (1559–1639) was intensely local, covering only a few miles between Florence and the nearby subject town of Pistoia. Although the two women met in person several times, the linchpin of their relationship was an intimate correspondence marked by frequent status inversions, in which the low-born Fontebuoni often took the lead as expert practitioner. Rather than marking the introduction of new commodities into a globalizing economy, their exchange maps the enduring power of magico-religious remedies and the healing power of consolation within the broader naturalistic framework of household medical practices.

Known by her Hapsburg title of archduchess, Maria Maddalena was a prolific correspondent. In her capacity as sister to the Holy Roman Emperor, she vigorously promoted Hapsburg religious and dynastic causes in letters to political dignitaries, diplomats, bishops, and aristocratic relatives scattered across Europe. She also cultivated contacts with nuns in dozens of locales, including her relatives in La Encarnación, the royal convent in Madrid founded by her sister, Margaret of Austria, in 1611. The archduchess fielded numerous pleas for favors from nuns across north-central Italy, ranging from the appointment of a preferred confessor to requests for the famed medicinal oils produced by the court pharmacy.[75] Nor was Maria Maddalena the first Medici consort to nourish close connections between court and convent through shared medical interests. Another pair of social opposites exchanged letters about health and healing: the much-reviled Bianca Cappello, an experimenter who tested remedies on herself, with the mystic Caterina de' Ricci, a skilled pharmacist in her own right.[76]

Nevertheless, the correspondence between Maria Maddalena and Fontebuoni stands out because of its longevity, intimacy, and highly performative nature. The thirty-three surviving letters written by Sister Orsola between April 1616 and January 1627 span the archduchess's late childbearing years, widowhood, and co-regency.[77] During this time, Maria Maddalena's husband, Cosimo II, was confined to his bedchamber for extended periods because of gout, fever, tuberculosis, and gastrointestinal problems that became increasingly acute after 1614. The archduchess drew on local spiritual resources as well as court medical expertise in tending his needs, directing monks and nuns across the dominion to perform special devotions on his behalf. In fact, Orsola's first surviving letter to the archduchess both assured her of the nuns' prayers and offered medical counsel about her husband's condition.[78] The two women not only discussed treatment options but also hinted at their inner emotional states when discussing health problems. Because she was illiterate, Orsola dictated her letters to a nun-secretary, which gave these missives a fluid, conversational quality. Although she retained a sense of deference, Fontebuoni frequently underscored the special nature of their friendship, singling out the archduchess as a "refuge" second only to God. Phrases such as "I'm writing in confidence" or "I tell you in secret" heightened the sense of privileged communication between intimates. The extent to which this Hapsburg noblewoman shared these feelings must be inferred from textual traces; only three letters she wrote to the nun have been identified to date.

Despite her lofty political connections, Maria Maddalena has attracted less critical attention than other Medici consorts. The co-regency she shared with her mother-in-law Christine of Lorraine from 1621 to 1628 conventionally has been seen as precipitating the decline of the Tuscan state. The two women supposedly set Tuscany on a downward path by showing a poor grasp of foreign policy and favoring courtiers over competent officials.[79] Recent studies have offered a different view of Maria Maddalena's state-building activities. The keen self-fashioning evident in theatricalized devotions, music-making, and decorative programs promoted a public image of the archduchess as an exemplary wife and widow advancing the Catholic Hapsburg mission within a just, pious Medici state.[80]

Sister Orsola was an unlikely confidante to this rich Austrian noblewoman. Born Camilla di Giorgio Fontebuoni, she became a serving nun in the Benedictine convent of San Mercuriale in Pistoia around 1570, when she was eleven years old. Her family probably lacked the means to make her a full-fledged choir nun.[81] By contemporary standards, the venerable convent she joined remained a small community, counting only twenty nuns in 1582, far fewer than the 180 women populating the Murate in the same years.[82] Limited endowment income forced the nuns to support themselves by boarding young girls, making medicines, spinning thread, and raising silkworms inside the convent complex. The latter was nasty work commonly performed by women and children living in custodial institutions throughout Tuscany. Orsola and her coreligionists thus formed part of a low-paid labor pool on which the Medici dukes and local elites relied to advance the local sericulture industry.[83]

Fontebuoni's convent also ran a commercial pharmacy serving the general populace. Although the lack of surviving business accounts hampers analysis of its activities, the shop was renowned locally for its remedies and medicinal confections. This enterprise was successful enough to warrant expansion in the early eighteenth century, during which time one of her successors compiled an extensive recipe book totaling 260 pages.[84] This *ricettario*—one of the few surviving convent formularies—included recipes developed by a range of medieval authors, local physicians, and nun apothecaries. In typical fashion, it juxtaposed recipes for unguents, plasters, syrups, and eye washes with instructions for making perfumes, inks, and disinfectants. Whether Orsola invented any of these secrets is unknown. The pharmacy remained a thriving business right up to the French suppression of Tuscan religious houses in 1808.[85]

Fontebuoni lived in the aura of sanctity within this tight-knit community for almost forty years. Her growing reputation for visions, ecstasies, and prodigious healing powers attracted a steady flow of visitors and patrons. Luminaries like the Spanish ambassador to the Tuscan court, the Archbishop of Florence, the Duke and Duchess of Mantua, Christine of Lorraine, and Maria Maddalena herself—once with three of her young sons in tow—flocked to the convent seeking her help.[86] Among her admirers was Galileo's daughter, Sister Maria Celeste, a talented nun-pharmacist at the Florentine convent of San Matteo in Arcetri. Writing to her father in 1630, Sister Maria Celeste reported that she kept "4 or 5 letters from [Fontebuoni] as well as several other very inspirational writings of hers."[87] Given their mutual medical interests, it is possible that the two nuns circulated practical information about remedies in letters that are no longer extant. Despite her reputation as a charismatic, however, Fontebuoni made no claims to inedia or other somatic proofs of holiness like the stigmata. The fact that her ecstasies and visions eluded verification placed her in a privileged but ambiguous position. These spiritual graces prompted her coreligionists to elect her abbess in 1618, when she was sixty years old; she served in this position for eleven years, having been reelected for several three-year terms. It was highly unusual for a serving nun to hold this office, especially considering the limited voice they exercised in chapter. Class frictions between well-born choir nuns and lower-class serving nuns were rampant in early modern convents, but apparently were overridden by the spiritual and secular favors Fontebuoni enjoyed.[88]

As a recognized charismatic, Fontebuoni stood at the nexus of Catholic reform initiatives and Medici court circles. She testified at the canonization hearings of the Tuscan holy man Ippolito Galantini in 1622–1624, where she confirmed his miraculous cure of one of her convent sisters as well as her own chronic gastritis.[89] Contemporaries also considered Fontebuoni to be the "spiritual mother" of Giovanni Visconti, another court intimate who ministered to the Medici chivalric Order of Santo Stefano. Early modern holy women often inverted conventional gender dynamics by inspiring devout men to greater goodness, although several of them later found themselves charged with simulated sanctity.[90] Solidifying Orsola's position within court circles was her kinship with her cousins, the painters Anastasio and Bartolomeo Fontebuoni. The latter served as a Jesuit missionary in Goa, India, and Ceylon from 1605 until his death in 1630, in which capacity he procured exotic gifts such as ivory, bezoars, and "a very beautiful Japanese print" for the archduchess.[91]

In tending the health needs of both convent and court, Fontebuoni seamlessly melded her spiritual gifts with naturalistic healing skills. An accomplished pharmacist, she was conversant with old standbys like quince paste as well as newer additions to the European medicine chest like sassafras, taken to relieve respiratory congestion or as an adjunctive therapy for syphilis.[92] We catch a rare glimpse of Orsola at work in the convent infirmary in February 1621, when she was busy dictating a letter to Caterina Medici Gonzaga, the Duchess of Mantua, while simultaneously administering a medicament she had prepared from some type of "china root" *(cina)* to a sick nun. This substance was likely the New World herb *Smilax pseudochina,* commonly used in Italy to treat respiratory illnesses since the mid-sixteenth century, rather than the famed "Peruvian bark" (cinchona) introduced into Europe by the Jesuits circa 1632 as a cure for malaria.[93]

In exchanging remedies with the archduchess, Orsola introduced her brand of healing knowledge into learned court circles. In November 1620, the nun sent the court a batch of fortified wine in which iron nails had been steeped *(acciaio stillato),* a remedy used to treat intestinal blockages. In return, Fontebuoni received a thin medicinal syrup made from powdered gemstones whose costly ingredients would normally be well beyond her reach. Continuing the exchange, Orsola returned the favor by providing the archduchess with "two ampules filled with the liqueur of S. Niccolò di Bari to rub on your knee." In the interim, Maria Maddalena had consulted the nun about ways to relieve the arthritis afflicting her left knee. The archduchess often compensated for this condition by shifting her considerable weight onto her other leg, causing her to fall periodically. Orsola constructed a two-stage treatment plan in response to her request. She recommended first applying her prized topical remedy to the painful joint; if the archduchess's condition didn't improve after repeated use, she should pursue the more aggressive measure of bathing her legs for a half-hour daily in leftover wine pressings. It is not clear where this therapy figured into a larger hierarchy of resort involving court physicians. Despite trying the immersion treatment the following autumn, Maria Maddalena's gait remained compromised over the long term.[94]

What amplified the healing power of Fontebuoni's naturalistic remedies was the inclusion of sacred elements. Early modern nuns were renowned for practicing "between heaven and earth" by putting sacred objects to therapeutic uses.[95] Working at the nexus of religion, magic, and medicine, Fontebuoni manipulated the power of relics, blessed bread, and the sacred word, especially when treating problems associated with pregnancy and female infertility. Jacqueline Musacchio has analyzed the role of sympathetic magic embodied in Renaissance images and objects ranging from birth salvers to fertility devices, which gave women and families greater control over the dangers inherent in pregnancy and childbirth.[96] Often scorned today as superstitious, these articles constituted part of the mainstream culture of reproduction in Renaissance Italy. Charms and prayers had a secure grounding in humoral medicine and were by no means an avenue of last resort.[97] For the most part, however, quasi-magical materials used in the birthing process were associated with midwives, who commonly wrapped birth girdles inscribed with prayers or scriptural verses around the bodies of women in labor or recited the legend of Saint Margaret, patron saint of childbirth, at their bedside. Occasionally, enterprising monks and friars were criticized for exploiting their access to the sacred by producing birth charms and textual amulets for profit.[98]

The Fontebuoni correspondence offers exceptional insight not only into how these practices were used at court, but also how they were mediated by religious women. Besides treating common physical complaints, Orsola also served as a birth consultant to several Medici court women, supplying them with magico-religious remedies as well as medical advice. In early June 1617, for instance, when the archduchess was heavily pregnant with her last child, Leopoldo, Orsola sent a soothing stomach poultice to relieve her discomfort. "I don't know if Your Highness will be pleased with it," she said modestly, "but it comes recommended by the Most Illustrious Signora Maddalena Nobili, for whom I made a similar one." This was only one of the many efficacy statements sprinkled throughout her letters, indicating that a remedy had been tried, tested, and proven effective. Characteristically, Orsola shifted assertions of proof to her noble female clients as a mark of humility. The nun assured the archduchess that "there's nothing in [the poultice] that can harm you," just "powdered wormwood, mint, cloves, cinnamon, nutmeg and musk." To these fragrant substances Orsola had added some "St. Nicholas Bread and other blessed things placed in an amulet," presumably slipped inside the silk wrapper itself.[99] The latter ingredient was simply bread marked with a cross and dipped in water, but it was renowned for having miraculously cured its titular saint after a prolonged fast. Orsola did not play with fire by using a consecrated host, but her intention of harnessing the sacred is unmistakable. Six days later, the archduchess wrote that she had found the remedy to be quite beneficial and asked Orsola to make two more poultices.[100] In reply, the nun advised her that "these poultices last and are effective for three or four months, but they should be re-heated each time" by steaming them over hot Malvasia wine. "When it's hot, put it on your stomach; that way it will bring greater comfort and benefit." Orsola also instructed her to burn the poultice when it lost effectiveness "because of the holy things inside it."[101]

As an afterthought, Fontebuoni scribbled the following message on a separate slip tucked inside the letter: "it occurs to me that a small relic of the Blessed Passitea [Crogi] placed on your stomach might help."[102] This recommendation not only tapped the sacred but reinforced the intensely personal female networks fostered by the Medici co-regents. Passitea Crogi (d. 1615) was a Sienese nun-mystic and prodigious healer whose magnetic personality had attracted many Tuscan noblewomen, including Christine of Lorraine and Marie de' Medici. Crogi played a similar role as confidante to Christine as Fontebuoni did to her daughter-in-law.[103] Orsola noted that "this little relic is kept by Signora Salviati," an administrator at the San Gregorio hospital in Pistoia, indicating the circulation of sacred objects within local female circuits. Personal ties to this holy woman must have made her relics particularly meaningful within these circles. Two weeks after Leopoldo's birth, on November 6, 1617, Fontebuoni followed up with yet another gift in the same vein. This time she sent the archduchess "a little heart with relics to place around the neck of the new little prince" to ward off illness and sudden death—a common prophylactic device given to infants. She concluded by saying, "what's more, I tell you that the tiny reliquary cross" included with the letter "has special divine powers for women in childbirth."[104]

Fontebuoni's manipulation of the sacred in reproductive matters is most apparent in her attempts to help Maria Maddalena's sister-in-law, Caterina Medici Gonzaga, Duchess of Mantua, conceive an heir. By 1618 the duchy of Mantua faced a succession crisis. In January 1616, Caterina's future husband Ferdinando assumed the ducal throne after having renounced his cardinalate; a month later, in February 1616, he secretly married the young noblewoman Camilla Faà. Despite the birth of a son in December of that year, the union was declared invalid. Federico's subsequent marriage to Caterina in February 1617 failed to produce immediate issue. To compound the problem, the only remaining Gonzaga heir from the main line, Federico's younger brother Vincenzo, was married to an older woman unlikely to bear children. Consequently, Caterina was under enormous pressure to ensure the Gonzaga succession.

As early as April 1617, Caterina's mother, Christine of Lorraine, sent the first of many fertility aids, this time in the form of "a [small] bottle of oil that is regularly made by a gentlewoman of the Capponi family, which is very good for that problem as demonstrated by many trials." As one scholar has argued, such remedies relied more on their perceived efficacy than on women's claims to greater competence in matters pertaining to the female body.[105] Still, the circulation of similar fertility aids within and across generations fostered a discernable female culture of gift-giving centered on reproduction. Although Christine, her daughter, and Maria Maddalena often used gift exchange in competitive ways, in this case all three women pursued the same objective.[106] Among the items sent to Mantua from the Medici court were "a belt of an animal that came from Poland" (probably a marten) to be worn around the waist; a "pregnant stone" *(pietra pregna)* that had helped the archduchess in birthing "all her children"; and assorted talismans that could be worn around the neck or applied to the body to preserve the duchess's health.[107] Pregnant stones, commonly known as eagle stones, enjoyed broad social appeal across Europe from antiquity to the nineteenth century as a means of facilitating childbirth. These hollow concretions containing loose pieces gained added legitimacy in the Renaissance as a prized natural wonder. The Sienese physician Pietro Andrea Mattioli likened eagle stones to a womb containing an embryo.[108]

After Caterina suffered the first of several miscarriages in June 1617, she widened her healing networks to include Fontebuoni. Between June 1618 and July 1624, the nun sent numerous fertility aids consisting of textual amulets (*brevi* or *brevini*)—some apparently blessed—accompanied by small images of the Infant Jesus or the Madonna. The use of amulets in healing had a long history dating back to early Christianity. Commonly placed around the neck, amulets were "writings worn on the body for protection." Whatever form they took, amulets derived their apotropaic power from the sacred words inscribed on them. These ranged from benedictions and short prayers to snatches of Psalms, lists of divine names, and invocations to saints. Like eagle stones, birthing amulets were meant primarily to relieve labor pains and facilitate safe delivery, but they could be used to treat female infertility as well. Many surviving examples formed part of what have been called "birthing kits"—a combination of amulets and devotional objects—passed down as heirlooms from mothers to daughters, much like family recipe books.[109] Tuscan *brevi* generally utilized short scriptural verses but could be customized by invoking a saint associated with a specific malady, such as an appeal to the plague saint Sebastian on one side combined with a general prayer or rhyming incantation on the other.[110] The more diminutive Italian *brevini* were inserted into covers made of lavishly embroidered silk or fine gold filigree. These objects enjoyed such currency in late Renaissance Italy that the sculptor Benvenuto Cellini discussed techniques for fabricating "cases for *brevi* to wear around the neck" in his treatise on goldsmithing.[111]

What made amulets particularly attractive as a gift of health was their endless variability. In constructing these remedies, Orsola might select one of countless biblical passages, combine them in new ways, or compose a short prayer herself for specific occasions or types of illness. She and her convent sisters compounded their power by packaging handwritten texts with cheap print images of Jesus or saints in small silk pouches, often embroidered with gold or silver thread. These were hybrid objects that blended newer printing technologies with time-honored manuscript forms and needlework techniques. Orsola's convent produced scads of such articles for distribution to generous benefactors as a sign of their "gratitude" and "affection." Fontebuoni dispatched an entire box of *brevini* to Caterina at the Gonzaga court when the duke fell ill in July 1624; similarly, she sent amulets to Maria Maddalena whenever the Medici children took sick.[112] The nun occasionally apologized to wealthy female patrons that she and her sisters had "nothing worthy" to reciprocate their valuable gifts. Like the Murate nuns, Fontebuoni rhetorically emphasized material disparities while responding with the stuff of spiritual healing.[113]

Enhancing the potency of these objects were prayer and "active viewing" by recipients. Contemporaries believed that the maternal imagination exercised a strong influence on conception and the child's physical form. Gazing at images of healthy little boys or simply contemplating the word *maschio* (male) could work a kind of sympathetic magic in conceiving an heir.[114] Using an image of the Infant Jesus in a fertility device, as Fontebuoni did, blended the best of material and magical worlds. To be successful, however, the viewer had to actively engage her imagination in the therapeutic process. Hence when the nun sent Caterina "some of our amulets" along with a "little Madonna to affix to your rosary" in June 1618, she reminded the duchess to "keep your intentions firmly in mind" while praying.[115] Bolstering this fertility quest were the constant prayers offered by Orsola's community. She assured the duchess in July 1619 that "not a day goes by" when the nuns failed to pray that "Jesus may satisfy your desire" to have children.[116] Here the nuns' direct spiritual plea to Christ rather than a saintly intermediary reflected the resurgent trope of Christ as physician in post-Tridentine religious currents. Fontebuoni herself declared on more than one occasion that "Jesus is our doctor and our medicine."[117] In sum, as both social and spiritual objects, these ancient devices proved responsive to new technologies as well as religious trends.

As Caterina's hopes for an heir were dashed repeatedly, Fontebuoni increasingly turned to the healing power of words. "Healing eloquence," as George McClure has called it, had a long therapeutic tradition dating back to Hippocrates. Consolation improved bodily health by healing the mind and steadying the emotions. Drawing on Stoic philosophers, early Renaissance humanists like Petrarch and Salutati debated the efficacy of words in mitigating physical pain, but they nevertheless agreed in principle that rhetoric had an important place in the healing arsenal.[118] Christian precepts added another layer to this philosophical foundation. Words stressing faith in God's will and the redemptive quality of suffering offered medicine for the spirit in the face of adversity. "We don't know what Jesus intends for us," the abbess consoled the still-childless Caterina in May 1620, while reminding her that "prayer is never wasted but always achieves something."[119] Much of this spiritual counsel was formulaic. Yet these simple phrases voiced by a venerated holy woman may have worked like a healing balm in bringing psychological relief and a sense of perspective. Recognizing the value of consolation, the duchess reported that she took great comfort from the letter her mother sent following her first miscarriage.[120]

Fontebuoni ministered a similar type of healing eloquence in virtually every letter she wrote to Maria Maddalena, whether about the archduchess' health problems or her husband's chronic ailments. Her letters *were* the gift: they formed a repository of consolation and practical advice to which the archduchess could turn over and over when spirits were low.[121] Running throughout all her prayers, Orsola said, was the hope that Jesus would restore the archduchess to "perfect health" and tranquility of mind, despite her many travails. This rhetorical repetition in both spoken prayers and written letters doubled its emotional impact. Fontebuoni constantly urged the archduchess to calm her spirit by taking a wide Christian perspective. "Keep the eye of the intellect always fixed on God, like the eagle does on the sun, and this internal view brings peace and sweetness to one's heart," she wrote. When Maria Maddalena was bedridden with fever or debilitating arthritis, Orsola reminded her that earthly troubles were transitory; she also offered immediate solace by stressing that the nuns remembered her in their daily prayers.[122] Enhancing the power of this verbal salve was Orsola's own suffering. The fact that this healer was often beset by fevers, respiratory ailments, and chronic intestinal pain deepened her credibility as a spiritual counselor. Fontebuoni rhetorically modeled suffering in her letters, saying that God intended it "as a purgation for my defects and sins." She wrote movingly of the ailments plaguing her convent sisters, whose pain she would gladly embrace as her own. Maria Maddalena remarked in one of her few surviving letters to Fontebuoni that the nun's steadfast faith and soothing words buoyed her spirits, especially when her husband was incapacitated.[123]

The resulting sense of trust enabled Fontebuoni to offer candid medical advice, especially about Cosimo's treatments. Her letters make clear that the archduchess often solicited her opinions on particular medical matters. On several occasions the nun disagreed with the course of action recommended by court physicians, thereby giving this Medici consort alternatives that were consonant with her pluralistic medical values. "I tell you this in secret," she wrote to Maria Maddalena in April 1616. "I would try to dissuade [Cosimo] from the doctors' [advice], saying that he's too weak, and that if he truly needs the spa, the waters can be brought to him." Stressing the importance of the non-naturals to recovery, she continued that "if he wants to take some air, he can go to the mountains or other similar places where there is good air that will benefit him."[124] Two years later, Fontebuoni was more circumspect in her criticism. "I think the doctors know what's needed," she remarked to the archduchess; "may the Lord give you the grace to choose which remedies are most advantageous." Mindful of social boundaries, she stepped back from challenging court physicians directly by saying "I have no experience [with them] and I don't know what things" they favor.[125]

However, Fontebuoni clearly objected to the frequent bleedings prescribed for the grand duke. "About bloodletting in these times when the fury of the illness has not yet subsided," she wrote in 1618, "it's not good to let [blood], especially in a body as weakened as his." The nun continued that "I believe he has more urgent need of building up [his blood] than letting it."[126] Although she did not spell out her medical reasoning here, Orsola clearly advocated supportive treatments to assist recovery instead of a purgative regime. Presumably what she had in mind were traditional restorative measures focused on diet and rest. Although these principles had a long history in humoral medicine, they began to coalesce as a distinct field of convalescent medicine only in the late Renaissance.[127] By framing therapeutic recommendations within intimate conversations, Fontebuoni gave the archduchess specific language with which to explore various treatment options with court physicians, while still leaving the final decisions to her. Through these epistolary exchanges, Fontebuoni both supplemented and challenged professional medical advice, guiding Maria Maddalena as she sorted through alternatives. Her sustained involvement with medical decision-making as both healer and health broker underlines the continued pluralism of household medicine at the Medici court well into the seventeenth century.

The arrival of plague in August 1630 brought a dramatic end to these exchanges. Cosimo's successor Ferdinando II commissioned his librarian Francesco Rondinelli to chronicle the course of plague, along with the measures taken to curb its spread. The Tuscan state instituted standard plague controls like quarantine, which slowed economic activity as well. Established notions of contagion meant that cloth and other materials suspected of harboring the "seeds of most certain death" were seized and burned. Apothecaries were prohibited from exporting medicaments outside the city to ensure adequate supplies and avoid price gouging.[128]

Mortality rates were not exceptionally high compared with Italian cities further north, but this plague episode, which sputtered on until 1633, dramatically revealed the economic malaise into which Florence had sunk. The decline of local textile industries brought chronic unemployment that only worsened living conditions among the laboring classes. Complicating matters was the fact that the urban poor were increasingly seen as vectors of disease. The street-by-street survey undertaken by confraternal plague officials to develop support strategies not only shone a spotlight on the faces of poverty; it also unmasked an urban environment "ruptured by rotting sink-holes, drowning in ordure, suffocating on its own noxious fumes and leaking like a sieve," as one historian has put it.[129] Still, the city apparently was spared the acute destabilizing effects of plague, such as the surge in homicidal violence that shook neighboring Bologna.[130]

Long experience with plague had produced an abundance of remedies ranging from homemade cordials and "all things acidic" to prepared syrups containing ground-up pearls and gemstones.[131] The most famous plague remedy was theriac, used as both prevention and cure. Subject to strict manufacturing controls, theriac's closely guarded recipe, complex preparation, and multiple ingredients (including opium) put its cost beyond the reach of the working poor, who were most at risk in early modern plague epidemics.[132] Consequently, household practitioners devised their own prophylactic remedies using inexpensive ingredients like vinegar and rue. Use of these homemade remedies underscores the extent to which Italians across the social spectrum proactively supplemented public health measures to increase their chances of survival. One such medicament was prepared by Sister Maria Celeste Galilei, who sent her father "two small jars of electuary for safeguarding against the plague" that she had made herself. One jar held a thick mixture "composed of dried figs, nuts, rue and salt," bound together with honey. She advised Galileo to take "a dose about the size of a walnut" every morning before eating, followed by some good wine, which "provides a marvelous defense."[133] Occasionally, Venetian families obtained patents for "secret" plague cures after demonstrating their efficacy in the city's plague hospitals.[134] This enormous variety of remedies nevertheless masks a persistent medical conservatism in treating plague throughout the early modern period.[135]

As the epidemic tightened its grip across Tuscany by late summer, Abbess Orsola began dispensing her own preventive remedy as a charitable gesture. This was a special potion made with water from the convent well, which reportedly had been blessed by the miraculous intervention of Saint Benedict. Focusing on prevention as the most effective tactic, Fontebuoni maintained that drinking the liquid "would save one from the plague," although she made no curative claims for it. Not surprisingly, large crowds hurried to obtain the potion. Maria Celeste Galilei herself put great stock in this remedy and managed to procure some for her father through monastic networks. Just a few days after delivering her own homemade medicament, she sent him "a small quantity of the healing potion made by Abbess Orsola of Pistoia." As she explained, "I was able to get some as a very special favor, since, as the nuns of that convent are prohibited from giving it out, whoever obtains any clings to it like a holy relic." She urged Galileo "to take it with great faith and devotion . . . as the most effective preventive sent to us by Our Lord," and further implored him "to put your faith in this remedy."[136]

Fontebuoni's expansive claims for her own tonic soon attracted the attention of the Holy Office, which opened inquisitorial proceedings against her on the grounds of feigned sanctity. By the early seventeenth century, ecclesiastical authorities had grown less receptive to ideas of female sanctity. In contrast to the numerous holy women recognized by the Church in the thirteenth and fourteenth centuries, the "ideal" Counter-Reformation saint was a male cleric of noble origin. Such an exemplar stood at the center of the ecclesiastical establishment rather than threatening it from the edges. As the post-Tridentine Church struggled to reassert its authority amid deep religious divisions, lower-class holy women like Fontebuoni presented a special challenge to the aristocratic church-state partnership taking shape on the Italian peninsula.[137] A host of female charismatics and healers like the abbess quickly found themselves charged with "pretense of holiness," which first emerged as a distinct crime in the early seventeenth century. The result was a flurry of investigations conducted by the Holy Office into persons thought to be "either deluded or attempting to perpetrate a fraud."[138]

The lengthy inquiry into Fontebuoni's visions, ecstasies, and healing activities was spearheaded by Alessandro Del Caccia, reforming bishop of Pistoia. During the trial, Maria Maddalena did nothing to assist her former confidante; nor did the abbess's clerical protégés come to her aid. Such a stunning failure of nerve was not uncommon. Noble protectors frequently abandoned those accused by the Roman Inquisition or even contributed directly to their prosecution if they felt betrayed by false spiritual claims. At the end of the proceedings, which ran from October 3 to November 28, 1630, Orsola was found guilty of simulating sanctity. A contemporary chronicler insisted that she had acted without malice, however, and speculated that the lack of a good spiritual director may have contributed to her "illusions."[139] Following the judgment, Fontebuoni was deposed as abbess and imprisoned in the convent jail, where she remained for the last nine years of her life. The Holy Office also ordered that "all of the writings" in her possession, "which were numerous," be burned immediately in the public square. Furthermore, it interdicted "all of the amulets, rosaries and images blessed by this nun" that had been distributed to "great outside personages whom she held in friendship."[140] When the eighty-year-old nun died nine years later (January 27, 1639), she was honored with a splendid funeral that attracted large crowds hoping for some miraculous sign. In this they were disappointed: as the same chronicler noted, "Her body gave no indication of sanctity whatsoever, as many had expected."[141]

Fontebuoni's downfall stemmed from a confluence of several factors—the gradual breakdown of Medici political favor, an emboldened local Church apparatus, deepening distrust of female sanctity, and economic woes exacerbated by plague—rather than a general loss of faith in magico-religious remedies. Certainly her requests to the archduchess had become bolder and more insistent over time, shifting what had been a traditional gift register into a more coercive vein.[142] In August 1618, several years into their relationship, for instance, she asked to see Maria

Maddalena's newest religious acquisition—the *Volto Santo,* a small painting replicating the Veronica veil in St. Peter's—which the archduchess generously dispatched via messenger.[143] It was not long, however, before Fontebuoni asked her to intervene in a local property dispute, help recover overdue debts, and sway ecclesiastical authorities so that convent novices could profess earlier than Tridentine norms allowed.[144] Although these favors were the very stuff of traditional patronage relations, the lack of a strong religious rationale suggests that the nun may have begun to overreach in the years leading up to plague. In addition, the failed outcome of earlier fertility quests probably weakened her reputation with Medici court protectors.

A decisive factor contributing to Fontebuoni's ruination was the political theatre surrounding the canonization hearings for another female healer, Domenica da Paradiso (d. 1553), taking place at the same time. The Medici court had avidly promoted the cult of this visionary nun for some years; several Medici princesses had sent votive offerings and prayed on their knees at Domenica's tomb long before plague broke out. In 1621, Christine of Lorraine placed one of her own daughters in private apartments attached to the convent of La Crocetta that Domenica had founded, and even testified on the holy woman's behalf.[145] In addition to this obvious Medici court backing, Domenica's cult also enjoyed great support from the Florentine populace. The most compelling testimonies of her healing powers issued from the current epidemic. Bread that had touched Domenica's exhumed body was boiled into a soup that miraculously cured many artisans and the plague-stricken poor. Reflecting the importance of household care routines, it was the mothers, wives, and daughters of the afflicted who organized the treatment regimen, which carried clear eucharistic overtones.[146] Nuns from La Crocetta further advanced the cause of their own venerable healer in this competitive thaumaturgic environment. Physicians who had examined Domenica's uncorrupt body lent additional weight to the inquiry. Like the later canonization hearings for Caterina Vigri in the 1670s, medical expertise played a crucial role in transforming miracles "from objects of faith into objects of knowledge."[147] When formal beatification proceedings for Domenica opened on November 15, 1630, Fontebuoni's trial was still in full swing and plague mortality at its peak. Supporting the charismatic claims of the now-suspect Fontebuoni threatened to derail these plans, in which the Medici regents had invested so much time and energy. In this contest between local female plague healers, the long-dead figure of Domenica da Paradiso won out as the safer, more established alternative. Despite being at the center of religious controversy in her own day, Domenica had become a scrim upon which different historical actors—female co-regents, opportunistic nuns, well-meaning relatives, among others—could project their own ambitions and political agendas.

Renaissance gifts of health spanned an enormous range, from globalized commodities and magical remedies to the priceless offerings of consolation. The two case studies presented here show that nuns and noblewomen used these gifts as a form of political currency to magnify their public presence and extend their social reach. Shared charitable investments deepened the interdependence of these two groups while integrating them more fully into existing circuits of power. As religious and charitable institutions, Renaissance convents simultaneously mediated the performance of aristocratic largesse and served as important distribution points for medical goods and products. The commodity exchange between Portugal and Italy examined in the first case helped develop the market for sugar and other imports on the Italian peninsula; these transactions lift up women's under-studied involvement as secondary distributors within global trade mechanisms. The "little things" Florentine nuns offered the Portuguese queen in return acted as both material and discursive agents in reproducing hierarchical social relations. Yet at the same time, their handmade counter-gifts of prayer books, talismans, and *naturalia* targeted specific health-related objectives. What made it possible to conduct this type of value-for-value exchange across different gift registers was a holistic conception of health uniting body and soul, on the one hand, and the sense of community created by shared religious tastes, on the other.

The second case offers a different angle on medical gift-giving that highlights the value of the intangible as well as the circulation of treasured objects within local female circles. By extending solace as well as therapeutic advice, Orsola Fontebuoni gained the trust of prominent female consorts, which affected not only their personal well-being but also the political fortunes of the Tuscan and Mantuan states. In offering her own brand of healing eloquence, the nun's words and letters became the gift itself. At the same time, Fontebuoni trafficked heavily in the unseen healing powers from which magico-religious remedies drew. The routine use of amulets and fertility aids at the highest levels of Italian society testifies to their integral place within the Renaissance healing armament. Fontebuoni refreshed some of these time-honored remedies by integrating print technologies and new religious currents into their composition—a repackaging that casts her as a knowledge-maker of some sophistication, despite her illiteracy. Her downfall also provides important insight into the structural vulnerability of some female alliances, which could unravel quickly in the face of ecclesiastical opposition and conflicting political demands. Taken together, these case studies give greater analytical purchase on how early modern Catholic women used gifts related to the pursuit of health to exercise agency, communicate affection, and create exchange networks of considerable cohesion and importance.

CHAPTER THREE

The Business of Health

Convent Pharmacies in Renaissance Italy

In recent years the history of Renaissance pharmacy has acquired a brand-new look. Moving away from organizational structures, recent studies have investigated the spaces, networks, and social identities associated with this healing art. Early inquiries in this vein depict pharmacies as vibrant nodes of sociability and exchange in Italian Renaissance cities.[1] On occasion, these same spaces became hotbeds of religious heterodoxy. The Venetian pharmacy of the Two Doves, for instance, hosted one of the most active centers of Protestant worship in the city, providing a gathering place for heretical preachers, physicians, and other medical men to discuss and disseminate reform.[2] Other studies have explored the distinctive consumer experience of shopping in pharmacy spaces marked by neat rows of jars and flasks, specialized cupboards and drawers, boxes of ingredients and heady medicinal scents. Late Renaissance apothecaries were among the first professionals to enhance this consumer experience by setting up chessboards, stocking cabinets of curiosities, and even playing music in their dispensaries.[3] Other pioneering research has explored the diversity of vendors populating the medical marketplace, ranging from itinerant mountebanks to flamboyant figures like Leonardo Fioravanti, the self-styled "professor of secrets."[4]

Recent studies also have established that early modern women were among the most avid and skilled practitioners of pharmacy. European women from different walks of life fabricated an enormous range of remedies in courts, convents, noble households, urban kitchens, convalescent hospitals, and custodial institutions. This abundant new evidence forces us to rethink conventional narratives about women's marginalization from early modern medical practice.[5] For the most part, however, scholars have explored these activities in relation to gift-giving, domestic duties, or charitable ventures. Their commercial dimensions, especially as part of local health markets, remain poorly understood. The extent to which female purveyors clustered in specific market sectors or engaged with new commercial practices and large-scale regulatory mechanisms has not been explored in any depth. In this chapter I take a first step in that direction by situating female medical agents within the business of health in late Renaissance Italy.

Early modern women retailed medicines through three main avenues: as wives and widows of licensed apothecaries who maintained the family business; as small-scale vendors who worked in the interstices of guild structures; and as religious women whose medical activities knit together charity and commerce. European guilds customarily extended corporate standing to apothecaries' wives and widows, who often possessed valuable managerial skills. In France and England, widows could inherit their husbands' apothecary shops, but were required to retain an experienced journeyman to supervise operations. A few widows trained new apprentices.[6] Early modern guilds increasingly took on the features of family businesses by transferring corporate privileges to daughters as well.[7]

Although the fluidity of occupational labels makes quantification imprecise, the number of female guild-affiliates probably expanded as early modern medical corporations brought new types of bodywork under their purview. The twenty-three women enrolled in the Florentine Guild of Physicians and Apothecaries between 1596 and 1610 represented an eclectic mix. They included hat sellers, hairdressers, and women who "washed hair in the piazza"; their inclusion among the rank and file underscored new hygienic practices. Guild membership extended to a servant who "sold powder for mange" as an approved sideline, along with female shopkeepers who retailed wellness products, distilled medicines, and confections.[8]

Working alongside them were small-scale female vendors who sold their goods within an economy of makeshift. Their numbers and products varied considerably, in keeping with life circumstances and local demand. As they move in and out of the market, their voices are heard most often in legal proceedings such as Inquisition records. The story of Maddalena, a middle-aged widow who ran a cosmetics business out of her home in early seventeenth-century Rome, could be multiplied many times over. Maddalena drew on much the same production methods and skill set as affluent female householders when making grooming products and love potions; she marketed these wares through wide social networks developed via other economic activities ranging from weaving to procuring.[9] A more regular presence in local markets was evidenced by the single women who supplied medicinal plants as part of an institutional healthcare matrix. The hospital accounts of Santa Fina in the Tuscan town of San Gimignano record scads of payments to local women working as "herb sellers" or "root dealers," some of whom provisioned materials in exchange for tenancy agreements.[10]

Within the Italian context, nuns were the largest, most visible group of female medical purveyors, and their numbers were not negligible. Italy had the highest proportion of women living behind convent walls anywhere in Europe between 1500 and 1650. One in every eight Florentine women in 1552 was a nun; more than half of Venetian patrician women had entered religious life by the 1580s, with numbers rising even higher over the next century. Local variants of this pattern took root in Milan, Bologna, Naples, and other Italian cities. With thousands of women sequestered in convents and enormous resources devoted to their maintenance, female monasticism became a defining feature of late Renaissance Italy.[11]

Renaissance nuns continued a long tradition in herbal medicine by preparing and administering remedies to members of their own communities. The Bolognese holy woman Caterina Vigri (d. 1463) reportedly "carried around a little box containing medications of her own" to treat various maladies. One of her convent sisters remarked that "hardly a day would go by" when Caterina was not "employed in treating now two, now three, now four" of her fellow nuns, "some in the hands, some in the ears, some in the mouth."[12] At times, nuns put their medical knowledge to probative uses that established the sanctity of their coreligionists.[13]

What was new about this therapeutic tradition was its commercialization after 1500. In the sixteenth century, Italian nuns seized on the burgeoning interest in medicinals to develop new revenue sources and enlarge their charitable scope. By making and selling remedies to the public, Renaissance religious women both augmented the medical resources available in Italian urban society and acquired roles of public significance beyond the spiritual realm. Convent pharmacies across the peninsula continued to grow in both number and reputation throughout the seventeenth century, at times outstripping the medical resources provided by monks and friars. A 1695 public record of "all the places that practice pharmacy in Rome" named thirty-four pharmacies run by nuns, compared with only eleven affiliated with male monastic houses.[14] In numerous rural locales, religious women constituted the backbone of public healthcare.

Studying convent pharmacies holds obvious interest for understanding the nature and extent of women's medical agency, on the one hand, and the production of specialized medical knowledge outside university and guild settings, on the other. These female-run businesses demonstrate the extent to which Italian women were deeply implicated in early modern medical discourse as well as commercial markets. Like other pharmacies of their day, convent apothecary shops served simultaneously as sales outlets, distribution points for medical advice, and production sites for making medicines. It has long been recognized that monastic houses figured prominently in local health markets and medical provisioning. In fact, they were among the first retailers to promote brand-name wares. Yet while pharmacies run by monks and friars have dominated the historical literature, records kept by Renaissance nuns—account books, chits, memoranda, letters—paradoxically offer more comprehensive information about products and marketing strategies. Especially valuable are Florentine convent sources, which open fresh insights into the pharmaceutical trade in a city renowned for commercial and technological innovation.

In this chapter I probe the commercial dimensions of Florentine convent pharmacies and the distinctive contributions they made to local medical provisioning. Why did nuns enter the medical marketplace around 1500? What conditions fostered their continued involvement in the business of health? Examining the spatial distribution of pharmacies in the early modern city shows that Florentine convents provided crucial healthcare resources to the urban periphery, home to dense concentrations of the working poor. An in-depth study of nuns' products, clients, suppliers, and the regulatory landscape in which they worked reveals that these women engaged in commercial innovation ranging from product development to marketing practices. Fiercely protective of their shops, Florentine nuns negotiated conflicts between reform ideals and market forces, on the one hand, and between powerful families and an emerging state apparatus on the other. In so doing, they dramatically shaped the business of health in the late Renaissance city.

Entering the Market

The commercialization of Italian convent pharmacy after 1500 was a complex response to external pressures and internal strains on convent budgets. Lacking sufficient income from their endowments, most nuns across the Italian peninsula had performed various types of market work to make ends meet since 1400. In fifteenth-century Florence, nuns played an essential role in the development of the silk industry by providing a cheap labor pool for reeling silk and spinning gold thread.[15] As convent populations ballooned after 1500, existing revenues were stretched even further, forcing nuns to look for openings in other economic sectors. Pharmacy was one such growth area, much as textile work had been a century earlier. Since diversification was a core principle of mixed convent economies, working in a different market sector buffered nuns against shifts in the textile industry, their principal source of wage earnings. These new investments in medicine proved to be well-founded, since income from embroidery and certain types of silk production declined after 1550.

A range of external factors also spurred nuns' entry into the marketplace. Changing disease environments, the crushing need for poor relief, and the renewed concern with healthy living all accelerated the global trade in medicinals. After 1500, Italian cities faced mounting challenges in the form of recurrent famines, foreign invasions, plague epidemics, and severe unemployment stemming from the collapse of traditional textile industries. Other developments inviting new practitioners into the marketplace included broad shifts in piety that emphasized social welfare over older forms of penitence. The charismatic Dominican preacher Girolamo Savonarola (d. 1498) highlighted medical charity as a spiritual duty, while the "new philanthropy" of Catholic reform placed charitable outreach at the center of spirituality.[16] In other words, the emergence of convent pharmacies after 1500 was a constituent part of the restructuring of social welfare mechanisms and the rationalization of medical resources occurring throughout the Cinquecento.

There certainly was room for new pharmaceutical initiatives across the peninsula in the early sixteenth century. In Florence, a relatively small number of apothecaries served an urban populace of 60,000 to 70,000 people. Apothecaries had declined steadily in number since the early 1400s, reaching a low point by the mid-sixteenth century, when a guild census counted "46 or more" apothecaries in the city—roughly one shop per 1,500 inhabitants.[17] This ratio was higher than in Venice, famed for its pharmaceutical culture; in 1565, seventy-one pharmacies served a population of roughly 172,000, or one shop for every 2,422 residents. These low ratios opened opportunities for other vendors to fill perceived gaps in terms of products or accessibility.[18] Renaissance Rome boasted the highest density of pharmacies—one shop for every 589 residents in 1526—owing to the huge influx of pilgrims and tourists streaming in to the papal city in need of medical assistance.[19]

Monastic pharmacies supplemented guild resources in all three cities. The Dominican friars of Santa Maria Novella in Florence began selling medicines as early as 1381 in response to recurrent plague episodes. However, they launched a full-scale commercial operation only in 1542, when they opened a new dispensary; even then, business remained modest. The pharmacy gained state support as part of an emerging public health apparatus only in 1612. Pharmacy specialties ranged from soaps, confections, and perfumes to plasters, unguents, and electuaries.[20] Similarly, the Dominican friars of San Marco became commercially active after 1450, selling various stomach tonics, herbal liqueurs, and the brilliant red elixir called alkermes, reportedly prized by Lorenzo the Magnificent. South of the city, the Vallombrosan monastery of Camaldoli marketed remedies out of its infirmary from the 1460s.[21]

Although these and other Italian monastic pharmacies developed robust reputations over the years, spotty business records make it difficult to reconstruct their activities during this time.

One important distinction between monastic pharmacies and female-run shops should be noted at the outset. Male houses frequently hired lay apothecaries to grow their businesses, whereas convents did not. In 1612, Santa Maria Novella paid the guild-licensed apothecary Simone Marchi a monthly stipend of three scudi to manage their shop. Marchi devised some new products that used imported medicinals but remained on-site for only one year; he trained an apprentice to continue his work. The monks and friars of Santa Trinita and Camaldoli had followed the same path in the sixteenth century.[22] Similarly, the monks of San Pietro in Modena hired the local apothecary Giovanni Francesco Migliarini in 1610 to operate their shop, paying him an annual salary of fifty ducats. Migliarini's contract stated that a monk had to be present when remedies were compounded and further stipulated that products had to be priced below other pharmacies—a major bone of contention with guildsmen. Finally, monks and the poor were to be furnished with medicines at no cost.[23]

By contrast, nuns were forced to be more self-reliant in making and marketing medicines. Because church and state officials tried to curb social interactions between nuns and laity, religious women never hired lay pharmacists to expand their enterprises but instead were left to their own devices. To be sure, they relied on outside suppliers to procure a wide range of materia medica and garnered technical information from friends and relatives. Yet, unlike their male counterparts, Renaissance nuns controlled the entire production cycle in order to protect reclusion. Making medicines also gave nuns greater commercial freedom than doing silk work, since selling remedies directly to the public eliminated mediation by guild brokers.

In some ways, it is not surprising that Renaissance nuns jumped into the medical marketplace after 1500. Religious women enjoyed a competitive advantage over itinerant empirics and charlatans who hawked their wares in the streets. They capitalized on business networks developed through the textile trades and book manufacture over the preceding centuries, which helped them scale up pharmaceutical production quickly. Moreover, the majority of convents claimed a large physical complex where medicines could be distilled in volume and stored on-site, along with raw materials. In other words, most Italian convents already possessed the infrastructure needed to succeed in this business sector. From the standpoint of human capital, convents housed a significant pool of literate women easily trained in this healing art, while their institutional culture guaranteed the transmission of craft knowledge. In addition, these communities boasted extensive patronage networks and a socially varied clientele. Above all, their religious status gave nuns legitimacy as healers that other female practitioners lacked.

At least nine Florentine convents scattered around the urban perimeter launched commercial pharmacies in the early sixteenth century (Fig. 3.1). Disparities in available sources—account books, convent chronicles, guild petitions—mean that some shops are better documented than others. All nine convents housed substantial populations, ranging from fifty to two hundred women. On the leading edge of commercial pharmacy were the Benedictine nuns of Le Murate, discussed in Chapter 2, and the lesser-known house of Sant'Agata. They were quickly joined by two Franciscan communities (Sant'Orsola and San Francesco), and five groups of Dominican nuns and tertiaries (Santa Caterina da Siena, San Domenico, San Jacopo, Santa Lucia, and San Vincenzo, called the Annalena). The prevalence of Dominican houses stemmed from the order's established role in pharmacy, their association with the Savonarolan movement, and internal practices. It was customary for Dominican sisters across Europe to pay their own medical bills, rather than being supported by the communal treasury.[24] Consequently, these nuns already engaged with local health markets more extensively than did women belonging to other orders.

In addition, several convents located outside the city walls, such as Monticelli and San Matteo in Arcetri—home to Galileo's daughter Maria Celeste—sold medicines to the public, distributed them freely, or included remedies in gift exchanges.[25] Nunneries in the neighboring Tuscan towns of Prato and Pistoia launched similar commercial ventures, with local hospitals following suit in the same decades. The hospital of Santa Fina in San Gimignano, built in the mid-thirteenth century, opened a pharmacy serving the general public around 1507, after absorbing the resources of two smaller institutions.[26]

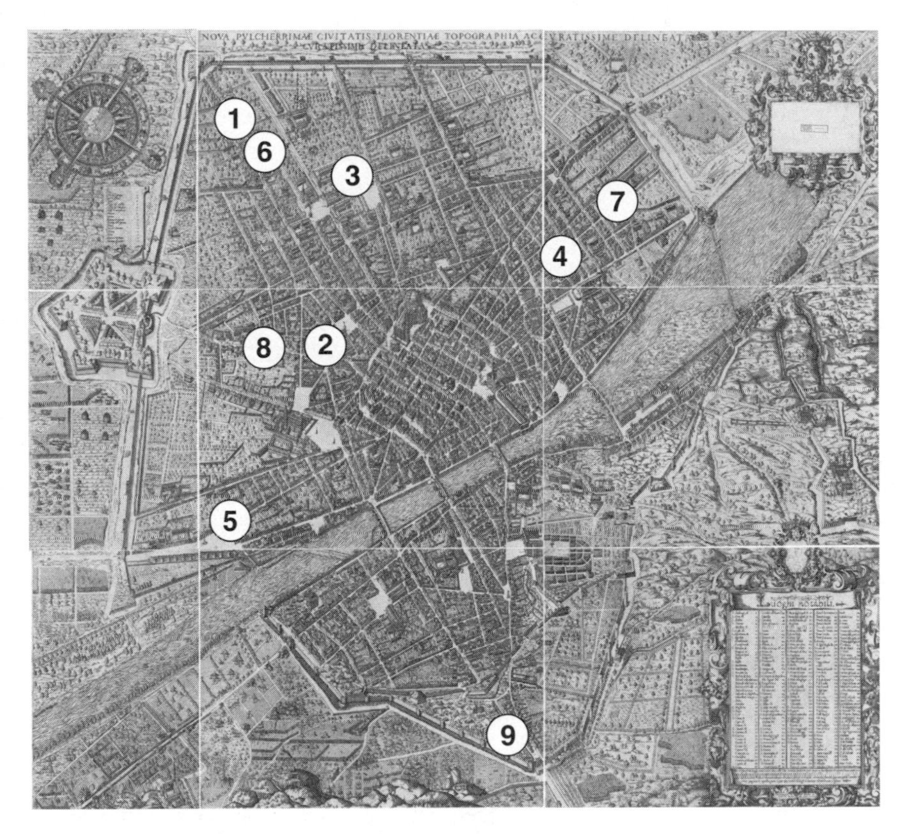

Fig. 3.1. Map of major Florentine convent pharmacies, circa 1584.
1. S. Agata. *2.* S. Caterina da Siena. *3.* S. Domenico. *4.* S. Francesco. *5.* S. Jacopo di Ripoli. *6.* S. Lucia. *7.* Le Murate. *8.* S. Orsola. *9.* S. Vincenzo, called the Annalena.
Courtesy of the Harvard Map Collection and the Emory Center for Digital Scholarship

Thinking about the city spatially helps explain why these particular convents entered the Florentine health market and how they adapted to localized conditions in competitive ways. The ducal census taken in 1561 maps the location of shops, trades, and residences with exceptional precision. In contrast to trades like tanning that were purposely sited on the periphery for hygienic reasons, pharmacies ideally would have been distributed throughout the city. That spatial pattern would have put apothecaries within easy reach of clients seeking urgent care. Yet, as Burr Litchfield has shown, more than three-quarters of the forty-five guild-registered pharmacies in 1561 were concentrated in the downtown business district ringing the Old Market; the rest (24 percent) were situated in an intermediate zone between the city center and urban periphery. Four shops stood within the Old Market itself, with nine more nearby. Another two stood on the street bearing the apothecaries' trade name, the Via degli Speziali Grossi, running directly off the Old Market. Others were located on streets leading to bridges spanning the Arno or at important intersections, where they were likely to attract foot traffic.[27] This dense concentration in the city center was largely a vestige of medieval guild life. As street names indicate, trades were highly localized in Italian medieval cities—a spatial practice that enabled efficient distribution of goods and services and reinforced a sense of occupational identity.

The clustering of Florentine pharmacies in town created real gaps in healthcare provisioning on the urban periphery. Although this area was poor in pharmacies, it was rich in convents, since numerous communities had been sited on the outskirts to escape urban hubbub.[28] Property values also were lower in neighborhoods between the second and third circuit of the old city walls, making them home to vibrant communities of day laborers, foreign migrants, and working poor. Prostitutes—often considered vectors for venereal disease—could legally ply their trade on certain streets in these zones after 1547.[29] Yet not a single guild-licensed pharmacy served the urban periphery in 1561, where healthcare needs probably were greatest because of poverty and the sex trade. This lack was particularly noticeable on the fringes of populous parishes like San Lorenzo, Sant'Ambrogio, and San Frediano stretching toward the walls. The entire working-class quarter of the Oltrarno, especially near the southern gate of Porta Romana, was drastically underserved.

Florentine convent pharmacies stepped into this spatial breach. The proximity of these institutions to prostitutes and the poor proved advantageous from both charitable and commercial standpoints. Convent decisions to invest in pharmacy not only embodied broader spiritual values, but also reflected specific management strategies for limited resources. Giving medicines freely to the poor required wealthy benefactors and regular clients to underwrite expenses. Not all convents possessed the resources or were prepared to risk precious funds, irrespective of location. Still, these pharmacies were well placed to attract paying customers as well as the poor. On the east side of town, the Murate served textile workers and migrants living in the parish of Sant'Ambrogio; the tertiaries of Sant'Orsola on Via Guelfa were perfectly positioned to dispense medicines to sex workers concentrated on that street. Several Dominican convents located on or near Via San Gallo—a major urban corridor—supplied medicines to the northern and western reaches of the city, as well as to the densely populated parish of San Lorenzo. Working-class residents of the Oltrarno apparently remained underserved, receiving medical help primarily from the Annalena nuns or from convents sited outside the walls.

These spatial patterns also help to explain why some convents, like the wealthy old house of San Pier Maggiore, had little incentive to develop a new business. In-town districts were flush with guild shops and remained legally off-limits to prostitutes. Under these circumstances, there was little incentive to invest precious resources in new business ventures. Moreover, old convents in the urban core were hemmed in by other buildings, leaving little space to add a pharmacy workshop rife with fire hazards. In mapping urban healthcare resources, it is worth noting that the major hospitals of Santa Maria Nuova, San Matteo, and Messer Bonifazio did not market medicines to the general public but dispensed them exclusively to patients. In sum, there was both a pressing need and a golden opportunity for convent practitioners to enter the medical marketplace after 1500.

The Murate nuns were probably the first to embark on commercial pharmacy (no. 7 on Fig. 3.1). Neither pharmacy ledgers nor general convent account books survive for this period, but the rich convent chronicle penned by Sister Giustina Niccolini in 1597 offers some clues to early activities and the disposition of work spaces. Using income from other cottage industries—sewing, spinning, copying, craftwork—the nuns constructed a new ground-floor infirmary in the 1460s.[30] This healing space for sick nuns anchored later commercial expansion. Once the infirmary was complete, the nuns added "the lower rooms of the pharmacy, [which were] fitted out for distillation and other necessary things, with the equipment and furnishings used in those endeavors." The affluent Benci family built private chambers over the still room as a spiritual retreat for female relatives, indicating that the initial production zone lay outside the designated areas of *clausura*.[31] Queen Leonora's generous gifts of sugar and spices discussed earlier probably were decisive in lifting medicinal production to the level of visible commercial exchange in the early Cinquecento. Further extensions of the business shared a common timetable and rationale with other Florentine convents. In the 1520s, Abbess Bonifazia Risaliti aimed "to improve her convent in ways that would be useful and profitable." After consulting with the convent confessor, Jacopo Manelli, the abbess elected to make "some improvements in our pharmacy," especially in "the earthenware vessels used to confect [remedies] and other pertinent things used in that craft." Additional funding to upgrade equipment came from the convent's longtime surgeon, Maestro Baccio, who endowed the pharmacy with twelve scudi annually to enhance healing efforts.[32]

A clearer picture of how and why Florentine nuns ventured into commercial pharmacy emerges from accounts kept by the Dominican tertiaries of Santa Caterina da Siena (no. 2 on Fig. 3.1). Founded in 1496 at the height of the Savonarolan movement and officially chartered in 1509, this convent is best known for its rich musical and artistic culture, especially as home to Sister Plautilla Nelli, the "first woman painter" of Florence.[33] From its inception, Santa Caterina was supervised by the friars of San Marco, situated across the square bearing their name. Their intertwined networks and shared charitable ideals proved decisive to the pharmacy's foundation. One of the principal goals in setting up shop at Santa Caterina was to make medicines that could be distributed to the poor at little or no cost in a circular moral economy. Ideally, expenses would be underwritten by spiritually engaged supporters. Savonarolans considered this kind of poor relief central to a righteous republic that cared for its most vulnerable members. Since the community initially owned little rental property, making medicines enabled the convent to generate much-needed revenue. The Santa Caterina pharmacy thus united commercial and charitable impulses into a single endeavor.

Guided by their confessor Fra Mariano Ughi, the nuns invested roughly ninety florins to launch their business in 1515. They purchased necessary stock and equipment from Bernardo Mini, an ardent Savonarolan and noted apothecary hailing from a long line of medical practitioners.[34] Within two years, the Santa Caterina nuns were selling herbal distillates. Within a decade, they had added a full line of ready-made items such as syrups, oils, electuaries, unguents, powders, pills, purgatives, and medicinal waters that traditionally stocked the shelves of Renaissance pharmacists.[35] Other committed Savonarolans financed the venture in its early years. Among them was Battista Landucci, whose father Luca was a well-known pharmacist, chronicler, and avid supporter of Savonarolan welfare projects.[36] One of the first women to train in this healing art at Santa Caterina (a process described in Chapter 4) was Sister Bartolomea Bettini, a young nun in her twenties. Not only was her father Piero deeply involved in the cause; five of her brothers also became friars at San Marco.[37] These Savonarolan connections continued to shape the development of the pharmacy throughout the sixteenth century, which in turn made the convent a model for other affiliated communities.

By reinvesting earnings from lucrative silk work, the Santa Caterina nuns quickly transformed the pharmacy into a profitable venture that supported living expenses and building campaigns. By 1525, the pharmacy was already generating 6 percent of earned income to a mixed convent economy; silk work remained dominant, accounting for roughly half of earned income. Monastic dowries still made up the bulk of revenues as the community accepted more than a hundred new members by 1520.[38] Spinning gold thread was more profitable than making remedies, but it was saddled with moral objections; by contrast, making remedies met a core charitable objective. It is impossible to gauge the full contribution of pharmacy production to convent budgets in the early sixteenth century without civic tax reports or visitation records, but it is clear nevertheless that this newly commercialized activity enabled the nuns to cover shortfalls, purchase property, and diversify their portfolio.

Over the next half century, the Santa Caterina shop grew in economic significance and expanded its scale of operation. In 1570, five years before enclosure was imposed on them, these Dominicans made an impressive capital investment in their apothecary business. That year, the nuns purchased a large house on Via Larga (now Via Cavour), contiguous with their complex, for 1,150 florins—the rough equivalent of twelve monastic dowries—to serve as the workshop's expanded premises. An additional 250 florins went toward remodeling the property for commercial use. This location ensured that customers enjoyed walk-in access to the shop from a major civic artery running north from the cathedral to San Marco.[39]

Siting convent pharmacies was tricky. Shops had to be readily accessible to customers, yet remain outside designated areas of enclosure. In 1577 the influential reformer Carlo Borromeo described the ideal location of the pharmacy within the convent complex. One of his principal concerns was the noise generated by collective labor in the workshop, which might interfere with the nuns' liturgical duties. Borromeo recommended that "the pharmacy should be located as far as possible from the church, the workplace, and in general from all parts of the monastery which should not be bothered by noise." He went on to explain that "the many mortars in a pharmacy, where herbs are pounded and medications prepared, together with the frequent conversations, make it anything but silent."[40] Each convent navigated business needs and clerical recommendations differently, depending on site considerations and level of ecclesiastical supervision, especially as Tridentine rulings took hold by 1600. At the rural convent of San Benedetto in Montefiascone near Viterbo, the nuns relocated the pharmacy in 1654 from its site in an outbuilding to the convent's ground floor to improve customer service. This move forced them to make costly structural modifications in order to accommodate lay visitors while preserving enclosure. Unfortunately, problems with this ground-floor site soon became apparent, as high humidity levels began to damage wooden pharmaceutical boxes and their contents.[41]

Pharmacy work not only raised revenues but also brought other rewards to Santa Caterina in the form of craft skills transferable to artistic production. Painting and pharmacy shared many of the same materials and techniques; dragonsblood, for instance, could be used as both medicine and pigment. In Florence, dealers who traded in materia medica also retailed pigments and other substances used to prepare canvases or make varnishes. In this respect, the Florentine commodities market differed from Venice, where pigment sellers *(vendecolori)* were organized into a separate trade.[42] Moreover, technical skills learned in the pharmacy gave nun-apprentices a deeper understanding of how certain materials behaved when mixed, pulverized, or manipulated by heat. This introduction to various processes grinding, boiling, mashing, straining, mixing organic and inorganic materials—worked in synergy with other crafts, such as making altarpieces, refurbishing frames, casting reliefs, and sculpting plaster.

Given its thriving pharmacy, it is not surprising that Santa Caterina also housed several important artistic workshops in the sixteenth and seventeenth centuries. Most notable was the workshop headed by Plautilla Nelli, known for monumental paintings of biblical and religious subjects. After her death in 1588, Nelli's disciples successfully sustained the workshop by working in various mediums. Their output ranged from sculpted figurines of angels and saints to illuminated manuscripts, small devotional paintings, and altar frontals, all marked by high-quality workmanship.[43] By the 1590s, the nuns of Santa Caterina had become the artists of choice for producing small religious pictures used by novices at the neighboring house of San Domenico.[44] The concurrent operation of these workshops not only points to an extensive, reliable network of suppliers and clients, but also indicates an internal management system of some complexity.

Other local convents encouraged similar synergies between workshops. In the 1460s, Filippa de' Medici shuttled between the scriptorium and pharmacy at Monticelli, where she put her knowledge of blending medicines, inks, and pigments to multiple uses.[45] As nuns' pharmacy became more commercialized in the next century, the interplay among artisanal activities intensified. By the late sixteenth century, the nuns of Sant'Agata staffed both a thriving pharmacy and a vibrant painting workshop whose production is just coming to light. The pharmacy directly supplied the noted convent artist Ortensia Fedeli with pigments and other materials for painting eight monumental lunettes in the convent church in 1612–1613. Assisting her were two other nuns, including a woman from the famed Buonarroti family. Fedeli may also have initiated the young boarder Arcangela Paladini, daughter of the painter Filippo Paladini, into the craft; Arcangela later matured into a talented painter and embroiderer at the Medici court in the 1610s and 1620s. Similar overlap among textiles, pharmacy, book production, and other craft workshops can be traced at San Domenico, the Murate, and the Annalena by the 1580s.[46]

These craft synergies shed new light on the occupational clustering of pharmacists, painters, and physicians within families—a practice that apparently encompassed female kin as well. Scholars have commented on the large number of artists' daughters among the ranks of Florentine nuns, but their concentration in particular convents warrants further attention.[47] Surely it is no accident that painters like Jacopo Ligozzi, Francesco Rosselli, and Monte di Giovanni, among others, enrolled their daughters at Santa Caterina, where they plied the family trade in convent workshops. Similarly, the Flemish painter Jan van der Straet (known as Giovanni Stradano) made his daughter Prudenza a nun in Sant'Agata, home to multiple craft endeavors. At these houses, daughters furthered their training in painting, sculpture, and fancy needlework under the tutelage of nun-artists whose skills intersected with pharmacy work. As discussed in Chapter 4, these familial networks furnished a key avenue for information exchange in the late sixteenth century, whether about medicines, pigments, or experimental techniques.

As Santa Caterina was developing its business, another convent pharmacy was springing up across the river. After the grueling siege of 1529–1530, the Annalena nuns established an apothecary shop marketing comparable wares (no. 9 on Fig. 3.1). Much of the nuns' property south of the city had been damaged or destroyed by foreign troops, depriving them of rents and foodstuffs in the midst of a subsistence crisis worsened by plague. Led by two of its members, Daniella di Battista and Maria Salome de' Lioni, this community strategically rebuilt its treasury and extended poor relief efforts via pharmacy. The convent also capitalized on the dearth of healthcare services in the quarter of Santo Spirito, which claimed only three apothecary shops in the 1480s despite being the city's second most populous district. That situation had not changed significantly fifty years later.[48] The Annalena pharmacy opened its doors in 1535 using stock purchased from Andrea Del Garbo.[49]

Deteriorating economic conditions throughout the 1530s and 1540s increased demand for affordable medicines, which could be expensive in the best of times. The Annalena nuns soon traded "in every sort of medicament needed for any kind of illness" and diligently compounded "every sort of medicine" in their workroom.[50] The business quickly became integral to convent finance, consistently generating 5 to 10 percent of earnings over the second half of the Cinquecento. Only textile production and boarding fees netted more income.[51] Although pharmacy revenues remained modest, they could tip the balance between solvency and debt because convents operated on such small margins. The pharmacy made an internal "loan" of thirty-five florins to the general convent treasury in 1562 to cover a shortfall, a figure equivalent to 2.5 percent of annual expenses. Similar loans followed in the 1590s, ranging from fifteen to fifty-five ducats each.[52] Monks and friars used pharmacy earnings in similar ways to subsidize in-house construction projects or offset internal deficits.[53]

Other Florentine convents quickly embarked on a similar path. The pharmacy run by the Benedictines of Sant'Agata (no. 1 on Fig. 3.1) was flourishing by the late Cinquecento, although little is known about its origins and development. Located on the major corridor of Via San Gallo, its proximity to several hospitals and convents, as well as underserved neighborhoods, probably accounts for its rapid success. By 1574, the Sant'Agata nuns reported that their apothecary generated almost 10 percent of earned income. Several nuns demonstrated their deep investment in the business by leaving permanent endowments as the century came to a close.[54] Similarly, the Franciscan tertiaries of Sant'Orsola (no. 8 on Fig. 3.1) claimed in 1561 that their pharmacy provided "their sustenance" and that they had "no enterprise more lively" than the shop. Unfortunately, there is scant information about how and why they had entered the medical marketplace.[55]

By the mid-sixteenth century, the success of Florentine convent pharmacies encouraged other regional communities to follow suit. In Prato, located fifteen miles northwest of Florence, the Dominican nuns of San Vincenzo (est. 1503) started marketing medicines in 1562. This venture stemmed from their Savonarolan ties as well as the influence of Caterina de' Ricci, the convent's most famous resident. San Vincenzo shared common religious roots with its "mother" house of Santa Caterina and was quickly colonized by the preacher's supporters. Supervised by the San Marco friars, the nuns of San Vincenzo continued to invigorate the Savonarolan movement in its waning days. Religious plays, devotional texts, artwork, and key clerical personnel circulated between the two convents, giving rise to a common literary and charitable culture.[56] Like its Florentine affiliate, San Vincenzo also supported a thriving artistic workshop renowned for painted figurines and small panels featuring religious subjects.[57]

Ricci was the prime mover behind the establishment of the San Vincenzo pharmacy. A mystic endowed with superb administrative skills, Ricci ran the community either as prioress or sub-prioress from 1553 until her death in 1590. As noted previously, she was an accomplished pharmacist in her own right. Hence Ricci was tasked with supervising the construction of a new pharmacy as part of extensive convent renovations. The house chronicle noted that the original pharmacy, dating from 1527, was situated "in a small, humid room inside the cloister." This unsuitable location limited production, damaged raw materials, and barred easy client access. The nuns replaced this older site with a "beautiful and roomy" addition overlooking the garden, at the steep cost of 1,190 scudi. Because the nuns voluntarily embraced enclosure even before Trent, the dispensary remained separate from the actual production site. Despite this inconvenience, the revitalized shop opened its doors to the public in 1562. Ricci also seized this opportunity to alleviate the high incidence of tuberculosis in the convent—the leading cause of death in the community, which killed three of her own sisters. Heeding the importance of ambient air, Ricci "decided to lower and enlarge the windows in the old cells, which were too high and small" in order to create a healthier living environment for the nuns.[58]

The reasons for embarking on commercial pharmacy differed according to circumstance, but shared a common rationale centered on medical charity. Marketing medicines enabled growing numbers of enclosed nuns to engage in new social welfare initiatives associated with Catholic reform. Retailing medicines not only provided crucial income to under-endowed convents but also broadened nuns' role as redemptive agents attending to body and soul. By developing greater expertise in pharmacy, nuns also improved prospects for their own health.

Products and Prices

Renaissance convent pharmacies retailed a wide variety of goods that ran the gamut from prescription medicines and ready-made wares to everyday wellness products. In the 1580s, the Murate pharmacy sold electuaries, syrups, and distillates as well as soaps, perfumes, and topical ointments.[59] Unfortunately, no inventories of stock comparable to those kept by the Speziale al Giglio have surfaced for Florentine convent pharmacies, nor have their business ledgers survived. Despite these evidentiary gaps, convent account books afford a look at the broad categories of products nuns made and sold.[60] Registers kept by Santa Caterina, San Domenico, and the Annalena are especially useful in this regard. When reconstructing product lines, it is important to note that Florentine convents had to meet the standards set out by official pharmacopeia.

Italian medical guilds had long established a conceptual divide between remedies taken by mouth and those administered externally. Given that medicines taken orally might do great harm, Renaissance guilds insisted on tighter licensing requirements for their makers. Yet this distinction between internal and external remedies was virtually impossible to maintain in practice. Traditional medicinals could be put to multiple uses, a practice known as polypharmacy. For instance, distilled rosewater—a staple marketed by both nuns and friars—might be ingested to clear phlegm from the chest or applied topically as an antiseptic to clean wounds. In addition, this multipurpose product was commonly sprinkled around interiors as a disinfectant, especially during plague; it also was used to dilute wine and furnished the base for other remedies and beauty products.[61] Other common remedies like eye drops defied easy categorization, since they treated organs with permeable membranes, which made them neither fully inner nor fully outer. Convent pharmacies regularly sold both internal and external remedies. However, they did not market wax—a consistent moneymaker over which guildsmen enjoyed a monopoly—which probably softened guild opposition somewhat. Nor did nuns market their own brand of theriac, a renowned cure-all, the recipe for which was jealously guarded by civic authorities. In fact, the friars of Santa Maria Novella began retailing a house brand of this panacea only in 1750.[62]

Ready-made items probably formed the bulk of merchandise sold by convent pharmacies, although account books overemphasize their prevalence in the absence of individualized prescriptions. Tightly sealed in jars, boxes, and flasks, these products enjoyed an extended shelf life of several weeks to several months, allowing inventory to be replenished as needed. Stockpiling these items enabled nuns to even out demand, which could be highly variable.[63] Convent remedies spanned a wide range. They included plasters, salves, and ointments to treat burns, wounds, and skin conditions; analgesic pastes to relieve gout or toothaches; distillates to eliminate worms or relieve digestive problems; syrups, electuaries, and pills to treat fevers and coughs; purgatives to eliminate peccant matter. Among their popular wellness products were brandies, cordials, and tonics that strengthened the body; perfumes that protected the brain from harmful odors; and confections and hygiene products that enhanced healthy living.

Probably their best-selling category of medicinal products was distillates and syrups—a popularity that paralleled sales by guild apothecaries. At the Giglio, for instance, medicines in this form constituted "the most important class of goods for retailing," accounting for 42 percent of sales circa 1500.[64] Some of these liquids could be made at home, but they often required lengthy preparation and proper equipment. The Sienese physician Pietro Andrea Mattioli mistakenly asserted that the art of distilling herbs was new in his day, but he was right in signaling its growing popularity, thanks in part to his own publications.[65] The surging interest in alchemy and chemical remedies by midcentury also boosted the popularity of distillates, which allowed practitioners to extract the "quintessence" or active ingredients of medicinal plants. As a process, distillation straddled both Galenic and Paracelsian medicine, despite their radically different understandings of the body. The influence of Paracelsian ideas on Italian medicine has been the subject of some debate. Antonio Clericuzio has argued that, although chemical medicine became widely diffused in Italy between 1550 and 1650, it never provoked the acrimonious confrontation with Galenism seen elsewhere in Europe.[66] Florentine convent records both support this interpretation and provide additional evidence for the inclusion of chemical remedies in local medicine chests. Nuns regularly incorporated chemical substances like gold and steel distillate into their wares, although the bulk of their merchandise remained firmly herbal.

Convent ledgers are littered with payments for "herbs to be distilled," but these materials and their resulting products were often lumped together in accounts. Still, the Santa Caterina pharmacy did itemize a few of its herbal products such as sorrel distillates, which physicians praised for treating fevers. Febrifuges were in high demand by clients needing urgent care, especially considering the prevalence of malaria in Tuscany. Other popular distillates included medicinal waters made from betony, which reportedly relieved toothache, improved respiration, and detoxified the humors. The same remedy could be used externally as an eye wash or mixed with oil or hot water to make a strengthening tonic.[67] The Santa Caterina nuns also manufactured large volumes of distillates and syrups using agrimony, chicory, maidenhair fern, melissa, mint, polypodium, and rhubarb. They decocted violets and roses in huge batches to obtain their essence, which could be used alone or in other products like medicinal oils, ointments, confections, and perfumes. Finally, they produced numerous liquors and brandies.

Convent pharmacies also excelled at making wellness products, which constituted a vibrant market sector. Renaissance paradigms of healthy living placed new weight on personal hygiene, which ranged from cleaning the teeth, ears, and beard daily to washing hands, face, and bed linens more frequently. Central to this evolving "culture of cleanliness" were differentiated soaps and cleansing agents.[68] In sixteenth-century Italy, soap became available in many varieties, grades, and price points, depending on the materials used in its production. Industrial-grade soaps, usually referred to as "red" or "black" soaps, played a crucial role in textile manufacture; the Genoese were acknowledged industry leaders in this area. Entrepreneurs across north-central Italy competed for state patents on industrial soaps made in huge batches weighing up to 500 pounds; most Florentine guildsmen specialized in this type of soap production.[69] Certain soaps could be used to disinfect goods suspected of harboring plague or even to cauterize plague boils. Indeed, the owner of a Lucchese soap factory, Girolamo Maccioni, wrote several short tracts extolling the virtues of special soaps, including one during the 1630 plague.[70]

By contrast, convents and other small-scale manufacturers capitalized on the growing interest in hygiene when developing products for a local clientele. The commercial potential for cleansing agents intended for personal use invited product innovation across Italy. These products could be made in small batches featuring an endless variety of shapes, sizes, and scents. As early as 1508, the pharmacy at Santa Maria Novella used special molds to form little balls *(pallottole)* of fragrant soap. Historians of consumption have argued that novelty itself could be attractive to consumers, but medical arguments about prevention further intensified product innovation in Italian Renaissance cities.[71] Small-scale producers like convents could respond nimbly to shifting taste and consumer demand by engaging clients directly about their preferences. Because soaps and other wellness products were intended for local consumption rather than export, consultations between nun producers and local consumers in the dispensary or parlor enhanced the likelihood of commercial success.

In addition, hygienic products could legitimately carry trademarks, which made them effective advertising vehicles for other pharmacy wares. The semiotic field of trademarks was especially rich in Italian Renaissance cities; producers branded their goods with recognizable signs that remained visually distinctive in some way. One ambitious Genoese soap-maker aiming to establish a new facility in Medici-controlled Pisa asked to register his trademark of "a Pisan cross inscribed in a circle" as part of commercial privileges requested in 1597. Nuns probably marked their wares with emblems similar to those placed on institutional goods like tableware and altar cloths. These emblems, incorporating convent initials, patron saints, or other religious signifiers, were distributed liberally throughout convent interiors and even functioned as public property markers on Florentine streets.[72] Marketing opportunities were enhanced by the fact that guilds focused quality controls on industrial-grade cleansers. Finally, locally produced soaps enjoyed a competitive price advantage because they did not carry import duties.[73]

One convent that aggressively pursued this product line was San Domenico, located on Via Venezia near San Marco and the Medici stables (no. 3 on Fig. 3.1). These Dominicans were relative latecomers to the medical marketplace, launching their pharmacy only in the 1560s. Although they sold a variety of syrups, distillates, plasters, sweets, and spring waters, their late entry pushed them into increasingly specialized niches at both the high and low ends of the market. These nuns wrapped high-end pills in thin sheets of gold leaf, which both eased swallowing and reputedly intensified active ingredients. Skills needed to cut and wrap gold leaf were already familiar to them from making metallic thread. At the opposite end of the market, the nuns invested heavily in preliminary production processes like pressing almonds that were the pharmaceutical equivalent of reeling silk: both processes entailed tedious, back-breaking work done in volume. In 1571 these women purchased a mechanical press for extracting great quantities of soothing almond oil, used in their famed rose unguent and fancy soaps; it also could be marketed as a stand-alone product to other vendors.[74] Purchase of this apparatus costing more than five scudi signaled both the convent's long-term investment in pharmacy and the diffusion of new technologies into female institutional spaces.

Among the first products the San Domenico nuns retailed were cleansing agents: soap for washing the head *(sapone da capo);* soap for washing silk *(sapone da seta),* presumably at home; coarser soaps for washing bed linens; and fragrant soaps for personal use scented with floral distillates made in-house.[75] This line of goods reinforced status distinctions within the convent, much as linen and silk production had done a century earlier.[76] Soap-making was a nasty, laborious business handled by lower-class serving sisters, with well-born choir nuns likely taking charge of distillation, branding issues, and product refinement. Some of these products remained controversial for consumers, despite the growing culture of prevention. Considerable anxieties swirled around washing the head in the later Cinquecento, since it might harm the brain or otherwise threaten health.[77] Despite being considered a health hazard by many, especially when done by pregnant women or the sick, this practice generated a new, specialized market. The nuns may have indirectly swayed their clients' medical views by washing their own heads (albeit infrequently) with convent products.

Comfits and sweets formed another expanding market after 1500—one in which nuns did a booming business. As noted previously, sugar was a key ingredient in both curative remedies and health-giving foodstuffs taken as part of a preventive health regimen. Bridging the medicinal and culinary, confections like quince paste were readily consumed by the healthy, the sick, and convalescents alike. The mystic Caterina de' Ricci willingly ate pennets and other "sugared things" prescribed by physicians during her numerous illnesses, although she refused other foods recommended for ailing nuns.[78] Both monastic and guild apothecaries constructed taste by retailing sweets as an important part of their business. Yet confections raised a host of moral issues, given the charitable rationale under which nuns worked. Despite the falling price of sugar after 1500, sweets remained luxury items beyond the reach of most city folk. The status value of sugary things was enhanced by elaborate confections used in aristocratic and court festivities.[79] By the 1590s, the Murate pharmacy, which sustained close ties to the Medici court, not only sold staples like quince paste; the nuns used pharmacy supply chains to also fabricate delicacies like candied chestnuts ordered by local noblewomen.[80] These confections consolidated important patronage relations vital to institutional health, but they represented a radical departure from initial charitable objectives.

Beauty products constituted another problematic category for nuns. As noted earlier, no bright line separated cosmetics from medicines in the Renaissance. Products that both cleansed and improved the appearance of bodily surfaces were seen as integral to good hygiene. Nevertheless, cosmetics raised concerns about creating a false, seductive appearance.[81] Given the nature of polypharmacy, nuns probably participated at least indirectly in this market segment. Their scented waters and distillates could be used as face washes; their rose unguents not only treated wounds but also softened skin and smoothed wrinkles; their toothpastes both whitened teeth and eliminated bad breath. As discussed later, the Murate nuns openly invested in these products around 1600, at great cost to the equilibrium of the house.

In keeping with Renaissance commercial culture, convent pharmacies developed reputations for certain specialties. These remedies did not necessarily involve exotic ingredients, but simply displayed some distinctive feature or demonstrated greater efficacy compared with other products. The medicinal oil marketed by the Paduan monks of Santa Giustina in the early sixteenth century, for instance, consisted solely of sixty-three different herbs macerated in good olive oil and left to steep in the sun. Still, the monks jealously guarded the recipe.[82] By the late sixteenth century, Santa Caterina had become famous for its syrups and distilled rosewater. Similarly, Sant'Orsola achieved notoriety for its "ducal pills" taken by Cosimo I, among others. The eighteenth-century antiquarian Giuseppe Richa claimed that these pills were still being made in his day by a physician using the original formula and marketed under the brand name "pills of S. Orsola."[83] Virtually every convent pharmacy in seventeenth-century Bologna boasted specialty products such as plasters, electuaries, and cherry cordials. Similarly, Neapolitan convents were so famous for certain "healthful" confections that the Tuscan jurist Giovan Battista Pacichelli compiled a detailed list of offerings while visiting in 1685.[84]

Pricing convent wares remained a major bone of contention between guild and monastic druggists throughout the early modern period. Guildsmen complained that monastic pharmacies wreaked havoc on prices because they consistently undersold licensed apothecaries or distributed medicines gratis. In 1593 the Medici state set prices on medicinals for the first time as part of a broader attempt to surmount guild economic power while safeguarding public health. Following a devastating famine, Ferdinando I established the fair market price for every conceivable type of medicament sold by apothecaries, citing growing imports from Rome and Venice that might stimulate price wars. However, these prices applied only to guild pharmacists, leaving monastic workshops untouched. Only one item manufactured by nuns—the nourishing chicken broth intended for convalescents—was assigned a fixed price, costing half of what guild apothecaries could charge.[85] Interpreting this price differential is complicated, but it can be seen as a sign of state approval for convent remedies, in keeping with other Medici public health initiatives, such as the establishment of the first Florentine convalescent hospital in 1588.

Regardless, prices for medicaments remained embedded in larger credit networks and patronage relations, making them subject to wide variation and different payment schedules. Monastic institutions probably determined the asking price based on the client's relationship to the house and perceived ability to pay. Moreover, benefactors often supplemented pharmacy purchases with alms to sustain other projects. Camilla Martelli, Cosimo's second wife, who took up residence at the Murate after his death, gave forty scudi to the house in 1574. She described this payment "as alms for the convent and in compensation for the pharmacy's efforts while she was ill."[86] Prices for convent remedies were not set in stone but rather reflected a wealth of social and spiritual factors.

Sources, Suppliers, and Botanical Innovations

Commercial pharmacy exposed nuns to commodity chains stretching across the early modern globe, mirroring the networks cultivated by large Florentine apothecary shops like the Speziale al Giglio.[87] Exotic ingredients such as bezoars, gemstones, resins, and guaiac could be procured only from long-distance merchants in Venice, Rome, Livorno, and intermediate hubs. Every convent pharmacy in late Renaissance Tuscany maintained accounts with several different brokers to ensure a steady stream of supplies. In the 1560s and 1570s, the Annalena nuns regularly bought large quantities of spices, foodstuffs, and medicinal items from the apothecary firms of Buonaccorso Pinadori and Bernabo Bicci, with whom they maintained extensive credit relations.[88] Ducal privileges further shaped supply chains for essential commodities like sugar. In January 1544, Duke Cosimo granted one-year rights to import sugar from the island of San Tomé to the consortium formed by Alessandro Antinori and two Salviati kinsmen. Two years later, he extended a fifteen-year privilege to Jacopo and Paolantonio Pinadori to refine sugar in Pisa, which amounted to a monopoly on sugar distribution within the Medici state. In 1558 Cosimo renewed this arrangement for another fifteen years and bundled it with rights over the traffic in scented Cyprus powders used to make perfumes and beauty products.[89] Consequently, the Pinadori firm had a virtual chokehold on local sugar supplies throughout the late Cinquecento.

By contrast, organic materials such as herbs and flowers had to be sourced locally in order to retain their freshness and medicinal "virtues."[90] Tuscan convents turned first to their own gardens to source common medicinals such as violets, roses, betony, borage, and chamomile. The ability to grow some of their own materials gave monastic pharmacies another competitive edge over lay druggists. The Annalena nuns cultivated simples in their extensive gardens and harvested saffron on their farmholdings for commercial purposes throughout the sixteenth century. San Vincenzo in Prato boasted "a spacious garden and orchard" that supplied its "noble pharmacy" by the end of the century.[91] The Murate grew medicinal herbs, fruits, and flowers in its celebrated gardens and orchards, which supplied both its apothecary business and burgeoning gift exchange. Two important cultivars grown there were pomegranate, used in remedies like eye washes for its astringent properties, and quince, the basic ingredient in the popular digestive aid called *cotognato* discussed earlier.[92] Praised for their beauty by Archbishop Antoninus in the mid-fifteenth century, the Murate orchards figured into Leo X's itinerary when he visited the house in 1516. A half century later (1574), the Florentine archbishop Alessandro de' Medici remarked that this convent "had the most beautiful orchard in the whole of Italy."[93] However, the large volume and variety of simples required for commercial production meant that convents also relied on other local vendors, much like the hospital of Santa Fina in San Gimignano, which regularly bought medicinals from local herb-women.[94]

The supply chains and brokerage arrangements at Santa Caterina warrant a closer look, both because these nuns were in the vanguard of commercial pharmacy and because their sourcing was somewhat anomalous. When Santa Caterina launched its pharmacy, circa 1515, the house was still a new establishment in the process of building its physical complex. Consequently, the nuns initially relied on outsiders for batches of manna, violets, cassia, and various herbs, as well as for huge casks of sugar and exotic plants. They obtained some of these commodities through their physicians and regularly bought in bulk from at least four apothecary firms. Their chief supplier, however, was the friary of San Marco, which used its own pharmacy as a trade conduit.[95] The San Marco friars also conducted all liturgical services at Santa Caterina and other convents under their tutelage, thereby deriving money, prestige, and influence from female affiliates. To compensate the friars for medicinal supplies and liturgical activities, the Santa Caterina nuns sometimes paid in kind with pharmaceutical items fabricated in-house. Some of these items were used by the friars themselves; other wares such as soaps and syrups were subsequently retailed through the San Marco pharmacy. Even though these products were made off-site by nuns, they still carried the prestigious San Marco brand. In managing convent wares, these friars simultaneously acted as suppliers, consumers, and vendors of nuns' merchandise.[96] These interlocking roles solidified San Marco's control over the convent's market production and made the pharmacy vulnerable to local political struggles within the Savonarolan movement and, later, with the duke himself. These provisioning arrangements were unique among Florentine convent pharmacies; shops affiliated with other monastic orders were not supplied or supervised in comparable ways.

Despite the importance of convent gardens to local health markets, these horticultural resources have never been considered in relation to the development of medical botany and scientific agronomy, both of which were vital to early scientific culture. Italian convents have been viewed as important sites of cultural production in needlework, the visual arts, and music-making, not as epistemic spaces in science and medicine. Yet archival evidence demonstrates that Renaissance convents were used as botanical testing grounds that produced commercially valuable knowledge. These institutions displayed many of the resources necessary for trying out new horticultural techniques. Their physical sites contained differentiated outdoor spaces that could be adapted to particular plant needs; their walls enclosed dense concentrations of literate women schooled in herbal medicine; and their residents enjoyed connections to local scientific elites. Both the scale and stability of convent spaces transformed their gardens into laboratories for naturalistic inquiry as well as key transfer points in larger experimental networks.

There was a complex feedback loop between the growth of Italian convent pharmacies and nuns' adoption of new horticultural techniques in the late sixteenth century. Demand for innovative products triggered botanical experiments both inside and outside convents. Advances in floriculture proved particularly significant for manufacturing perfumes and scented items, which were considered health-giving products that strengthened the brain. By 1550, perfumed pastes were incorporated into a stunning array of objects, such as scented buttons and gloves, pomanders attached to rosaries or belts, earrings that diffused a steady stream of fragrance around the head, and perforated perfume burners that scented domestic interiors and sometimes doubled as hand warmers.[97] The popularity of scented items in the late Renaissance reflected an ambitious campaign by urban elites and the middling classes to replace the foul with the fragrant in the pursuit of health.

Two innovative practices became pivotal to the expansion of convent pharmacies in late Renaissance Italy: the technique of forcing potted plants to bloom out of season, and the cultivation of cold-sensitive plants. Both carried enormous commercial and scientific potential for increasing production, while also elevating the status value of cultivars introduced into Europe by early globalization. To meet the growing demand for flowers by midcentury, Tuscan nuns mastered the technique of forcing, which significantly extended the natural growing season. Forcing involved planting specimens in pots that could be nestled in glass cases—the precursors of modern-day greenhouses—or brought into protective structures like cold sheds during the winter. Similar results could be achieved by plunging pots into special beds composed of tanners' bark or horse manure that generated natural heat. Successful use of this labor-intensive technique required detailed attention to variations in timing, temperature, humidity, light levels, and the spatial requirements of particular cultivars.

The Dominican friar Agostino Del Riccio discussed this method at some length in his three-volume manuscript titled *Agricoltura Sperimentale* (1595), which drew on thirty years' experience as curator of the Santa Maria Novella gardens. Del Riccio (1541–1598) was a key figure in Florentine scientific circles. A frequent visitor to the Medici laboratories, he maintained close friendships with Jacopo Ligozzi, Stefano Rosselli, and Niccolò Gaddi. Among his various works were treatises on the ideal garden, the history of gemstones, and the memory arts.[98] In his massive exposition of scientific agronomy, Del Riccio maintained that forcing techniques had been in use for some years by "our nuns of S. Domenico, who live near S. Marco in Florence." Recognizing the commercial value of this practice, this master gardener boasted that the San Domenico nuns "garnered more than fifty scudi per year" by artificially extending the growing season.[99] This sum represented twice the annual rent (twenty-seven scudi) the nuns realized from leasing an adjacent garden to Don Luis de Toledo, brother of Duchess Eleonora.[100] According to Del Riccio, other local nuns engaged in similar practices, including "the shrewd and prudent Florentine women living outside the Porta alla Croce" (current day Piazza Beccaria)—a likely reference to the nuns of San Giovanni Evangelista resident at San Salvi, just east of the city. Thanks to these innovations, these nuns "have very many violets, carnations, cornflowers and other flowers from which they make plenty of money every year." These communities reportedly exported fruits and flowers cultivated out of season to Bologna, Ferrara, and other Italian cities. The friar broadened the significance of these activities by linking them to the political economy of the Medici state. Del Riccio stressed that nuns' botanical exports contributed to the balance of trade, thereby strengthening the Medici fisc. The commercial value of these practices led the Medici dukes to extend their use. When the Venetian entrepreneur Marco di Fugeni petitioned for exclusive rights on using heated stoves to force flowers in 1596, the grand duke granted the privilege, despite opposition from "various convents that tend these flowers, along with many peasants."[101]

It is not clear from existing records how and when the San Domenico nuns acquired this horticultural expertise. Certainly they were no strangers to craftwork: by the 1490s, they were casting plaster figurines of saints and angels and painting rosary beads in a vibrant workshop setting.[102] Although new horticultural practices were becoming more widely diffused among monastic institutions by midcentury, the arrival of a new confessor in the 1560s, Fra Domenico Meninconi of Perugia, probably tipped the balance. A polymath, Meninconi introduced the nuns to various innovative technologies, including a new hydraulic apparatus that he designed and installed in the main garden. This device collected rainwater from the cloister roof and used its flow to turn a new type of water wheel, which funneled water into irrigation troughs. This apparatus successfully watered large areas under cultivation and subsequently was adapted for other convent locations.[103]

Cultivating flowers out of season was a smart strategy for San Domenico, and the nuns promptly reinvested earnings from experimental floriculture in their pharmacy. In 1569 they purchased a copy of the newly revised Florentine *ricettario,* which set the latest standards for making medicines. They also bought a large lead distilling bell and commissioned twelve pharmaceutical jars *(albarelli)* marked with the convent emblem from a noted manufacturer in Montelupo. Over the next few years, the nuns invested in a mechanical press and purchased new mortars, flasks, beakers, jars, and other equipment. These upgrades permitted the apothecary to scale up quickly in order to retail larger volumes of more varied products. Like its counterpart at Santa Caterina, the convent deepened its expertise by appointing a head pharmacist, Sister Maria Maddalena Asini, who served in this capacity from the 1570s until her death circa 1584, when two of her apprentices took over the business. Part of Savonarolan networks, the nuns maintained accounts with two apothecary firms but obtained most of their supplies through San Marco.[104]

The second significant innovation for Italian convent pharmacy concerned the cultivation of cold-sensitive plants, especially oranges and other citrus. Bitter oranges had been grown in Italy since antiquity, but the new, sweet varieties imported from distant locales in the early Cinquecento "could not tolerate the cold," as the Florentine plant expert Giovan Vettorio Soderini observed.[105] Grown in pots and clipped into round or conical shapes, dwarf orange trees emitted a luscious scent while giving sculptural form to Renaissance pleasure gardens. The very susceptibility of ornamental orange trees in colder climes north of Naples only added to their appeal and status value. As their popularity skyrocketed in sixteenth-century Europe, citrus became both a commodity and a curiosity that stimulated new horticultural techniques and proprietary knowledge. The Spanish friar Juan del Pozo tried to curry favor with Philip II of Spain in 1578 by offering to share his "secret" of how to protect citrus plants from the cold without needing to cover them.[106] Soderini advised owners to protect precious specimens from cold damage by one of several methods: wrapping the plants in cloth, which was unsightly when done on a large scale; building temporary protective structures around them; or moving them into permanent outbuildings during the winter. Oranges also could weather a mild frost if placed against a south-facing wall that retained heat.

By 1550, techniques for cultivating citrus and other tender plants were in widespread use at monastic houses as well as Medici gardens. The ducal secretary reported to Cosimo in March 1551 that "the sweet orange trees" at Castello "have suffered a little," although he hoped that with the advent of spring "they [would] soon sprout new leaves."[107] In procuring precious cultivars, the duke sometimes used local convents as both suppliers and way stations for plants destined for his gardens. In January 1564, Cosimo instructed his factor to obtain eighty potted orange trees and twenty-five potted lemon trees from Santa Felicita and Santa Brigida del Paradiso, along with any similar plants he could find at other convents. The duke urged him to act quickly in order to obtain "the largest ones that can be found" and to exercise appropriate care in their handling.[108] The number of plants involved in this single transaction is impressive in its own right, considering that plant-loving English aristocrats usually ordered about six orange trees at a time, owing to their high transport costs.[109] Sources are silent about where these specimens were grown at convent sites, but it is clear that Cosimo—a plant expert himself—believed the nuns had the necessary know-how to tend them in the dead of winter.

While Cosimo prized citrus specimens primarily for their status value, the nuns of San Domenico cultivated them for commercial purposes. Fresh orange blossoms were ideal for distilling into fashionable perfumes. Florence was unrivalled in the art of perfumery throughout the sixteenth century. Both the Medici court laboratories and small-scale producers expanded the sensory repertoire of Renaissance elites by incorporating new floral fragrances as well as exotic animal scents like musk and civet into scented goods. In developing this product line, the San Domenico nuns profited by growing oranges on espaliers inside the cloister at little cost. Del Riccio observed that the "beautiful and fresh" appearance of blooming citrus had the added bonus of lifting the spirits of enclosed nuns. Capitalizing on homegrown materials, nuns made floral distillates and scented waters that could be used for washing hands, sprinkled in rooms to combat bad smells, or mixed with other liquids as a dessert beverage.[110] All of these practical uses fed a burgeoning culture of prevention. Once again, however, convents compensated activities aimed at high-end clients with charity. According to Del Riccio, Dominican nuns regularly delivered branches of flowering citrus to the Florentine city prison as an act of medical charity. Echoing contemporary concerns about the quality of ambient air, these fragrant branches placed "in the closed rooms with little air" made prison living quarters a little healthier.[111]

These findings demonstrate that late Renaissance convents became incubators for horticultural advances in tandem with commercial pharmacy, embedding nuns in experimental networks as well as commercial markets. The practical knowledge Renaissance nuns gained as a result augmented their activities as purveyors of medicines, cosmetics, and perfumes within local health markets. Their engagement with innovative techniques underscores the extent to which early modern knowledge production was a networked enterprise extending beyond the male worlds of court laboratories, guild workshops, and university lecture halls.

Clients

The financial success of convent pharmacies raises important questions about the customers who bought their products. Because secondary shop ledgers have been lost, pharmacy clientele must be reconstructed from general convent registers, which do not always name paying customers. Despite this source problem, it is clear that nuns built an extensive clientele via predominantly female networks—relatives, friends, boarders, neighbors, tenants, spiritual allies—with whom they enjoyed multilayered associations. The high degree of interconnectivity linking convents and identifiable customers confirms the observation made by Shaw and Welch that the Renaissance apothecary trade was "intrinsically social in nature." Given the nature of the evidence, it is more feasible to reconstruct the identities of what these scholars have called "habitual" clients—mainly elite consumers who purchased sweets and perfumes as well as medicines—rather than "crisis" clients, such as poor artisans, laborers, prostitutes, and servants seeking immediate relief.[112] Although these groups displayed different consumption patterns, extant sources do not permit a detailed analysis of their purchases. It is likely, however, that these groups hailed from different urban locales. Charity clients seeking urgent care probably came from the immediate vicinity, whereas habitual clients must have been drawn from citywide networks that mirrored convent recruitment patterns in place by midcentury.[113]

Buying goods from convent pharmacies can be seen as an everyday act of patronage. Purchases not only supported these communities financially but also fostered their charitable mission. In this sense, customers advanced medical charity by using convents as spiritual brokerages, much as did Queen Leonor of Portugal. Patrician women who bought "medicines, syrups and other remedies" from the Annalena pharmacy also commissioned embroidery and sewing projects from the nuns.[114] Many of them had a long history with the house, making it easy to extend credit or otherwise adjust payment terms. Other steady customers at the Annalena included pupils and boarders of all ages, since nuns embraced a duty of care for all convent residents. For instance, the nuns used purgatives made in-house to treat their teenage boarders Caterina and Maddalena Ridolfi in 1557, and kept a running tab for the "medicines, syrups, and other medicaments . . . from our pharmacy" given to the boarder Margherita Bonini "when she was ill" in 1587. This latter account went unpaid when Margherita left the convent soon afterward to marry, forcing the nuns to sue for compensation. Elderly widows and annuitants residing in the convent often stipulated in their residential contracts that medicines be freely provided during their stay—a typical feature of Renaissance custodial arrangements.[115]

As in guild apothecaries, nuns made business decisions that were deeply bound up with social relationships. Among the habitual clients patronizing the Santa Caterina pharmacy were nuns' relatives and longtime benefactors like Bernardo Gondi, whose family associations with the house extended back to its foundation. In 1607 this Florentine nobleman bought "several medicines" on credit for his ailing mother, worth ten florins. A few years later, he perpetuated these institutional links by placing two of his daughters in the convent as nuns.[116] These reciprocal social relationships not only conditioned nuns' willingness to extend credit, but probably also influenced the development of customized products. Other grateful pharmacy patrons occasionally earmarked pious bequests for the business, which both strengthened its financial health and augmented charity to the poor. Conversely, pharmacy patronage might stimulate other kinds of economic relationships. As noted previously, one of the earliest patrons of Sant'Orsola's pharmacy was Maria Salviati, for whom the nuns concocted special tonics in the 1540s. This connection soon blossomed into doing specialized textile work for the Medici court, including burning down metallic thread to recoup the precious metals.[117]

Over the course of the sixteenth century, Florentine convent pharmacies developed thicker commercial ties to institutional clients such as convents and hospitals. The Santa Caterina pharmacy was particularly active as a medical supplier for the nearby Dominican convents of Santa Lucia and San Luca.[118] Selling medicines to female religious houses was a lucrative market, since their numerous inhabitants experienced frequent health problems: Santa Lucia housed about 130 nuns in the late Cinquecento.[119] Santa Caterina also retailed some bulk commodities in smaller quantities, echoing the distribution system used by lay apothecaries to streamline marketing. In this respect, the convent not only functioned as the female analogue to San Marco, but also demonstrated the extent to which these businesses were integrated into coordinated marketing networks by the late Cinquecento. This intermonastic commerce added yet another strand to existing social and commercial transactions between female communities, complementing the purchase of small paintings and other artwork from Santa Caterina in these years.[120]

One such transaction is recorded in two itemized bills of sale written by Giovanna Ginori, head pharmacist at Santa Caterina. The first is signed "I, Sister Giovanna, *spetiala* of S. Caterina." It is tempting to read a sense of occupational identity into this statement, considering her long service and esteemed reputation. Dated July 29, 1567, the receipt lists the large volume of medicinal waters provisioned to Santa Lucia every few days between June 9 and July 11, totaling sixty-two flasks of *acqua del tettuccio* and twenty-five flasks of *acqua della porretta* (Fig. 3.2).[121] A best-seller, *acqua del tettuccio* was a local spring water from Montecatini often taken as part of a comprehensive regimen; the papal physician Paolo Zacchia praised it as an efficacious digestive aid for the sick and convalescents.[122] The price tag of four and a half florins was consonant with current wholesale prices, so that Santa Caterina made little profit on the transaction. Nevertheless, this expense for a month's worth of medicinal waters alone illustrates how costly basic medical care could be.[123]

The prominence of female clients—some of whom reportedly came into the shop themselves—underscores a significant difference between convent pharmacies and guild apothecary shops. Monica Green has argued that women's medicine in medieval Europe was not strictly women's business. Instead, women utilized a wide range of medical practitioners rather than relying solely on other women for healthcare services—an observation that holds true for Renaissance Florence as well.[124] Commercial transactions never followed a simple gendered divide but were conditioned by class, purchase value, and other circumstances. Indeed, guild shops like the Speziale al Giglio in late fifteenth-century Florence catered to a mixed clientele. That said, most account holders there were men, which should not be surprising in view of women's lesser means, their financial dependence, and social norms that frowned on excessive visibility. Since guild pharmacies were staffed by male shopkeepers and apprentices, these venues can be seen as essentially masculine sites.[125]

By contrast, convent pharmacies were vital nodes of female sociability, much like convents themselves. Respectable women could shop there in person without jeopardizing their reputation, since pharmacy visits could be easily integrated into routine social exchanges with nuns or attending mass at convent churches. In this respect, convent pharmacies enjoyed a competitive advantage over guild shops. Sociability also facilitated the transfer of practical medical information. Waiting in the dispensary, female clients might confer directly with nun apothecaries, who reportedly were generous with their knowledge, or share health advice with other customers, including medically informed women who ran girls' shelters and tended to the needy on the streets. Other convent spaces such as parlors served similar functions. At Santa Caterina, where

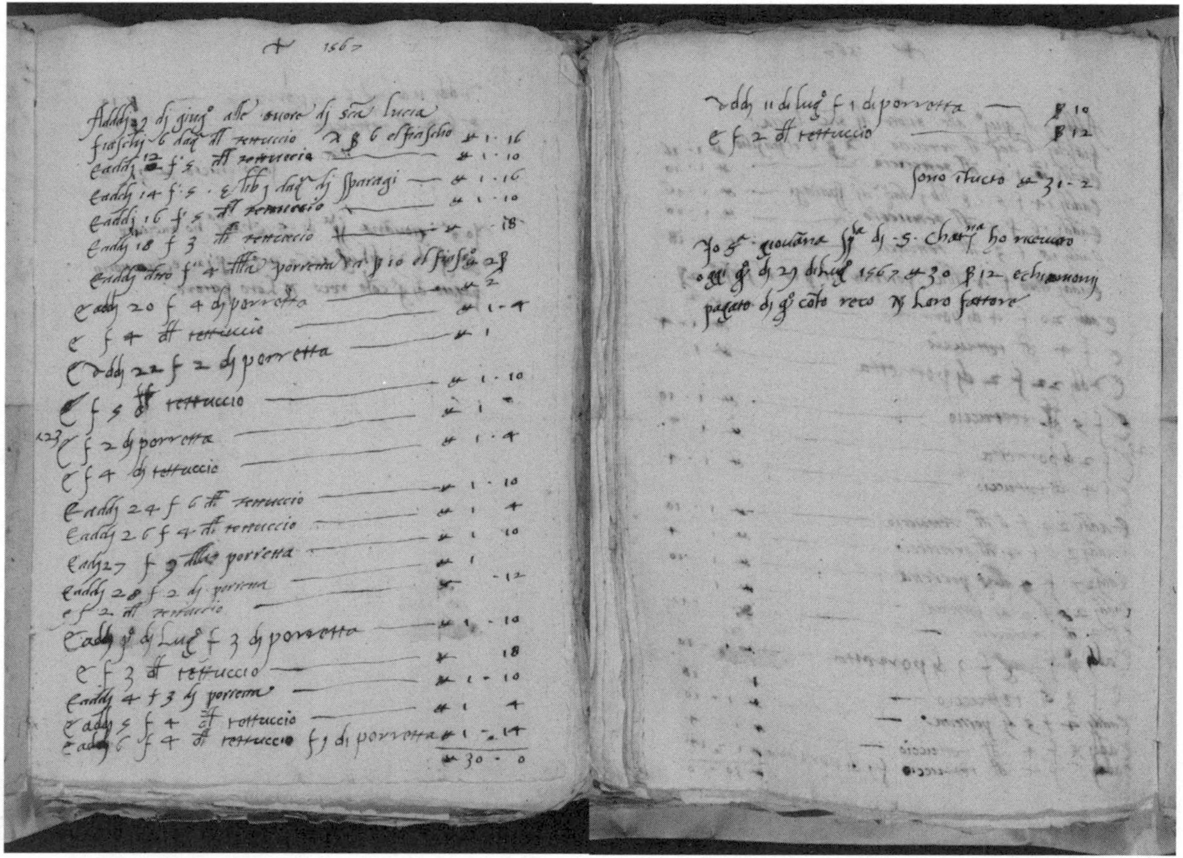

Fig. 3.2. Pharmacy receipts in the hand of Sister Giovanna Ginori, 1567. ASF. CRSGF. 111. Vol. 67, unfol. By permission of the Ministero per i beni e le attività culturali

open reclusion was the norm until 1575, laywomen involved in social welfare initiatives visited frequently, sometimes staying overnight in the house to pray and sing hymns. The Savonarolan proponent Lorenza Ginori Rucellai, kin and confidant to head pharmacist Giovanna Ginori and her sister Maddalena, reportedly was "always living, associating, and discussing confidential matters with them."[126]

The Santa Caterina pharmacy in particular acted as a linchpin for local Savonarolan activists. Among the charitable projects they sponsored were the Incurabili pox hospital (est. 1521) and La Pietà, a shelter for abandoned girls founded by two affluent laywomen in 1555.[127] Between 1515 and 1523, the apothecary Bernardo Mini furnished Santa Caterina and the Incurabili with their initial stock of medicinals, while the wealthy merchant Simone Ginori donated 300 florins to outfit the hospital pharmacy.[128] Although financial backing by male benefactors was crucial, especially in the early stages, these initiatives were sustained by grassroots support from pious laywomen hailing from the Rucellai, Gondi, Nucci, Tornabuoni, Landi, Altoviti, Giugni, Bettini, Buondelmonti, Lapini, and Mini families. For these local activists, patronizing the Santa Caterina pharmacy was a tangible way of demonstrating membership in the elect nation. The pharmacy continued its charitable mission throughout the seventeenth century by partnering with two other convents—Santa Lucia and San Luca—to provision syphilis remedies to the Incurabili.[129] As their Savonarolan origins faded, these convent pharmacies were folded into the public health apparatus of the Medici state.

Regulating Convent Pharmacies

After 1500, female-run apothecary shops added new complexities to state regulatory efforts concerned with quality control and fair pricing. Because they fell under ecclesiastical jurisdiction, convent shops escaped attempts by Italian states to consolidate control over local medical resources through guild licensing and health boards. Although convent ventures posed unmistakable competition to guildsmen, they were widely recognized as serving the public good. These political and commercial tensions were exacerbated by competing visions of female monastic life that pitted reclusion against an active apostolate. Even after the Council of Trent (1545–1563) seemingly decided the issue in favor of enclosure, the effectiveness of regulatory efforts by both church and state varied by locale. The result was a complex, often contradictory, regulatory landscape across the peninsula throughout the sixteenth century.

Having entered the marketplace in the early Cinquecento, Florentine convent pharmacies figured among the most highly supervised shops a century later. Much of that regulatory framework was built by Cosimo I, whose sweeping institutional reforms in the 1540s and 1550s brought poor relief, hospitals, convents, and guilds under the purview of a centralizing state.[130] Within this developing bureaucracy, two magistracies assumed responsibility for regulating convent apothecaries. The first was the Magistrato sopra i monasteri (est. 1545), which supervised practical aspects of convent life, from budgets to repair work. Building on earlier administrative structures, this magistracy functioned as a web of four lay supervisors assigned to every convent in the Medici dominion, who in turn reported to a central body comprising three deputies.[131] The establishment of the Magistrato created the potential for serious jurisdictional conflicts with the Florentine archbishop; each claimed the right to authorize visits to convent parlors and pharmacies via their own licensing mechanisms. This scenario was largely avoided during Cosimo's reign, thanks to a series of weak or absentee bishops who held the see between 1524 and 1567. Consequently, the state held the upper hand over convents for almost a half century. Only with the election of Alessandro de' Medici in 1574 did Florence develop a strong church-state partnership. Even then, the new Medici archbishop did not officially claim the diocese until 1583.[132]

The Guild of Physicians and Apothecaries was the second regulatory body that oversaw Florentine convent pharmacies. Cosimo reshaped the official practice of medicine in the duchy through a series of legislative reforms between 1547 and 1559. Among the key changes were the formation of the prestigious College of Physicians and the mandate that all apothecary shops be inspected triennially to ensure that their wares were "good, pure and not fraudulent in any way."[133] The guild reserved the right to "inspect and condemn" apothecary shops throughout the dominion, even those holding a ducal privilege. Inspectors examined the weights, measures, and instruments used to fabricate medicines, as well as the quality, labeling, and arrangement of products; they also maintained the right to destroy old, spoiled, or poorly made remedies. Resulting fines bypassed the guild treasury and instead supported the Medici fisc—another example of Cosimo's state-building techniques.[134] Owing to a weak local Church, monastic pharmacies located in Medici territory did not escape this oversight. Guild inspectors were empowered to certify that medicinal wares fabricated by nuns and friars met official standards—an unusual situation among fragmented Italian states. Monastic pharmacies in sixteenth-century Rome, for instance, were exempt from annual inspection, despite ongoing prohibitions against the practice of pharmacy by non-guild members.[135] Within this regulatory environment, however, only nuns answered to two different magistracies when marketing medicines.

This dual oversight led to frictions between supervisory bodies that could be exploited by different parties. Periodic calls by guildsmen to close convent pharmacies met resistance from convent magistrates charged with protecting nuns' livelihood. Similarly, guild attempts to maintain high standards via routine inspection clashed with attempts by convent magistrates to maintain equally high standards of enclosure. Conflicts arose even within the guild itself, since physicians who partnered with convent pharmacies had vested financial interests in maintaining these outlets. Nuns themselves influenced specific decisions by exploiting tensions between magistracies and appealing to ducal protection. These competing claims resulted in a haphazard regulatory scene running counter to professionalizing aspirations. Generally speaking, the duke was inclined to protect the healthcare resources these businesses represented, but he did so on his own terms, sometimes undercutting the authority of his own magistrates in the process. Cosimo adjudicated conflicts in ways that extended his personal influence over state functions and increased dependence on ducal favor. This complex regulatory environment gave Florentine nuns some latitude, but it also transformed convent pharmacies into yet another site of Medici state formation.

Convent petitions to the duke illustrate both the ad hoc nature of Florentine pharmacy regulation and the strategies developed by various players. In October 1561, the Sant'Orsola nuns asked Cosimo to allow "any person . . . to enter their convent" without needing to obtain an "additional license from the [archbishop]." Their stated rationale was "to preserve their apothecary shop," which they maintained was "their sole means of support." In requesting a single license issued by the state, the nuns explicitly recognized the primacy of state authority over ecclesiastical supervision. Obtaining this first permit from the magistrates had already slowed business, they complained, since the "inconvenience and bother" reportedly drove "many persons" away. Because the nuns "freely gave away the things they made in the pharmacy," both the sick poor and cash-paying customers were forced to line up "at every hour" to obtain entry permits from the Magistrato. Requiring customers to secure yet another license from the archbishop would prove absolutely fatal to business. Since these nuns were Medici clients, Cosimo instructed the magistrates to seek some kind of accommodation. After making inquiries, they confirmed that the new policy did indeed harm both the pharmacy and its clients. As a workaround, they proposed "that all persons who either have a prescription from any Florentine physician or who get permission from the abbess or pharmacist of that convent can freely enter the premises without additional license." This concession gave the Sant'Orsola nuns firm control over convent spaces and real latitude in dealing with their clientele. Still, Cosimo noted that his decision should not be interpreted as setting a precedent for other convents or giving special support to this "pharmacy business."[136]

Having satisfied these magistrates, Sant'Orsola still had to pass muster with the guild, which conducted pharmacy inspections for the first time the following year (1562). After making the rounds, guild inspectors had nothing but praise for the hospital dispensary of Santa Maria Nuova, "in which they found an abundance of drugs and compounds" maintained "with a rare orderliness and ministration." The same inspectors commended the pharmacies at Sant'Orsola and Santa Caterina as being "well-run and diligently managed" and "found nothing there to fault." The inspectors arrived at a very different conclusion regarding the San Marco pharmacy, however, run by a certain Fra Baccio. That dispensary reportedly "displayed few medicines, the majority of which were badly made and managed." Guild officials proposed that the friar be barred from selling medicines in future, but lacked authority to enforce this recommendation.[137]

Other nuns quickly followed suit in petitioning the guild for modifications. In December 1562, six months after Sant'Orsola submitted a plea, the Santa Caterina nuns asked guild officials to amend the inspection process. Guild statutes stipulated that "religious foundations that sell or dispense medicines outside" their premises should be inspected periodically by two physicians and four respected apothecaries. The abbess and other convent officers, including the pharmacist Giovanna Ginori, humbly argued that convening such an esteemed group was unnecessary, considering "the little bit of pharmacy they practice, above all for their own poor community." They suggested instead that two men—their physician Francesco Gamberelli and the court pharmacist Stefano Rosselli, brother to three of the nuns—inspect "the compositions produced in their pharmacy" as needed, rather than on a fixed schedule. Both men enjoyed solid professional reputations as well as a vested interest in the shop. Since recent guild inspection proved that the nuns made "good medicines, diligently kept," Cosimo granted the request, thereby replacing corporate oversight with a small, self-interested group.[138]

Next to petition were the Annalena nuns (1563), who affirmed the wisdom of guild inspections. In fact, the nuns reminded Cosimo that prior guild review had shown their pharmacy to be "furnished with good, well-compounded medicines." Their central concern was the requirement that "every time they wish to compound medicines, they are obliged to call these deputized inspectors." Here the Annalena nuns skillfully played on heightened expectations about enclosure to strengthen their case. "Since the nuns have their pharmacy in the middle of the convent," their petition stated, "it displeases them greatly that these lay visitors have to pass through almost the entire cloister" to do their job. To avoid this situation, they proposed that the convent physician serve instead as the designated inspector. Once again the nuns successfully narrowed the regulatory circle through the duke's good graces.[139] Ten years later (1572), guild officials eked out a small victory. New restrictions allowed the nuns of Santa Caterina and the Annalena to dispense prescriptions only when their physicians and deputized pharmacists were present, although they made no mention of ready-made products.[140]

Such piecemeal measures continued to mark state regulation of convent pharmacies under Cosimo's sons and successors. Neither Francesco nor Ferdinando developed coherent regulations but instead granted favors to one group or another opportunistically, in line with ducal interests. This pragmatic approach was characteristic of both republican and ducal regimes, which preferred to manage anomalies on a case-by-case basis rather than eliminate them. In the 1580s, Francesco adamantly refused to support guild proposals requiring that all pharmacies be licensed, thereby maintaining a more open medical marketplace. His brother Ferdinando actively supported convent ventures by helping to rebuild the Murate pharmacy after extensive flood damage in 1589.[141] Patronage considerations aside, both rulers recognized that these female-run businesses provided valuable healthcare resources to the local populace.

Catholic reform initiatives complicated this regulatory picture but never succeeded in shuttering convent pharmacies. Tridentine enclosure provisions slowed sales somewhat by the late 1580s and probably necessitated architectural modifications to dispensaries. Still, contradictions within reform initiatives allowed convents to continue offering their products to a more limited public after Trent. The notion of monastic self-sufficiency offered a powerful corollary to *clausura;* consequently, Catholic reformers like Carlo Borromeo encouraged nuns to produce their own medicines in order to reduce dependence on outsiders. In fact, Borromeo urged every convent to maintain its own pharmacy so that nuns could manage simple ailments on their own. Caring for convent sisters in this way had the added bonus of reinforcing a sense of community, which became increasingly fragile as convent populations exploded. Borromeo also recommended that convents cultivate "useful plants" like medicinal herbs, rather than purely decorative ones. In 1601, the Florentine archbishop Alessandro de' Medici (later Pope Leo XI) actively encouraged pharmacy as a path to financial autonomy, advising nuns to increase both the production of textiles and herbal remedies.[142] Backed by high-level churchmen, convent pharmacies flourished across Italy and Catholic Europe throughout the seventeenth century.

The confection of sweets also caught the eye of Tridentine churchmen, leading to increased regulatory efforts. Sugar-based confections played a major role in the Renaissance gift economy, especially as presents given at life-cycle events like weddings and baptisms. Increased demand for these items as well as tighter medical oversight pushed nuns to invest more heavily in their manufacture. Yet the production of sweets remained dogged by complaints about excessive contact with laity and misspent convent funds. One anonymous reformer grumbled in 1563 that aristocratic Neapolitan nuns wasted time and money making "sugared things, little soaps and other culinary products" for people who frequented the premises under the guise of devotion. According to this critic, Neapolitan houses that hosted these crowds "did not appear to be enclosed convents of women, but rather public markets."[143] New ecclesiastical ordinances passed in 1589 vigorously tried to limit the production of "sugared things or those in syrups to be sold or given away."[144] Despite repeated attempts at controlling production, the social currency of sweets was too deeply embedded in Neapolitan society to be curtailed that easily.

Tensions persisted in church circles around the moral implications of marketing medicines openly. As Tridentine rulings gained greater purchase around 1600, Florentine nuns exploited interstices within and between ecclesiastical strictures to retail remedies. One way to skirt opposition was to request donations from clients rather than selling items directly. Taking this approach, the Annalena nuns kept making medicines "for their benefactors," thereby sustaining a workshop that accounted for 7 percent of all convent revenues in 1602–1603. Between 1629 and 1632, a period encompassing a massive plague outbreak, that intake rose to 10 percent of earned income, which kept the convent treasury afloat via a series of loans.[145] Still, marketing medicines, or simply giving away surplus goods under the guise of "friendship and familiarity," rankled some church officials, who rarely spoke with a single voice. In 1637 the Congregation of Bishops and Regulars tried to prevent Roman religious pharmacies from selling remedies for profit, and their remit was understood as applying to all religious houses across Europe.[146] Florentine nuns largely eluded these bans by dispensing their wares gratis or by asking a "just price" in return. By skillfully adapting church regulations and embedding medicines in a gift economy, the pharmacy at Santa Caterina continued to generate income as late as 1805, when local convents were suppressed.

Similar strategies enabled convent pharmacies across the peninsula to operate throughout the early modern period. In Bologna, the apothecaries' guild protested repeatedly to the state medical board that monastic pharmacies presented unfair competition. After lodging complaints in 1697 and 1699 against "the monasteries of both friars and nuns who sell internal and external remedies," the guild took its campaign to the pope himself. Forceful lobbying finally brought the desired outcome. In 1737, Clement XII prohibited "all regular clergy of both sexes" from practicing "the art of the apothecary in Bologna unless for their own internal use," and specifically banned them from selling to the public.[147] Nevertheless, efforts to uproot such long-standing practices often backfired, since they disrupted entrenched patronage patterns and threatened to eliminate still-vital healthcare resources. Given the fragmented politics of the Italian peninsula, local battles might be won but the war dragged on. When the Venetian health board finally brought all monastic pharmacies in the territory under its purview in 1769, the subject city of Padua still counted twenty convents running active apothecary shops.[148]

Despite being overshadowed in the historical literature by monastic pharmacies, apothecary shops run by Renaissance religious women significantly expanded the range of medical resources available to the Italian populace. By commercializing their knowledge, nuns both generated crucial revenue for their communities and developed a recognized public presence as medical agents. Their entry into the medical marketplace was facilitated by established business networks and longstanding familial connections. While Savonarolan motives loomed large in launching the Santa Caterina pharmacy, the glaring need for poor relief in the 1530s and 1540s led other nuns to follow suit. As convent apothecaries grew in number and importance, they fostered synergies between related crafts such as painting, sculpture, textile work, and pharmacy, which shared common materials and production techniques. Consequently, girls belonging to the families of pharmacists, physicians, painters, and artists tended to cluster in particular convents, where they continued to ply the family trade in convent workshops. At the same time, the spatial distribution of Florentine convent pharmacies broadens our understanding of sixteenth-century urban experience by demonstrating the paucity of healthcare resources available to poor or marginalized residents living on the urban periphery.

To be competitive in the medical marketplace, Florentine nuns developed a full line of products that crossed the conceptual divide between internal and external remedies. Ready-made wares—distillates, syrups, soaps, sweets—probably formed the bulk of convent business. Goods aimed at hygiene and well-being offered a huge growth opportunity after 1550, as the culture of prevention gained traction. Capitalizing on this expanding business sector, nuns introduced new products and pursued aggressive marketing strategies such as trademarking wares. They drew on an already expansive network of business connections to supply materials, but they also invested in horticultural experiments to gain a competitive edge. Archival evidence convincingly positions convents as important epistemic spaces for horticultural experimentation. While this area warrants further research, it is clear that Italian religious women played a larger role in early modern empiricism than has been recognized.

Essential to the success of convent pharmacies were steadfast female clients seeking new charitable avenues as well as reliable medical products. Certainly nuns' kinsmen and male allies sent business their way. Yet the ability of convent pharmacies to accommodate gender restrictions, adjust prices, and extend credit within known female circuits gave these businesses a competitive advantage over guild apothecaries. Female patrons probably seized on the sociability afforded by convent dispensaries to extend their social networks while exchanging medical advice. Even within the limitations of available evidence, convent pharmacies emerge as significant transfer points for medical knowledge as well as lively retail outlets.

Professionalizing trends made limited headway in the sixteenth century, in part because the early Medici dukes protected convent pharmacies as recognized assets within the local healthcare infrastructure. State regulation had an opportunistic quality that invited negotiation by interested parties, ranging from the duke and his magistrates to nuns and their friends. Still, the fact that Florentine convent pharmacies came under state regulation at all marked an anomaly on the Italian peninsula—one that signaled the weakness of the local church in the mid-Cinquecento. Although marketing medicines remained a bone of contention between nuns and churchmen, Renaissance nuns successfully adapted Tridentine reforms to their own ends. Consequently, these businesses and the women who ran them remained an integral part of a pluralistic medical landscape well into the eighteenth century.

CHAPTER FOUR

Agents of Health

Nun Apothecaries and Ways of Knowing

SOME YEARS AGO, a noted medical historian remarked that we know more about Renaissance pharmacy than about Renaissance pharmacists.[1] That picture has begun to change recently in conjunction with a broader reassessment of empirical knowledge and the material turn in early modern scholarship. As skilled artisans, Renaissance apothecaries exemplify the ways in which textual knowledge and craft skills informed each other in the early modern period. Throughout the sixteenth century, guild apothecaries engaged in both medical reading and hands-on training to meet increasingly stringent standards of quality control. These medical artisans can be viewed as "hybrid experts" who worked along a "differentiated continuum of forms of knowledge."[2] Apothecaries not only mastered traditional ways of working but also capitalized on market trends, promoting new wares through mechanisms such as advertising. Even if some products represented only slight variants on old standbys, they nevertheless indicate the productive nexus that existed between market and laboratory.

Recent studies of artisanal practice and early modern craft culture raise important questions about how skills and knowledge circulated within and outside commercial workshops, which integrated female labor to varying degrees. Little is known about the training of small-scale retailers who marketed wares from their homes, or the knowledge base that wives and daughters of guild apothecaries developed while running the family business. The convent pharmacies discussed in Chapter 3 provide an exceptional opportunity to assess how early modern women working outside guild settings acquired, deepened, and transferred medical knowledge and technical skills. One reason for nuns' success as commercial pharmacists is the training they shared with guildsmen that was rooted in apprenticeship and medical reading. Like guildsmen, Renaissance nuns experimented with new recipes and modified old ones in order to remain competitive. These similarities support the view that Italian commercial pharmacy was distinguished more by varying skill levels than by distinct ways of knowing. The porous nature of convents also meant that new medical thinking filtered into these institutions through books and personal interactions, transforming convent workshops into important sites of medical exchange. By tracking nuns' training and experimental activities—the focus of this chapter—we can map the flow of craft knowledge in early modern cities.

Examining convent workshops also highlights the materiality of making medicines—its pungent smells, fierce heat, distinctive tools. Italian religious women quickly discovered that retailing medicines in large quantities required specialized equipment, dedicated space, and start-up capital, in addition to broad technical expertise. A close study of the Florentine convent pharmacy at Le Murate in the late sixteenth century takes us into the heart of a female-run workshop, where available technologies affected production strategies. The experimental nature of the convent "kitchen" contrasted with the performative space of the dispensary, where issues of social trust loomed large in the retail process. This analysis also reconsiders nuns' relationship to the Renaissance culture of "secrets." Ranging from practical household tips to industrial trade secrets, these hard-earned bits of knowledge pitted open exchange against proprietary interests. Some scholars have argued that Italian women were conspicuously absent as authors of secrets compared with their northern counterparts. The three case studies presented here challenge these views; they situate Italian religious women squarely at the intersection of scientific discourse, an emerging consumer culture, and the new information economy sweeping late Renaissance Italy.

The Making of Convent Healers

Medical Apprenticeships and Embodied Knowledge

In commercializing their products, Renaissance nuns built on the knowledge and skills developed over generations in tending health needs within their own communities. Central to the transmission of the healing arts were traditional modes of apprenticeship. This centuries-old arrangement remained the principal avenue for training guild apothecaries as well as transmitting assorted craft skills within the convent. Recent studies of artisanal practices have stressed both the embodied nature of early modern handwork and commonalities between types of craft production. Indeed, the sixteenth-century German noblewoman Anna of Saxony proudly referred to her medical remedies as "handiwork."[3] As Pamela Smith has argued, handwork required active, perceptual engagement with the world. Whether making medicines, copying manuscripts, or embroidering silk, practitioners became acutely aware of the materials in their hands as well as the surrounding environment. The essence of skill was attending to material in the moment: learners developed perceptual focus through observation, trained their hands through repeated practice, and improvised using intuition.[4] This way of knowing created a deep understanding of craftwork from start to finish, whether in pharmacy, painting, or fancy needlework. The act of making gave special insight into the finished product—what Bruce Moran has called a "maker's knowledge."[5]

Preparing medicines involved a high degree of sensory apprehension. Written texts and oral instruction shaped this learning process, but its core remained careful observation and the judgment born of experience. Practitioners relied heavily on all five senses to gauge when mixed ingredients had achieved the right color and consistency, made the right sounds while being processed, smelled burnt, or tasted off. Becoming an expert practitioner in pharmacy thus required careful attention to detail and technique. Both manual dexterity and disciplined attention were portable across certain crafts to some extent. As noted previously, grinding pigments for altarpieces or making inks required much the same skill set as preparing certain types of medicine. Similar transfers among types of handwork prevalent in convents—painting, sculpture, needlework, copying, illumination, pharmacy—conditioned workshop production.

Convent pharmacies also paralleled guild workshops in their organizational structure. The art of making medicines involved tacit knowledge best imparted by working alongside a practiced teacher. Consequently, convent workshops divided duties between an expert head pharmacist and one or more apprentices.[6] The head convent pharmacist *(spetiala maggiore)* was responsible for a wide range of administrative and practical duties. Among these tasks were compounding complex preparations, supervising production, maintaining accounts, taking inventory, training new assistants, staying current with medical knowledge and market trends, and coordinating efforts with the convent nurse and abbess. Working along a continuum of knowledge and expertise, some of these women became connoisseurs of herbals and other materials; others focused on refining traditional recipes. The financial success of the pharmacy depended to a high degree on their creativity, skills, and dedication.

Assisting the head pharmacist was an apprentice or assistant *(spetiala minore),* who carried out basic procedures such as pounding herbs or preparing binding materials for pills and ointments (Fig. 4.1). This arrangement prepared apprentices for more advanced work by introducing them to specific manufacturing techniques and developing their firsthand understanding of materials and processes. Collaboration was essential, both for improving skills and dividing labor efficiently. Apprentices also performed physically taxing work beyond the capabilities of older, expert practitioners. When the Chiarito nuns petitioned the Florentine archbishop in 1585 to admit a new serving sister "to assist in their pharmacy," they explained that "there are three nuns who have practiced the art of pharmacy there for many years." However, two of them were now "incurably ill" and unable to handle their former duties. The nuns ended their plea by stating that "this pharmacy is extremely useful and necessary to the convent, and we don't want [it] to crumble."[7] Nuns probably began these apprenticeships in their twenties, after gaining sufficient discipline and manual skill through silk work, which formed the lifeblood of Renaissance convent economies. The need to

Fig. 4.1. Dominican nun preparing rose unguent, seventeenth century. Biblioteca comunale dell'Archiginassio, Bologna. MS. B. 4231. Tavola 20. By permission.

master different kinds of craft work made them slightly older than young boys training in artistic workshops, who began apprenticing between ages eleven and eighteen.[8]

This system is typified by the career of Anna Capponi, a Franciscan nun at the Florentine convent of Sant'Orsola. Capponi began training with Benedetta Bettini—a renowned monastic practitioner in her day—and then succeeded her as head pharmacist upon the latter's election as prioress in 1550. Persistent health problems made it difficult for Capponi to handle the demanding workload, although she reportedly managed to complete her initial three-year term through sheer grit. When Bettini was elected prioress again a few years later, Capponi took over pharmacy operations. This time, however, she entrusted much of the heavy work to her "beloved apprentice and companion," Evangelista dal Borgo. Despite Capponi's chronic ill health, she continued working in the pharmacy until her death in 1574. Over the years, the partnership between master and pupil fostered a mutual sense of affection. When Capponi's apprentice died unexpectedly a few weeks after her, the nuns decided to bury the younger woman in the adjacent grave as a way to honor their collaboration. Upon opening the gravesite, they reportedly discovered that Capponi's body "gave off no stench whatsoever"—a sure sign of sanctity in the Catholic world.[9]

Launching a commercial pharmacy must have encouraged Renaissance nuns to consult lay experts about various technical matters, from equipment to economies of scale. The Florentine nuns of Santa Caterina likely got advice from their supplier, the apothecary Bernardo Mini, when setting up their still room in the 1510s. Similarly, the Annalena nuns probably conferred about materials and equipment with their supplier Andrea Del Garbo two decades later. These instructional partnerships circulated pharmaceutical knowledge through existing social channels, mapping information circuits onto existing social relations. Knowledge transfers assumed a more formal institutional footing in seventeenth-century Dijon, where the main civic hospital launched a commercial pharmacy in 1643. Initially the governing board sent two master apothecaries to instruct the hospital nuns three times a week in "the method of working and the necessary preparation of medicaments."[10] Once the nuns had acquired the requisite skills, however, they began training new apprentices themselves.

Other locales in early modern Europe relied on similar collaborations. Nuns in southwest Germany worked extensively with local apothecaries to create coordinated networks of care in the eighteenth century. Postulants who showed aptitude in this field trained in local town pharmacies for three years before taking vows, a period commensurate with German guild standards for apprenticeship. Afterward, the newly trained nuns entered the convent and served an additional stint as "journeymen" under the head convent apothecary. They continued to update their knowledge in order to provide expert care to local residents, especially in the countryside.[11] These modes of craft training remained deeply entrenched in pharmaceutical practice throughout the early modern period. Only in the nineteenth century did European pharmacy become a profession requiring academic training, which closed off opportunities for women's participation.

Because it took time to master this healing art, most Renaissance nun apothecaries remained in office longer than other convent officers, who usually rotated positions every one or two years. Agostina Rinuccini ran the Murate pharmacy for fifteen years before being elected abbess in 1572, while Giovanna Ginori served as head pharmacist at Santa Caterina for almost four decades prior to her death circa 1579. The nuns of San Vincenzo in Prato utilized this same model; Maria Angela Segni (d. 1606), daughter of the humanist Bernardo Segni, distinguished herself in the apothecary for twenty years.[12] The Annalena nuns took a slightly different approach when developing proficiency in this area. Rather than concentrating expertise in the hands of one person, they struck a balance between building skills and distributing medical knowledge more broadly throughout their community. The head apothecary there normally served for three to five years before rotating out of office, often becoming the next head nurse, where she put her know-how to related uses. Consequently, Orsola Del Riccio ran the Annalena pharmacy from 1558 to 1563 and again from 1566 to 1570; she had just begun another term in 1579 when she died unexpectedly. Similarly, Lena da Catignano fulfilled this office multiple times between 1563 and 1587. Nuns who displayed special aptitude for one or both positions rotated back in for additional terms.[13]

Given their depth of experience, knowledgeable healers were consulted years after leaving office. The Dominican serving nun Andrea

(d. 1613) reportedly attained great proficiency in making medicines at San Jacopo, despite being illiterate. According to the convent necrology, she distinguished herself in the pharmacy so well that "even though she didn't know how to read, she knew and understood everything about that office so well that after she finished her turn, all of the other nun-apothecaries always consulted her and never made an electuary or other important preparation unless she was present."[14] Her example highlights the extent to which many contemporaries considered Renaissance pharmacy a branch of practical knowledge open even to unlettered experts. By transmitting skills and knowledge from one cohort of nuns to another, Florentine convents developed a self-sustaining cadre of personnel who circulated between pharmacy and nursing. They still hired other practitioners—barber-surgeons to pull aching teeth, physicians to treat serious conditions—but these competencies meant that nuns could manage some of their own health needs while serving the public good.

Convent records afford a rare glimpse into the life and work of Giovanna Ginori, whose expertise proved crucial to the growth of the Santa Caterina pharmacy detailed earlier.[15] Giovanna (secular name Ginevra) entered the community at age twelve and took the veil four years later (1517). Three of her sisters became nuns at the affiliated house of San Vincenzo in Prato soon after; one experienced a miraculous cure by a vision of Savonarola himself. Their father, Simone di Giuliano, was a noted Savonarolan, and many of their kinswomen were known for charitable work and devotional self-confidence. In 1541 Giovanna's relative, Leonarda Barducci Ginori, founded a shelter for abandoned girls that typified local social welfare initiatives.[16] This shared family interest in helping the helpless probably accounts for Giovanna's early placement in the pharmacy, where she apprenticed for six years—a term roughly comparable to the three to eight years required of guild aspirants in Siena and Rome.[17] Ginori reportedly "quickly became so expert at the job that the other nuns decided amongst themselves not to confer any other convent office upon her."[18] Hence she ran the pharmacy for the next thirty-seven years, developing new products, expanding its client base, and building its reputation.

As she matured in her profession, Ginori became "extremely knowledgeable about medicinals and botanicals," according to her fellow nuns. By the mid-sixteenth century, she enjoyed a widespread reputation as an expert practitioner. Florentine physicians and other apothecaries reportedly maintained that "anyone wishing to take advantage of good medicinal products and be provisioned with conserves, juleps, and other substances to purge and maintain good health" would be well-advised to consult her. In order to perpetuate this legacy, Giovanna trained two "excellent apprentices" who shared her dedication and spiritual values. Over the years, these apprentices mastered Ginori's "way of working with great understanding." Every practitioner developed personal techniques and tips for making standard products, which were integrated into the training process. Knowledge transfers between masters and apprentices were not static but encouraged learners to improvise in response to unexpected situations. The process of learning by doing fostered intuitive leaps and invited experimentation that kept products alive to new approaches or market trends. After Ginori died, circa 1579, one of her apprentices—fifty-year-old Raffaella Gondi, whose relative Marietta Gondi had founded another girls' shelter in 1554—took over the pharmacy.[19] The other apprentice, Raffaella's older sister Gostanza, became the convent nurse. In turn, these Gondi nuns trained the next generation of convent apothecaries, who continued their healing work after 1600.[20]

Despite commonalities with guild apprenticeships, nun pharmacists faced different occupational expectations. These women still participated fully in the liturgical life of their communities, placing heavy demands on their time. It was difficult to balance these activities with commercial imperatives to test new products, keep the books in order, train new apprentices, and stay abreast of current knowledge. In addition, nun apothecaries often assisted the convent nurse in treating ailing members of their community. To meet these demands, Ginori reportedly stayed up "eight nights in a row with little or no sleep" caring for sick nuns. Similarly, her counterpart at the Murate, Beatrice Benci, supposedly "remained on her feet day and night without sleeping" during her many years as head apothecary. Other descriptions of convent pharmacists stressed compassionate virtues such as "charity and loving kindness."[21] Such reputed selflessness anchored nuns' public personae as charitable healers, which built social credit with clients. Consumer trust in convent remedies was grounded not only in perceived skills but also in the moral virtues of their makers.

In fact, the pharmacy became an important proving ground for top administrative positions within the convent. Running the shop melded compassion, specialized knowledge, and practical skills with sound judgment and an eye for public relations. All were highly valued traits when choosing convent leaders, especially in light of the tumultuous religious landscape of late Renaissance Italy. It is not surprising that nun apothecaries were elevated to top leadership positions more frequently than other convent officeholders such as sacristan or novice mistress. Two of the earliest known nun pharmacists, Cecilia Michelozzi at Santa Caterina and Maria Salome de' Lioni at the Annalena, were elected to the highest governance positions, as was Orsola Del Riccio later in the century. When the Murate transferred five of its members to launch the new convent of Santissima Concezione, circa 1590, the nuns chose the "prudent and judicious" sixty-year-old Umiliana Lenzi to serve as the first abbess. Among the women she chose to accompany her were two nuns who had worked beside her in the pharmacy for many years. Both women—Laura Aldobrandini and Laudomina Malatesta—were "known to be extremely suitable" when dealing with practical matters; both served lengthy terms as head apothecaries at their new convent before being elected as abbess there.[22]

The career path from still room to chapter room is perhaps best exemplified by Benedetta Bettini, who mixed technical skill with deep compassion for the sick. This trusted apothecary helped develop the "beautiful pharmacy" at Sant'Orsola much favored by the Medici, who relied on her proprietary pills and curative compounds. Writing from firsthand knowledge in 1580, the Franciscan chronicler Fra Dionisio Pulinari described Bettini as "a calm, quiet woman of humane words and excellent deeds." As a young nun, she had apprenticed in the pharmacy with Paola Signorini and took over the workshop when the latter woman became prioress. Like her Dominican counterpart Giovanna Ginori, Bettini cultivated the pharmacy business so successfully that one guild apothecary reportedly told the friar, "had Sister Benedetta been able to run a licensed pharmacy, all of Florence would have flocked there."[23] Bettini served a total of eighteen years as prioress of Sant'Orsola from 1550 until her death in 1577, shuttling between the apothecary and the head office. Her convent sisters reportedly were so satisfied with her mode of governance that they tried to retain her permanently as prioress, rather than electing a new head officer every three years as required. During these stints, Bettini relied heavily on her assistants to keep the pharmacy running at peak production. Grand Duke Francesco and his first wife, Giovanna of Austria, developed close personal ties to Bettini; the grand duchess commissioned a commemorative portrait at her death, and a similar likeness was ordered by the new convent prioress to honor her legacy.[24]

By contrast, the professional formation of guild pharmacists focused more narrowly on their intellectual training and, to a lesser extent, on their ethical qualities. The 1567 edition of the Florentine pharmacopeia was the first to articulate an official vision of the pharmacist and his shop. This text stated that "a good apothecary must have skill, an agile body, and good habits." As a tradesman, "he should be diligent, faithful and not greedy." Sound training and ethical comportment went hand-in-glove for any Renaissance professional, but the stakes were particularly high for apothecaries, whose sloppy or fraudulent practices might do irreparable harm. Upon entering apprenticeship, the good apothecary should be trained "in the knowledge of simple and compound medicines." Part of his knowledge derived from reading canonical texts. The 1567 formulary noted that a licensed pharmacist needed "to know enough Latin that he is able to read Dioscorides, Galen, Pliny, Serapione, Mesue, Avicenna and other authors." If an apprentice lacked Latin learning, he might "be instructed by an intelligent master and apply himself to reading the moderns who had translated these works or written about this subject in the vernacular."[25] Still, the 1567 pharmacopeia insisted that books and workshop experience were not sufficient in and of themselves. The apprentice also needed firsthand exposure to medicinal plants in situ. This recommendation reflected two new developments in mid-sixteenth-century Italy: the foundation of botanical gardens designed for teaching purposes in university venues such as Florence, Pisa, and Padua; and the passion for exploring the book of nature among naturalists, botanists, and natural philosophers.[26] By the late 1560s, Italian guilds were beginning to adjudicate complex relationships among forms of knowledge—Latin and vernacular, empirical and learned—while relying on a corporate ethos to shape everyday behaviors.

Medical Reading in the Convent

Although nun apothecaries were not destined for guild membership, their training conjoined experiential learning and medical reading in similar ways. These dual sources of knowledge were given powerful visual expression in the portrait of Pierrette Monnet (1557–1628), chief pharmacist at the Burgundian civic hospital in Beaune. Painted by the Dijon artist Nicolas Quentin in 1624, when the nun was sixty-seven years old, the portrait is still conserved at the hospital today. Known for her proprietary remedies and assiduous care of the sick, Sister Pierrette is depicted with a mortar and pestle—the signature tools of the trade—in one hand; her other hand rests on a book. This rare visual representation of a nun healer can be read as validating the ways that textual knowledge and artisanal skills informed each other.[27] The nature and extent of nuns' medical reading warrants further research, but it is clear that some religious women with inquiring minds educated themselves about the workings of the body and the natural world. In 1617, the erudite Bolognese nun Diodata Malvasia recalled that, following her appointment as convent nurse, "it pleased me to read something of medicine in order to carry out my duty as best I could." Like the humanists of her day, this autodidact went back to original sources, selecting Galen's treatise *On the Temperaments* to further her theoretical understanding of the body.[28]

In evaluating medical reading within convents, it is important to distinguish between reading books and owning them. Despite the flood of print materials after 1500, books remained costly, especially large, illustrated volumes like herbals. Inventories of Italian convent libraries do not necessarily give a full picture of what nuns read, since religious women often borrowed books from relatives and friends and then circulated them within the community. The Florentine polymath Fiammetta Frescobaldi (1523–1586) relied heavily on her brother and other relatives for the loan of reading materials that included published letters, travel accounts, and studies by natural philosophers. Culling information from these works, Frescobaldi penned an ambitious chronicle of the world as well as a history of her own community of San Jacopo. She also made learned compilations for the convent library, wrote abridged histories, developed a Spanish grammar containing useful phrases for nuns, and translated other writings.[29] As Elaine Leong has shown, these varied reading practices—selecting portions of texts, assessing their informational value, classifying extracts to form original, freestanding works—were not simply passive modes of reproducing knowledge, but instead consolidated broad-scale knowledge transfers about the known world.[30] In Frescobaldi's case, her engagement with borrowed texts also created unique treasuries of knowledge tailored for convent spaces. Not all nuns possessed equally inquiring minds; nevertheless, sustained exposure to contemporary intellectual currents through books helped integrate early modern convents into larger communities of medical practice.

A closer look at the medical texts housed in Italian convent libraries circa 1600 reveals a wide range of printed and manuscript materials written in both Latin and Italian. Most inventories that have come to light were compiled at the request of ecclesiastical authorities interested in ferreting out suspect works listed on the Index of Prohibited Books. The majority of convents reporting were small rural houses far from major printing centers. Unfortunately, no inventories of convent libraries survive from important printing hubs like Venice.[31] Although spiritual readings predominated, these lists reveal a substantial number of medical books with a practical bent. One Clarissan convent in the provincial hill town of Narni registered "eight books belonging to the pharmacy," five of them in Latin. These works included an edition of the *Antidotarium Romanum* (1585); two vernacular recipe books culled from Galen and other sources; Pietro Andrea Mattioli's short Latin treatise on distilling herbs *(De simplicium preparatione)* published in 1569; Manlio de Bosco's *Luminare maius,* a popular herbal recipe book first printed in Venice in 1494; and two major antidotaries, including the medieval compilation by Messer Nicolao.[32]

In addition, convent libraries across the peninsula housed texts that taught nuns how to fabricate medicines, care for the sick, and make informed decisions about healthcare. Library inventories listed numerous health manuals, plague treatises, books of secrets with a medical inflection, and at least one treatise on the medicinal value of foodstuffs. In contrast to the standard medieval formularies noted above, these vernacular texts reflected more recent medical thinking. Among the works owned by Italian convents was the health manual titled *Il perché,* first published by the Bolognese physician Girolamo Manfredi in 1474 with numerous reprints. This book had enough staying power as a reference work to be kept on-site at the Medici ducal pharmacy in 1591.[33] Other recent vernacular works listed in convent inventories included Leonardo Fioravanti's book on surgery (first issued 1561, definitive edition 1582), and the *Giardini de segretti* (1586) authored by Giovanni Vitrario (called "il Tramontano"), a Roman surgeon and distiller.[34] In contrast to the paucity of medical manuscripts found in medieval convents, Renaissance nuns clearly availed themselves of medical print as a way to expand their health literacy and artisanal capabilities.[35]

No inventory of the library at Santa Caterina survives, but the experience of Giovanna Ginori there permits some informed speculation about the role of medical reading in training convent apothecaries. As Smith has noted, the relationship between the written word and experiential training remains problematic. How-to manuals could only point to production techniques and their underlying bodily activity, rather than be fully formative of them.[36] Still, the emergence of civic pharmacopeia after 1500 reflected early institutional attempts to establish common standards for preparing medicaments. By listing the texts licensed pharmacists should own, the 1499 Florentine *ricettario* set the benchmark for what was considered foundational knowledge.[37] These texts included Simon of Genoa's *Clavis sanationis,* a botanical dictionary based on Greek and Arabic authors printed several times in the fifteenth century; Matteo Silvatico's *Liber pandectarum* and Abulcasis' *Liber Serapionis;* and the famed medieval antidotaries compiled by Mesué and Messer Nicolaio. Vernacular versions of all of these works, along with diverse commentaries, were readily available in both manuscript and print by Ginori's day.

Given Ginori's extensive connections and the pharmacy's business reach, it is reasonable to assume that she had access to standard reference works as well as recent publications. She almost certainly had read Mattioli's commentary on Dioscorides' *De materia medica,* which sold over 30,000 copies in the sixteenth century alone. Reissued several times between its first appearance in 1544 and the definitive edition of 1568, this best-seller quickly became the gold standard for Renaissance pharmacy. The appendix on the art of distilling herbs was later published separately; its brevity and affordability diffused technical information about making distillates even more widely. Similarly, practical manuals like Prospero Borgarucci's *La fabrica de gli spetiali* (1566) informed readers about recipes and equipment that could be utilized by both household and commercial practitioners.[38]

Moreover, the 1567 Florentine *Ricettario medicinale* functioned more like an instructional manual than had earlier editions issued in 1499 and 1550. The guild required all licensed shops to keep this civic pharmacopeia on hand to ensure that products conformed to its standards. In fact, this updated text was one of the first purchases the San Domenico nuns made when launching their pharmacy in 1570. Not only did this recently revised formulary reiterate the proper way to dry herbs (in a south-facing room free of moisture, dust, and smoke); it also added an entirely new chapter on distillation, including an illustrated guide to building a proper furnace. Another new feature was a long discussion of local medicinal waters like *acqua del tettuccio,* which formed an important part of Santa Caterina's brokerage business.[39] Nuns profited still further from the ongoing popularization of canonical medical works, whether as patients or practitioners. Ginori's successors at the workshop surely capitalized on the 1581 publication by Fra Serafino Razzi—confessor at San Vincenzo in Prato—of an Italian version of the famed Salernitan health manual *(Regimen sanitatis salernitanum),* which he translated for the benefit of "all those who can read, including women."[40] Medical books with a practical bent formed the bulk of nuns' medical reading, but like the literate public, they enjoyed growing access to the theoretical foundations of Galenic medicine through vernacular print.

Understanding nuns' training in pharmacy not only requires asking what books they read, but how they read them. Although we can only infer what titles Ginori may have read, these books almost certainly were subjected to intensive reading. We know that early modern Englishwomen perused their herbals methodically, going over individual entries one by one, making selective notes and sometimes culling the most useful information into stand-alone extracts. Acquiring knowledge through medical reading required active engagement with the text. After repeated scrutiny and consultation with other trusted knowers, genteel household practitioners both synthesized content from these useful books and tested it empirically.[41] The often idiosyncratic organization of herbals, recipe books, and health manuals lent itself to the process of selective reading and rereading. Similar reading habits in religious houses were facilitated by proximity to reference materials stored in convent workrooms. Benedetta Bettini stashed various well-thumbed business ledgers and vernacular works "in a cupboard in the pharmacy," where she could consult them as needed.[42] Intensive handling of working texts helps account for their high wastage, much like the everyday devotional objects that were literally consumed by pious users.

Information Networks

Hands-on practice and medical reading gave Renaissance nuns a solid platform for making medicines, but it was their rich information networks that kept them abreast of current medical thinking and market trends. Recent studies have emphasized the social dimensions of early modern experimental activity, noting that frequent interactions among artisans, physicians, botanists, and other scientific enthusiasts promoted innovation and risk-taking.[43] Nuns cultivated a wide range of medical informants to supplement their skill set. Throughout her tenure, Giovanna Ginori and her apprentices at Santa Caterina remained in regular contact with numerous pharmaceutical experts. These included the apothecary of San Marco; the convent's business associates at the Mini and Pinadori apothecary firms; and the convent physician Francesco Gamberelli, member of the prestigious College of Physicians, for whom Ginori compounded prescriptions. Other conduits for medical exchange included the nuns' lay supervisors, one of whom was the Medici court apothecary Stefano Rosselli. This botanical enthusiast also advised the Murate pharmacy in the 1580s, as discussed below. In this dual capacity, Rosselli widened the informational reach of both houses while drawing their businesses into closer contact. Other pathways for brokering knowledge were provided by intermediaries like Fra Serafino Razzi, who frequented Dominican pharmacies in Florence and Prato. With his easy access to convent precincts and his deep commitment to medical charity, this monastic go-between could transport product samples as well as technical tips among communities.

Employing the same physician also helped circulate medical information between religious houses. In the 1570s, Alessandro Bencivenni was simultaneously salaried by the nuns of the Annalena, San Niccolò Maggiore, and San Paolo; the latter ran a thriving hospital pharmacy. Physicians like Bencivenni were often deeply immersed in the business side of things, either directly or as members of professional medical families, giving them added impetus to share technical knowledge with nuns. The extended Bencivenni family enjoyed an ongoing relationship with San Paolo throughout the dramatic jurisdictional changes that transformed it from convent to hospital. Piero Bencivenni, who owned an apothecary shop on Ponte Vecchio, was the nuns' principal supplier of materia medica from 1548 to 1566. A few years after he died, his relative Alessandro was appointed physician there, while his kinsman Domenico became the hospital warden the same year (1570).[44] Alessandro Bencivenni's involvement with the Annalena was similarly complex. Besides serving as convent physician, he boarded his daughters Margherita and Gostanza there as pupils between 1577 and 1593. Bencivenni allocated part of his salary, some of which came from pharmacy revenues, to offset their boarding expenses, creating an internal financial loop.[45] These experiences point to the continued appropriation of Florentine religious institutions by local families in the late sixteenth century. At the same time, they illuminate how medical and technical information—some of which represented the latest experimental currents—flowed in and out of convents with relative ease.

As this last example suggests, one of the most striking features of nuns' information networks is the extent to which they mapped onto kinship relations. The centrality of kinship to Italian life needs no rehearsal here, but it is worth reiterating that nuns' relatives always enjoyed easier access to convent spaces, irrespective of enclosure provisions. Familial ties gave nun apothecaries exceptional connectivity to local medical networks, especially since painters, physicians, and pharmacists bunched daughters in the same religious houses. These patterns predated the clustering of health-related occupations analyzed by Sandra Cavallo for eighteenth-century Italy.[46] Moreover, many nun pharmacists hailed from the same lineages, such as the Bettini, Ginori, Gondi, Rucellai, Rosselli, de' Lioni, and Del Riccio.[47] Patrilineal networks are far easier to trace because of a shared surname, but conduits for medical exchanges included female relations as well. Nuns' kinswomen who were involved in social welfare work, especially those with Savonarolan affiliations, shuttled between convents and girls' shelters, creating additional pathways along which medical information circulated.

The case of Orsola Del Riccio illustrates how family ties exposed Florentine convents to wider empirical networks. An expert practitioner, Sister Orsola served several rotations as the Annalena's head apothecary between 1558 and 1583, during which time the pharmacy reached peak productivity. Between turns in the dispensary, she served as convent nurse before being elected prioress in 1591, when administrative responsibilities increasingly absorbed her attention.[48] Throughout these decades, this medical artisan cultivated contacts with local naturalists and experimenters through the mediation of her younger brother, Fra Agostino Del Riccio, the botanical expert noted in Chapter 3.[49] A voracious polymath, he frequented Medici experimental circles at the Casino di San Marco, the city's major clearing house for scientific and technological activity. The Casino pooled the technical know-how of men like Stefano Rosselli, the naturalistic painter Jacopo Ligozzi, the collector Niccolò Gaddi, the botanist Giovan Vettorio Soderini, the hospital surgeon Giovan Battista Nardi, the priest and glassmaker Antonio Neri, and other experimental enthusiasts.[50] Through her brother's mediation, Orsola probably was exposed to some of the latest medical, scientific, and technological advances that surfaced at the Casino, whether innovative methods for sheltering plants or new distillation techniques. By gathering useful facts, selecting practical information, and assessing its use value for the convent, Orsola could act as a knowledge broker in her own right. Communication between other family members must have extended the information networks linking the Annalena to other convents. One of their sisters staffed the Sant'Orsola pharmacy at midcentury; two other kinswomen joined the Murate in the 1580s and 1590s, although their role in that pharmacy is unclear.[51] Since all these convents engaged in pharmaceutical experiments of their own, it is not unreasonable to think that practice-based knowledge flowed out from them in multiple directions as well.

This clustering of medical and experimental interests along familial lines is further illuminated by the career path of Benedetta Bettini, discussed earlier. This longtime Franciscan apothecary shared a deep interest in pharmacy with her kinsman Giovanbattista Bettini, one of the three Florentine noblemen tasked with revising the Florentine pharmacopeia in 1567. His appointment to this position almost certainly reflected ducal perceptions that his views bridged both established and innovative medical thinking. Besides setting new civic standards for the production of medicines, Bettini contributed to a larger web of medical exchanges that fed into convent knowledge transactions. He was a devoted benefactor of the local pox hospital, in which capacity he recommended hospital nurses and other personnel until his death circa 1589. Crossing the often rancorous divide between mendicant orders, he also advised the Dominican nun Giovanna Ginori in the Santa Caterina apothecary as a lay convent supervisor in the 1570s and 1580s. In fact, Bettini had been appointed to this supervisory position in part because one of his older female relatives, Sister Bartolomea Bettini, had helped launch the pharmacy some decades earlier. Still another female relative allied through marriage became head apothecary at San Vincenzo in Prato.[52] This was Vincenza Nardi, daughter of the humanist Jacopo Nardi and his wife Elena Bettini, who reportedly had been healed by a portrait of Savonarola in the 1520s. Vincenza supervised the pharmacy along with her two sisters from its opening in 1562 until her death in 1570. Other Bettini nuns scattered across Tuscany staffed convent pharmacies, infirmaries, and painting workshops in both Franciscan and Dominican houses, while at least five Bettini friars served in comparable roles at San Marco. Similar patterns held true for the Ginori and Roselli families, whose numerous nuns and friars became experts in healthcare as well as the production of religious imagery.[53] Clearly nun apothecaries were a constituent part of occupational clustering and the creation of professional medical dynasties; these women cultivated familial networks across mendicant affiliations as a principal axis for producing, gathering, and disseminating health-related information.

Other opportunities to exchange knowledge about techniques and materials were created by the admission of Spanish and Portuguese women into Florentine convents after 1550. Sponsored by an increasingly international Medici court, these daughters and dependents of foreign courtiers breathed new life into local practice. One of the most active herbalists at the San Domenico pharmacy in the 1570s was a Portuguese nun renowned for her distillates. Her ways of working quickly spread to the nuns of the nearby Chiarito convent, who used their female gardener and the factor's wife as go-betweens in purchasing remedies and other wares.[54] Other windfalls energized the exchange of pharmacy practices. In 1577 the guild apothecary registered "at the sign of the Moor"—one of the city's leading enterprises—enrolled his niece in the convent of San Clemente, which was just opening a commercial pharmacy. This established guildsman provided the equivalent of 400 scudi in medicinals from his own stock in lieu of a cash dowry to launch the new venture.[55] Consequently, this shop started with a competitive business capacity, its shelves already stocked with an array of guild-approved materials and ready-made goods. Animated by kinship, charity, and common financial interests, nuns' information networks reveal the extent to which the circulation of medical knowledge mapped primarily onto well-traveled avenues of social exchange, but also moved beyond them.

Spaces, Materials, and Technologies

Convent records provide valuable insight into the spaces used to make and market medicines, as well as the objects that filled them. Like guild-licensed shops, Renaissance convent pharmacies typically featured two or three separate rooms: the dispensary for displaying and selling products; the actual production site or "kitchen" where herbs were distilled and medicines compounded; and sometimes an adjacent area for storing medicinals and ready-made products. Even the famed Medici ducal pharmacy conformed to this spatial arrangement.[56] Underpinning this disposition of space was the need to protect raw materials and finished products from the intense heat of the workroom in order to preserve their medicinal properties. In mentally designing an ideal convent complex in 1577, the reformer Carlo Borromeo recommended that convent pharmacies should annex a small storage room to the main workroom, "where the distilled water and other jars with ointments or medications" would be kept. Ideally the convent also would maintain "another cool room in which the herbs and distillation jars" would be stored.[57]

Other sources provide a glimpse into the public showroom of the convent dispensary, where customers ordered prescriptions, purchased ready-made products, and consulted with nun apothecaries in informal acts of "counter prescribing." Recent studies concerned with consumer experience have focused on the distinctive retail spaces of Italian Renaissance pharmacies, whose material furnishings promoted trust in both the efficacy of remedies and their makers. Dispensaries were performative spaces where issues of trust were enacted in the retail process. Maintaining a sense of orderliness was imperative for both medical and commercial reasons: apothecaries needed to be sure that they dispensed the right medicines, while customers wanted to be assured that they received the right remedies. Evelyn Welch has argued that Renaissance dispensaries relied on strong visual elements such as clearly marked storage jars and neat shelves of ready-made wares (Fig. 4.2), in order to inspire consumer confidence.[58] These considerations probably figured into the decision made by the Chiarito nuns in 1599 to invest five florins in painting the cabinets and shelves lining their retail space.[59] Given the limitations imposed by Tridentine enclosure provisions, it is possible that serving nuns who had not taken solemn vows handled face-to-face transactions, or that nun apothecaries were screened off from public view in some way. These retail spaces also could be situated outside of officially designated areas of enclosure.

Despite site challenges, it is clear that convent pharmacies participated in essential aspects of market culture that allowed them to compete successfully in a crowded medical marketplace. For instance, a number of Italian convents commissioned the beautiful maiolica jars for which Renaissance pharmacies are renowned. Unfortunately, only a few such collections have survived. One of the largest extant collections was commissioned by the Roman nuns of Santa Cecilia in the early seventeenth century. These women outfitted their commercial pharmacy with high-quality wooden cabinets and some 200 maiolica vases for display in the sales room. The entire contents of the pharmacy, including scales, mortars, recipes, documents, and medicaments, was transferred to the Vatican in 1870 in a perfect state of conservation, but remains off-limits to the public.[60] Similarly, the nuns of San Benedetto di Montefiascone near Viterbo commissioned a full range of maiolica wares for their dispensary in 1652, when they were expanding their business. Conserved at the Palazzo Venezia in Rome, many of the vessels in this magnificent collection depict the jovial figure of their patron saint Benedict.[61] The prominent display of monastic emblems and patron saints on storage containers (Fig. 4.3) not only helped to sacralize the medicinal preparations inside but also served an important advertising function. Exquisite maiolica jars lining dispensary walls visually distinguished phar-

Fig. 4.2. An apothecary instructing an apprentice in his shop, 1505. Courtesy of Wellcome Images

macy wares from competitors and promoted an early form of brand awareness within a crowded semiotic field.[62]

Other furnishings in convent dispensaries built consumer trust by linking convent healers to the sacred. Prior to the devastating flood of 1557, the ground-floor dispensary at Le Murate displayed a sculpted marble bas-relief of the Madonna and Child on a counter opposite the

Fig. 4.3. **Pharmacy jars depicting the saints John the Baptist and Lawrence, Venice, 1501.** Courtesy of the Science Museum, London

door "where the preparations and medicines were arranged." Little is known about its manufacture in the 1450s, either by Donatello or Desiderio da Settignano, or about its acquisition by the convent. However, its iconography, which explicitly invoked the cult of the Virgin, may have been particularly significant for the nuns and their female patrons. Having survived the flood, this image miraculously cured one of the nuns a few years later. The head apothecary, Agostina Rinuccini, reportedly spent many hours venerating this image during the fifteen years she spent managing the shop.[63] Dispensing medicines and "counter prescribing" in the presence of a miracle-working icon could only have enhanced the perceived efficacy of convent remedies, while reminding clients of the religious rationale animating nuns' pharmacy work.

Convent sources provide special insight into the materiality of pharmaceutical production in late Renaissance Italy, including the organization of workspaces and the technologies women used in fabricating remedies on a commercial scale. Recent explorations of the workrooms and medicinal inventories of early modern European noblewomen have offered a valuable glimpse into their everyday working conditions and personal work habits. Because their apothecaries were organized for familial and charitable purposes, however, these sites generally operated on a smaller scale than convent pharmacies and remained relatively free from market pressures.[64] Since Renaissance nuns had to petition multiple supervisors whenever they wanted to alter their physical complex, they indirectly described the material conditions, spatial constraints, and ordinary hazards implicit in making medicines on a large scale. In an undated letter, for instance, the abbess of Corpus Domini in Bologna—one of the city's most renowned healing centers—complained bitterly that their pharmacy was seriously inconvenienced by high humidity levels most of the year. The situation had become so severe that the nun pharmacist laboring in the "kitchen," where various distillation procedures took place, "was in danger of fainting, as has happened several times." Adding to their woes was the fact that the ground-floor workroom was plagued by ants in the summer, owing to the "sugary products" stored there. Hemmed in by adjacent buildings, the convent successfully proposed that a cooler underground room be excavated beneath the pharmacy.[65]

Evidence from the Florentine convent of Le Murate permits a rare look at the evolution of production spaces as well as their adaptation to new technologies and market trends. The entire pharmacy had to be rebuilt several times in the late sixteenth century because of flood damage. Although these episodes obviously disrupted business, they also created opportunities to update equipment and make other technical decisions affecting production. By 1500, the "kitchen," store room, and dispensary were all clustered together on the ground floor. This location allowed customers easy access from the street but also made the rooms vulnerable to flooding from the Arno River, since the Murate was situated in a low-lying urban area. Indeed, the epic flood of 1557 inflicted massive damage on the house, completely destroying the dispensary, infirmary, all of the medicinals, other storerooms, portions of the church, and much of the convent archive. Reeling from this blow, the nuns prioritized the building of a new storage room on an upper floor "to keep the medicinal items safe" as part of recovery efforts, despite the logistical inconvenience. Other pressing construction projects had to be postponed because of limited funding. Both the sequence of rebuilding and the significant capital outlay it required attest to the pharmacy's perceived value to the house. Completed in 1562 at the cost of 100 scudi, this spacious "new room" was realized through the outreach efforts of Faustina Vitelli, who solicited funds from her relatives to pay for construction. In turn, head apothecary Massimilla Salvetti assumed financial responsibility for repairing the distillation equipment.[66] This kind of mixed funding—corporate, private, familial—became increasingly common in sixteenth-century Italian convents, further deepening the privatization of these communities by local elites.

The Murate nuns capitalized on this disaster by investing heavily in special equipment that would help them satisfy growing consumer demand. To remain viable in the market, convent pharmacies had to stay current with the latest ideas and production methods. Under the direction of Salvetti's successor, Umiliana Lenzi, the nuns installed a massive brick apparatus called a pyramid in the ground-floor still room, at the cost of another 100 scudi. Lenzi undertook this project after conferring with the Medici court apothecary Stefano Rosselli, whose technical advice often proved decisive to convent decision-making. This innovative device probably resembled one of the furnaces popularized by Mattioli in his printed works on distillation (Fig. 4.4). Fully operational by 1581, this furnace allowed dozens of glass alembics or lead distilling bells to be placed over a single fire, thereby dramatically increasing the volume that could be obtained from a single process. The total number of vessels typically used at the Murate is unknown, but the convent chronicle reported that the nuns reserved the contents of two alembics to treat their own sick and either sold or gifted the remainder. Once again, this venture required joint funding by the nuns and committed patrons. The noblewoman Leonora Cibo, who enjoyed numerous ties to the Murate, pitched in sixty scudi to build the new furnace; the balance was paid by pharmacy earnings.[67] The adoption of this commercial-grade apparatus in a female institutional setting underscores both the diffusion of innovative technologies and women's firsthand exposure to early scientific culture. Unfortunately, this new apparatus was operational for just a few years before being flattened by yet another flood in 1589, which damaged other parts of the complex as well. This time, Grand Duke Ferdinando helped the business get back on its feet. At his request, Stefano Rosselli's son Francesco estimated the value of the "electuaries and other

Fig. 4.4. Pyramid furnace for distillation. Pietro Andrea Mattioli, *Del modo di distillare le acque da tutte le piante*. Venice, 1604. Courtesy of Special Collections, University of Wisconsin Library

medicinal items" lost in the flood at 131 scudi. Ferdinando paid this sum out of pocket and structurally reinforced the pharmacy storage room to avoid further damage.[68]

These recurrent episodes obviously required a longer-term solution. Supplementing the Medici replacement capital with another fifty scudi, the nuns rebuilt the distilling pyramid on the same site, but this time raised the floor level to prevent future flood damage. This proved to be an expensive solution, totaling 195 scudi in construction costs, because it also required raising the floor above, where the scriptorium was housed. Apparently the nuns considered this project to be a worthwhile investment. By 1589, when the second major flood hit, the pharmacy was netting about seventy scudi annually from "the sale of electuaries, distillates and similar things." Once again the Murate drew on its extensive social networks to gather the necessary capital. The convent chronicle noted that the entire project was conceived and supervised by head pharmacist Beatrice Benci, who contributed her "industrious and diligent earnings from the pharmacy" to bridge funding gaps.[69] These business decisions not only testify to the psychological and financial investments the Murate nuns had made in pharmacy by the late Cinquecento; they also demonstrate how commercial pressures drove the adoption of innovative technologies outside traditional scientific settings.

In the wake of these disasters, other technical decisions had to be made to maintain commercial viability. Mattioli had argued that, although lead distilling bells produced greater volumes, the resulting liquid often had a smoky, burnt taste that nauseated the sick. He advised using glass alembics instead to obtain higher-quality distillates—a recommendation that simultaneously supported the glass-making industry then developing in Tuscany.[70] Because glass was notoriously fragile, however, this practice led to higher replacement costs for commercial producers. When taking stock at the San Marco pharmacy in 1603, Fra Giuliano da Falgano observed that the workshop owned so "many glass vessels . . . of different sorts that one cannot count them exactly, because every day some of them break and [others] are bought."[71] Since using glass alembics cost more and diminished production, the Murate nuns faced an important business decision when rebuilding their still room after the 1589 flood. The head pharmacist, Umiliana Lenzi, reportedly consulted with Stefano Rosselli, who advised her that "herbs distilled in glass conferred greater benefit to the body than those distilled in lead." Echoing Mattioli, this rationale persuaded her that it was worth investing a hundred scudi—the equivalent of half a monastic dowry at the house—in glass alembics and other new manufacturing equipment.[72]

Mattioli also recommended using a bain-marie (a type of double boiler) as the best way to distill medicinal waters considered to be "superior in goodness and clarity." This device produced high-grade distillates that could be further concentrated and refined. By the late sixteenth century, the bain-marie had become the preferred method for making luxury items such as scented waters and perfumes extracted from roses and orange blossoms, as well as superior skin lotions decocted from lemon juice and tiny white snail shells.[73] The drawback to this production method was the limited volume obtained from each batch. Since the 1589 flood had also destroyed the convent's bain-marie, it too was rebuilt in conjunction with the pyramid furnace. These two business decisions—using glass alembics and replacing the bain-marie—had important consequences for the commercial trajectory of the Murate pharmacy. Both manufacturing methods shifted production to smaller volumes of high-end goods, moving the nuns away from their original charitable objectives and increasingly toward the luxury market. As Medici court patronage intensified after 1590, these decisions had critical consequences that are examined below.

Supplementing narrative accounts of pharmacy spaces is the visual evidence provided by a rare portrait of an unknown Florentine nun pharmacist. Painted by Alessandro Gherardini in 1723 and currently conserved at the Museum of Fine Arts in Budapest, the portrait has received scant critical attention and awaits fuller study[74] (Fig. 4.5). Although the circumstances surrounding its commission are unknown, the sitter's Dominican habit—black veil, white tunic, work apron, prominent rosary—places her in the same monastic tradition as Giovanna Ginori and her successors at Santa Caterina. Among the portrait's most striking features is its monumental size—some 1.72 meters tall—making it slightly larger than life-size. This imposing scale suggests that the painting probably had a commemorative function and that it was meant for public viewing in one of the several Dominican convent dispensaries still active at the time. By 1600, it had become commonplace to depict the illustrious heritage of a monastic pharmacy by displaying portraits of distinguished apothecaries who had worked there. These commemorative images simultaneously bolstered a sense of communal identity and of consumer trust. The public waiting areas at the San Marco pharmacy were lined with portraits of revered friars who had run the apothecary, along with images of generous pharmacy patrons. Among the latter was a portrait of the seventeenth-century noblewoman Eleonora of Montalvo, painted by Cristofano Allori. Montalvo's relationship to San

Fig. 4.5. Alessandro Gherardini, Portrait of an unknown Florentine nun pharmacist, 1723. By permission of the Museum of Fine Arts, Budapest

Marco is unclear, but she enjoyed a prominent reputation as the founder of a conservatory for noble girls that included instruction in household medicine.[75] Similar portrait displays mixing memory and marketing could be found in northern Europe, such as the pharmacy of St. John's hospital in Bruges run by religious women.[76]

Although many questions remain about the Gherardini portrait, it offers an important counterpoint to more familiar depictions of Renaissance laboratories and craft rooms. Paintings by Giovanni Stradano, Giovanmaria Butteri, and Alessandro Fei in the *studiolo* of Francesco I used the collective hubbub of all-male workshops to signify invention and industriousness. Often disseminated as prints, these works enhance our grasp of Renaissance craft activity by articulating the disposition of space, illustrating technical equipment, and conveying the sensory aspects of hand work. Yet by representing workshops as exclusively male spaces devoid of even allegorical female figures, these images claimed new forms of technological activity and ingenuity as distinctly masculine preserves.

The Gherardini portrait contrasts sharply with these representations, owing in part to its probable commemorative function. This sweet-faced, smiling healer welcomes visitors into her surroundings by extending her right hand toward the viewer. The tranquility and intimacy of her workspace create an impression of orderliness and personalized care, in contrast to the bustling, cavernous workshops referenced above. Still, Gherardini invests the sitter with a strong sense of legitimacy by means of carefully assembled details. The charcoal gray curtain in front of which she stands not only dramatizes the sitter's light-filled face, but also shields medicinals in the cupboard behind from damaging exposure to light. The position of the sitter's left hand on a Chinese-style pharmacy jar, fashionable at the time, visually reinforces the connection to her craft.

Amid this cluttered workspace are the books, tools, and other paraphernalia that traditionally signified skill in the healing arts. On the left side of the portrait stands a large brass mortar and pestle—the distinguishing sign of an early modern pharmacist. The large chest on which it rests is labeled "alabaster," a substance commonly used to make an analgesic ointment for swollen legs. The chest's massive size suggests a large quantity of material, perhaps indicating that alabaster figured in a prominent convent remedy. Also stationed on the chest is Mesué's hefty antidotarium, a standard reference book all pharmacists were required to own. The unidentified book on top with its leaves aflutter afforded a convenient vehicle on which Gherardini could sign and date his work. To the nun's right is a round table covered with a well-worn, gold-fringed cloth. On it rest a brass balance scale for weighing ingredients and a silver inkpot with quill for scribbling work notes, which are visually referenced by the various papers tucked into her books. Two more bound volumes sit atop the table. The bottom one, a text on medicaments, was authored by the eleventh-century Arab physician Iohannis Serapion the Younger, and was issued frequently from the sixteenth century on. The top volume appears to be a massive dictionary, perhaps a reference work on materia medica.

Lining the armamentarium are numerous glass jars containing powdered substances; several labels are still legible. Despite their cost, glass vessels became increasingly commonplace in Italian monastic pharmacies as local glass-making industries flourished after 1600. Among the drugs on hand were sarsaparilla, a New World sudorific used as an adjunctive treatment for syphilis and fevers; polypodium, a traditional Galenic purgative and vermifuge; and medicinal rhubarb, which experienced growing popularity in the early modern period.[77] Finally, hanging from the cupboard were the nun's tools of the trade. These included a long, thin spatula for smearing and blending ingredients; a double-ended silver spoon for measuring doses; and a small, fluted bain-marie for distilling and reducing liquids in small quantities. The amassed visual evidence stresses traditional, time-honored ways of healing and knowing, rather than foregrounding new wonder drugs or the latest medical apparatus. That may have been part of the point. By surrounding this practitioner with recognizable visual idioms, Gherardini conjured a sense of trustworthiness rooted in a long healing tradition that remained viable in his own day.

"Secrets" in the Convent

No assessment of nuns' medical knowledge would be complete without considering their involvement with the "secrets" tradition. Secrets have sparked tremendous interest in recent years, despite the difficulty of defining them. Scholars have argued that these bits of proprietary knowledge, often won through trial and error, accelerated the emergence of the "new sciences" between 1400 and 1600 by creating powerful synergies among the arts, crafts, and sciences.[78] Part of the new information economy transforming sixteenth-century Europe, books of secrets rolled off Italian presses in record numbers, giving readers access to practical information that ran the gamut from vaunted cure-alls to everyday remedies. In their written form, secrets resembled modern culinary recipes. Part of their attraction lay in conveying information in an accessible, straightforward manner, including detailed lists of ingredients, measures, and instructions. By trying out recipes for themselves, readers culled valuable insights into the nature and properties of artisanal materials while gaining helpful tips for household management. The vernacular reading public showed an insatiable appetite for these books. Alessio Piemontese's blockbuster book of secrets, first published in Italian in 1555, was translated into multiple languages and reprinted seventeen times in the late sixteenth century alone. Self-promoting physicians like Leonardo Fioravanti capitalized on the popularity of this genre to expand their professional reputation and clientele.[79]

Early modern secrets also functioned as a form of intellectual property fraught with tensions. On the one hand, secrets had little social or commercial value unless they could be shared openly; on the other, divulging secrets indiscriminately cheapened their value.[80] Easing these tensions between openness and secrecy was the development of state-issued patents that gave owners short-term privileges over their creations. In the sixteenth and seventeenth centuries, Italian civic governments ushered in an international patent system in their quest to simultaneously incentivize innovation and capitalize on it. By 1550, the value of patented secrets was so widely recognized that Italians commonly sold such privileges on the open market, used them as dowries, and transmitted them as legacies to their heirs.[81]

Despite surging interest in Renaissance secrets, the relationship of Italian women to this knowledge culture remains murky. Tessa Storey's recent survey of two major manuscript collections has uncovered few compendia of secrets written by Italian women.[82] The most renowned compilation remains the one assembled by Caterina Sforza, discussed in Chapter 1. It is well known that only one printed book of secrets was attributed to a female author in sixteenth-century Italy: Isabella Cortese's *Secreti,* first published in 1561 and reprinted eleven times in the next century. This popular work actually may have been compiled by three men, rather than being a female-authored text.[83] Meredith Ray has argued for a more capacious view of this genre by contending that Moderata Fonte's *Worth of Women* (1600), a dialogue commonly associated with the *querelle des femmes,* can fruitfully be read as a book of secrets.[84]

Greater insight into female knowledge-making arises when considering the interplay among manuscript, print, and social practices. Secrets were transmitted in multiple, often interdependent ways, especially in the early days of print. Family recipe books were never intended to be published but were handed down as handwritten texts, which allowed each generation to add new layers of useful knowledge.[85] Many Italian recipe collections, like those kept by the Medici and Bardi, incorporated secrets that either originated with female householders or were modified by them; yet their collective authorship was ascribed to men, in keeping with patrilineal record-keeping practices that were especially pronounced on the Italian peninsula. The public nature of print further obscured the nature and extent of Italian women's collaboration in developing family secrets. Scholars have documented numerous instances in which wives manufactured proprietary remedies while their husbands took responsibility for marketing them.[86] All of these social practices exert a strong masking effect on the written legacy of secrets produced by Italian women.

Pharmacy work gave nuns various points of entry into the Renaissance culture of secrets, as shown in the following three case studies. Each case reflects a different type of engagement with the secrets tradition. The first situates Italian nuns as creators of proprietary remedies; the second explores nuns as brokers and consumers of secrets; the third reveals their role as custodians of secrets left in their safekeeping. These cases not only shed new light on forms of female knowledge-making but also showcase the multiple sites of early modern empiricism.

Case 1. Semidea Poggi and the Magic of Plants

The first case examined here centers on the medical activities of the Bolognese nun Semidea Poggi (d. after 1637), a talented poet, musician, and plant expert. Born in the 1560s to two of the city's premier aristocratic families—the Pepoli and Poggi—this polymath is best known as the author of the lyric verse collection *La Calliope religiosa,* published in 1623.[87] Her elite social status, as well as the latitude permitted by her monastic order, enabled her to participate extensively in local literary and political culture. Renaissance Bologna was renowned for its educated noblewomen who capitalized on both the presence of a university and the absence of a dominant first family to throw themselves into public life as writers, intellectuals, and shapers of civic imaginaries.[88]

Little is known about Poggi's formal education; even the date of her entry into the convent of San Lorenzo is disputed.[89] Still, we know that she served in the convent infirmary in the late 1580s, shortly after taking vows as a Lateran canoness. This nursing experience quickly introduced Poggi to contemporary healing practices, but she likely honed her practical skills in the convent pharmacy. Her community of San Lorenzo was one of many Bolognese convents that sold remedies ranging from foodstuffs to syrups, unguents, and plasters for mending broken bones. Among the convent's specialties was elegant quince paste molded in small dishes, whose delicacy reportedly was "greatly esteemed by lords and ladies."[90] Another convent specialty was a nutritious capon broth marketed to convalescents and the elderly who had difficulty digesting other foods. There was considerable demand for this product in the late Renaissance; the papal cook Bartolomeo Scappi included several different recipes for it in his 1570 cookbook.[91]

As a young nun, Poggi became embroiled in a scandal that has been explored by Craig Monson at some length.[92] In 1584, local inquisitors discovered that the convent was a hotbed of magical practices. Members of her community allegedly used ordinary "finding" magic that relied on the manipulation of sacred words and natural objects to locate lost things. The nuns reportedly deployed "magic" in other ways, such as reciting prayers or incantations when gathering simples to give them added potency. More serious accusations focused on the transgressive use of love magic by Poggi and others. Insults exchanged between nuns flew fast and furious as the inquiry proceeded, attesting to deep animosities within the house. Woven throughout these disputes were institutional battles over monastic reform that turned increasingly ugly after Trent. Although inquisitors reprimanded several nuns, everyone tried to keep matters quiet in hopes of avoiding a public scandal. As Poggi matured, she renounced some of her previous magical activities in a spirit of repentance, while vigorously attacking female vanity and occasionally castigating other women.[93] Despite her apparent conversion, Poggi never abandoned her longstanding belief in the power of nature or in her own ability to exploit it.

When Poggi surfaces again in the historical record several decades later, it is as a proficient, self-assured healer trafficking in secret remedies. In the early 1620s, she showcased her expertise in a brief exchange of letters with Cardinal Scipione Caffarelli Borghese, the papal nephew who had amassed a huge fortune under the protection of his uncle, Paul V. Best known as a patron of the arts, the cardinal built the famed Villa Borghese in Rome, where he took special interest in developing its extensive gardens. Borghese had served briefly as absentee archbishop of Bologna from 1610 to 1612, governing the diocese through a suffragan. Although Poggi had never met him, she confidently picked up her pen on September 18, 1621, to share her special cure for bladder stones—a painful condition that prevented Borghese from urinating normally.[94] The fact that she was privy to his personal health problems speaks to her extensive connections as well as to the difficulty prominent figures had in keeping their health status private. In sharing her secret remedy unbidden, Poggi probably hoped to secure a "favor in waiting" from this contender to the papal throne, possibly concerning the anticipated publication of her poetry.[95]

Poggi opened her first letter to the cardinal with what scholars have called an "efficacy phrase," which asserted that a remedy had been tried, tested, and proven useful. Such phrases were ubiquitous in early modern medical recipes and were meant to reassure distant readers that a remedy had demonstrated positive results.[96] After empathizing with the "atrocious pains" Borghese was suffering, Poggi announced that relief was now at hand. "This remedy is certain," she told him; after taking the remedy, "you will no longer feel the pains of being unable to urinate." Although she had not tested the remedy on herself—always the ultimate testimonial—Poggi repeatedly stressed its efficacy. If the cardinal followed her instructions about how to make and take the cure, "there is no doubt that with the help of the Lord God the stones will break up, and you will urinate without pain of any sort and remain completely healed." Exuding confidence, Poggi boldly ended her letter by counseling the cardinal to heed her advice and "not trust those physicians who will do you harm instead of good."[97]

At first glance, Poggi's cure seems to be little more than a decoction of violets grown and gathered out of season. However, her careful indication of the exact type of plant to be used and its optimal harvest time conveyed a clear mastery of contemporary herbal practice. This pungent liquid was neither the standard herbal remedy prescribed by Hildegard of Bingen, nor a Paracelsian remedy utilizing animal ingredients such as dried goat's blood and dove droppings.[98] Rather, Poggi's secret almost certainly issued from continued experimentation with the natural world. In her eyes, the little flowers at the heart of her remedy embodied great curative power—if one knew how to tap it. In contrast to the "finding" magic she had practiced earlier, she now stressed the close observation of naturalistic details necessary to healing. After carefully describing the botanical features of the violets in question, she called Borghese's attention to certain "little green vesicles" containing the potent seeds that were the key to the decoction. To prevent misidentification, she included a packet of seeds with her letter for comparative purposes. Sharing botanical samples in this way was commonplace among early modern plant enthusiasts.[99] By engaging in these practices, Poggi marked herself as an expert practitioner and participant in a wider community of knowledge.

Apparently this signature cure was a great success. Although the cardinal's original reply has not surfaced, a jubilant Poggi wrote to Borghese again six weeks later, expressing delight that her remedy had brought the promised relief.[100] Although the cardinal never ascended the papal throne as many had anticipated, he probably rewarded her in some way. Failure to do so would have ignored established ethics of exchange governing scientific and social relations in early modern Europe. Perhaps Poggi suggested that Borghese use his influence to publish her poetry. While there is still much to learn about her medical pursuits, Poggi's secret remedy placed her squarely on the map of contemporary empirical practice.

Case 2. The Medici Secrets: Teaching Household Alchemy

While Poggi's example highlights the production of secrets, the second case explores their circulation among family members of both genders. Sharing secrets was an expression of trust that cemented social bonds within communities of practice, whether they comprised alchemical adepts or gentlewomen concocting home remedies.[101] In artisan families, trade secrets commonly circulated between and among fathers, brothers, uncles, cousins, and in-laws sharing allied occupations.[102] The most visible traces of these exchanges were recipe books, such as the one Stefano Rosselli compiled in 1593 for his sons, who followed him into the pharmacy business.[103] The oral transmission of secrets within familial networks is more difficult to document, especially when that exchange crossed gender lines. Still, it is not unreasonable to think that Rosselli family recipes informed the development of the pharmacy and artistic workshops at Santa Caterina, where the clan had long-standing ties. Rosselli's female forebears were among the first women to join the community; following them into convent life were his sister and daughter Fiammetta, who became a successful marble sculptor in the convent workshop.[104]

Some of the most striking evidence for the circulation of family secrets comes from Le Murate. By 1590, this community was experiencing dramatic changes in its social composition and religious tenor. From its foundation in the early fifteenth century, the convent remained a socially inclusive house with a pronounced appetite for asceticism. Among the 146 nuns living there in 1550 were descendants of carpenters, barbers, masons, weavers, shearers, pharmacists, and other artisans, who rubbed shoulders with local and foreign-born elites.[105] Under the influence of Christine of Lorraine and husband Ferdinando, however, the Murate was transformed into an extension of the Medici court. Ducal gifts and favors, such as the restoration of the pharmacy in 1589, gave the court a strong sense of ownership over convent affairs and personnel. That sense of ownership was amplified by the admission of four Medici princesses to the house between 1596 and 1610, for which Christine obtained special papal permits.[106]

A gift of Medici secrets made in 1598 further integrated the convent into court networks. The key protagonists in this transaction were two Medici cousins: Caterina Eletta de' Medici (d. 1634), the first Medici princess to be admitted to the community; and Don Antonio de' Medici (1576–1621), the natural son of Francesco I, and his mistress Bianca Cappello. Despite her lofty origins as granddaughter of the first duke Alessandro, Caterina's early life remains obscure.[107] In August 1596 she became a boarder at the Murate, eventually taking monastic vows there on December 8, 1604. This prolonged interval suggests that her marital destiny was still being decided by her relatives. During this period the princess lived well, supported by a live-in chambermaid and hefty annual boarding fees of 100 scudi.[108] Her father Giulio's death in 1600 apparently diminished her marriage prospects, consigning her to a life of forced celibacy. Upon profession, she brought a huge dowry of 1,000 scudi—five times the norm—and took the religious name Caterina Eletta, which combined her birth identity with an elevated sense of having been "chosen" *(eletta)*.[109]

On the other end of this transaction was her cousin Don Antonio, an avid experimenter and alchemist. Precluded from dynastic succession by illegitimate birth, this Medici prince pursued his scientific passions at the Casino di San Marco, home of the ducal workshops by 1576. Combining spacious laboratories, an extensive library, and a luxurious private residence, the Casino became one of the most vibrant sites of scientific culture in late Renaissance Tuscany. A wide array of products was developed and tested there, including medicaments, glass, porcelain, munitions, metalwork, porphyry sculptures, and clocks. Numerous local experimenters collaborated with Antonio on technical projects.[110] Don Antonio also maintained a close personal relationship with his half-sister Marie de' Medici, the future queen of France, a perfume enthusiast with whom he swapped recipes. Other correspondents included Caterina Medici Gonzaga and Eleonora Medici Gonzaga, both of whom expressed keen interests in medicine and botany in their roles as Duchess of Mantua. Unfortunately no letters exchanged between Don Antonio and Caterina Eletta have surfaced. He may have relied on intermediaries to communicate with her or capitalized on religious festivities to visit the convent in person.[111]

At the Casino, Don Antonio developed thousands of secrets ranging from analgesics to aphrodisiacs. To date, scholars have focused on two recipe collections representing the fruits of his labors, both of which reveal strong Paracelsian influences. The better-known of the two works contains more than 6,000 secrets, mostly medicinal in nature.[112] Apparently Don Antonio had prepared this ambitious work for publication by the Casino's in-house press in 1604. However, only the title page was printed; the rest remains in manuscript. Divided into four separately indexed volumes, this massive compendium includes not only his own remedies but also some attributed to contemporary physicians like Mattioli, Girolamo Mercuriale, and Baccio Baldini, along with earlier practitioners. Given the vast number of recipes, spanning some 3,000 pages, it is doubtful that Don Antonio and his assistants tested each one individually.[113] A related Paracelsian work authored by the prince was printed at the Casino the same year.[114]

The other, lesser-known collection is much shorter, covering a mere twenty-six folios. Written on vellum in Don Antonio's beautiful hand, it includes a few dozen recipes that may represent the prince's earliest experimental efforts.[115] As the title *Segreti sperimentati* suggests, these undated recipes probably were subjected to repeat testing. They convey more detailed measurements, more numerous ingredients and more exacting production processes than those inscribed in the larger *Apparato*.[116] These hand-tested recipes can be viewed as working notes that could be shared with other experimenters possessing abundant resources. Certainly the prince's recipes for everyday hygienic products lay beyond the financial reach of most household practitioners. His secret for teeth whitener, for instance, included twenty different ingredients ranging from spices to expensive corals, and took three days to prepare.[117]

Don Antonio can be securely associated with a third recipe collection that has escaped scholarly notice, despite its importance for the Medici genealogy of experimentation. Dated 1598, this manuscript covering 324 folios is housed at the Wellcome Library; it probably represents a midpoint between Don Antonio's shorter, undated book of secrets and his magnum opus of 1604. The prince devoted himself wholeheartedly to experimentation after taking up residence at the newly remodeled Casino in 1597.[118] It is this third collection that tracks the transfer of family secrets from Don Antonio to cousin Caterina. The hundreds of recipes it contains probably were extracted and copied by Caterina herself from the prince's copious working notes. Certainly the frontispiece leaves little doubt about the recipes' origin and their circulation within the ruling family. Writing in the first person, Caterina declared these recipes to be the "secrets of Don Antonio de' Medici, prince of Tuscany," which "were given and donated to me, Caterina Medici, in the Florentine convent of the Murate, in the year 1598." Because Caterina had not yet taken vows, she did not identify herself as a nun in this notation. The title also stressed the proven nature of the recipes, which purportedly had been "tested by [Don Antonio's] own hand" and found to be "most true"—a double confirmation of their value.[119]

Virtually all the secrets in this collection concern the fabrication of cosmetics, soaps, perfumes, and other hygiene products. Caterina displayed little interest in copying recipes for ordinary ailments like sciatica or worms. Yet this was not a standard herbal recipe book in the tradition of Mattioli. The frequent use of substances such as mercury, litharges, rock alum, sal nitrate, and various sublimates places the work firmly in local Paracelsian circles of alchemical enthusiasts, including Christine of Lorraine. As Sheila Barker has shown, Medici family interests in alchemy stretched back to Cosimo il Vecchio, Caterina Sforza, and Cosimo I.[120] Caterina Eletta must be counted among these Medici adepts, for she seemed completely at ease with both the components and processes discussed in the manuscript. Each recipe is accompanied by detailed instructions and practical tips designed to develop understanding and proficiency. For instance, she assiduously copied her cousin's advice against adding scented water when making perfumed balls, which might dissolve the entire mass. Similarly, Caterina transcribed his explanation that variegated ambergris was preferable to the pure black variety when fabricating perfumed beads because the latter "smells too much like civet and evaporates quickly."[121] The recipes also suggest ways to vary the same product by tinkering with scent or texture. Among the recipes listed are twelve ways to make rosewater, ten methods of concocting perfumed pastes to burn indoors, and fourteen variations on the delicately scented water known as "acqua d'angeli" popular at both the French and Tuscan courts.[122] In short, Don Antonio was teaching his cousin "household alchemy," in which basic procedures were dressed up as chemical thinking.[123] By closely following this practical guide, Caterina could become a prince-practitioner on her own.

This "copious treatise" containing Medici family secrets was divided into two roughly equal parts. The first part taught "the way of making all cosmetics that don't irritate the skin, and every water, oil and other sorts of beautifier that make the skin white, soft, and beautiful; and smoothing agents for the hair, beard and body hair."[124] Indeed, the very first recipe featured a "miraculous" talc-based face powder distilled from snail shells and other materials that took twenty days to prepare. Although some recipes harkened back to the beauty products developed by her forebear Caterina Sforza, these secrets added greater complexity in terms of materials and working processes. Other recipes in the collection called for using reflective agents such as ground pearls, powdered silver, and mercury sublimates when making facial cosmetics.[125] The 195 recipes for cleansing, lightening, brightening, and clarifying the complexion attest to the importance of the skin as an aesthetic marker in Renaissance court circles. A similar book of cosmetic secrets from the Tuscan court showcased a face wash that left the skin as "clean and bright as a mirror."[126] Even outside court circles, however, skin remedies loomed large. Daily environmental irritants and cutaneous diseases made skin remedies the single largest category of medicines (22 percent) sold by Italian empirics between 1550 and 1630.[127] Caterina's claim that these beauty products did no harm provided welcome assurance in view of the toxic nature of some Renaissance cosmetics. Because beauty products remained morally controversial, their manufacture and use linked her to Medici court values centered on artifice and display.

These Medici beauty secrets were aimed primarily but not exclusively at women. Scattered throughout her collection were numerous recipes for grooming products intended for men eager to maintain good health while keeping up appearances. One focal point was the beard, a key signifier of Renaissance masculinity that demanded specialized products such as oils and dyes. Douglas Biow has explored the complex motives driving the shift in Italian men's self-presentation from being clean-shaven to wearing a beard. By 1600, beards had become so commonplace among all social classes that they functioned as a means of personal identification in civic travel passes.[128] Italian books of secrets in both manuscript and print circulated surefire recipes for dyeing beards that enhanced a youthful look and sense of virility. The Bardi family's handwritten book of secrets opened with a recipe for a grooming product that promised to "blacken men's hair and beard or give it a reddish tint." Similarly, Stefano Rosselli disclosed the secret for darkening hair developed by the Medici court barber. In keeping with these trends, Caterina copied several dozen recipes for coloring and grooming men's hair and beard, along with remedies to combat baldness, which she may have intended as gifts.[129]

The second part of Caterina's recipe book focused on perfumes and scented articles. She noted that recipes in this section "taught the ways of making waters, oils, balsams, scented products, perfumes, aromatic pastes, pomades, perfumed creams, scented Cyprus powders," perfumed gloves, and various styles of aromatic soaps.[130] The popularity of scented goods reflected both a holistic view of health and a concern with maintaining "good" ambient air around one's person.[131] Printed health manuals and books of secrets offered some guidance for making aromatics, but the first comprehensive treatise on the art of perfumery was published only in 1555. Written by Giovanventura Rosetti, this tract was addressed to "all virtuous women" and contained over 300 recipes for making fashionable scented products aimed at affluent householders.[132] One of the reasons that aromatics became such a growth industry in the late sixteenth century was that they were easily customized according to season, fashion, price point, or personal taste. Growing demand for highly differentiated products expanded consumer options while boosting local economic activity, especially if the articles enjoyed an international reputation, like the fine scented soaps made in Bologna and Genoa.[133] Customized wares also gave producers great latitude in creating and marketing signature scents bearing their own trade names. Since perfumes generally fell outside of corporate regulations, small-scale producers (including nuns) could work in the interstices of guild structures, much as nuns had done a century earlier when manufacturing gold thread for silk entrepreneurs.[134]

The products Caterina described in this section were not affordable luxuries for middling householders but high-end goods meant for aristocratic consumption. Most of the items—scented gloves, fragrant sachets, beads made from perfumed paste—belonged to the world of court fashion. Some of these Medici family secrets featured extremely expensive ingredients; the musk used to scent different types of hand soap, for example, cost more than one hundred times per unit weight than the finest grade of soap itself.[135] Other recipes in this section gave detailed instructions about how to perfume gloves in the Spanish style, fabricate aromatic powders for stuffing sachets, and extract essential oils from flowers. A lengthy set of recipes elaborated the art of making beads from perfumed paste that could be strung as belts or bracelets, incorporated into rosaries, or worn around the neck as pomanders. Working techniques combined basic chemistry with elements of cookery. In distilling a wide variety of botanicals and making pastes of different textures, the practitioner was required to grind, heat, reduce, dry, and filter materials to achieve the right consistency. Don Antonio had provided precise measures and step-by-step instructions, but still left much to the maker's discretion. The inclusion of numerous practical details indicates that he anticipated his cousin would put the recipes to good use. By leaving room for personal judgment, however, he invited Caterina to experiment on her own.

Don Antonio seemed eager to share these secrets with his female relatives, despite being immersed in all-male circles at the Casino. In fact, it appears that he had compiled another, unattributed book of secrets two years earlier for Christine of Lorraine. This 1596 manuscript, also conserved at the Wellcome, was written in Italian but bore the French title, "Recette pour la baute."[136] Written on vellum, this elegant volume containing fifty-two folios shows striking similarities to both the Murate collection and Don Antonio's known Florentine works. His distinctive script is immediately recognizable, while the structure is identical to Caterina's recipe book: the first part is devoted to cosmetics, the second to perfumes. Similar Paracelsian influences are common to all four manuscripts, in terms of both working methods and specific ingredients. Other clues point to Don Antonio's authorship. Following every recipe is a brief explanation of the remedy's "virtues" or healing powers, a feature also seen in his 1604 published tract. Finally, small decorative elements like the French fleur-de-lis crosses penned on the opening folios establish visual connections to the grand duchess. Given the strong parallels between the two London manuscripts and their proximity in dating, Caterina may have seen this text while living at court in 1596, and asked to make a similar compilation when boarding at the convent.

This unattributed book of secrets assumes that the reader was a more sophisticated practitioner, like Christine herself.[137] The recipes gave the maker great latitude of judgment in preparing products and took more complex considerations into account. For instance, the reader was consistently invited to gauge when a substance had acquired the right texture, consistency or dryness before proceeding to the next steps.

The author also signaled that external conditions like room humidity had to be considered when making such decisions. Numerous recipes, such as those for perfuming leather gloves, left the quantities of materials used "to your discretion."[138] The practical tips in Caterina's recipe book read like a primer by comparison. Moreover, this collection includes several secrets more appropriate for a mature female sovereign, such as recipes for making the famed Medici poison antidote, removing stretch marks from the abdomen, and tightening the vagina. Several recipes stress their proven empirical value obtained through testing. Taken as a whole, this work established a respectful working dialogue between skilled practitioners, with precious secrets mediating the exchange.

It remains to be seen how Caterina de' Medici utilized this gift of secrets. By 1598, the Murate nuns were already experimenting with new products in their renovated workshop. Capitalizing on the Medici secrets would have significantly bolstered their growing line of luxury products. Yet Caterina kept her distance from the community, living in separate quarters even after she took vows in 1604. From that time until her death thirty years later, she held no convent offices, occupying herself instead with theatrical productions and other courtly pursuits under Christine's tutelage. The admission of three other Medici princesses—Caterina's cousins, the illegitimate daughters of Don Pietro de' Medici—between 1606 and 1610 fractured the community so deeply that a cadre of devout nuns tried to secede from the house in 1615. Irreconcilable differences between this group and the haughty princesses caused relentless turmoil over the next two decades.[139]

The extent to which Caterina pursued her training in "household alchemy" must remain speculative for the moment. She and her associates may have fabricated products for the Medici court, using their expansive living quarters as a production site. One of the other Medici princesses, Sister Maria Vittoria, ventured into the related craft of confectionary, making delicate sugar-based confections on demand for Christine and other court ladies.[140] Caterina's interests certainly facilitated the circulation of Paracelsian ideas, since visitors shuttled frequently between court and convent. Whatever the case, possessing the Medici secrets gave Caterina an entry point into understanding chemical processes and a means of engaging scientific discourse.

Case 3. Secrets, Ownership, and Print

The final case examined here illuminates nuns' engagement with the culture of secrets from yet another angle. Renaissance religious women benefited not only from information flowing into their communities but also from secrets entrusted to their care. Placing valuables in religious houses for safekeeping during political crises or prolonged absences was commonplace in Italian cities. Urban magistrates frequently utilized these institutions as safe houses for storing movable goods while the estates of deceased citizens were appraised or contested. Both their sacrality and physical security made monastic enclosures ideal repositories against thieves and impatient creditors. In addition, nuns maintained meticulous inventories of goods placed in their care.

Hence it is not surprising that the Florentine military captain Orazio Tornabuoni entrusted a medical secret to his nun-aunt in 1614, while shuttling between postings in Milan, Bologna, and his garrison near Grosseto.[141] This custodian is not identified by name in relevant correspondence, but she probably lived at San Domenico, home to a long string of Tornabuoni nuns and a thriving pharmacy. In the sixteenth century, the Tornabuoni had made a name for themselves in scientific circles. Niccolò Tornabuoni had sent the first tobacco seeds to Florence around 1570 while serving at the French court, giving this medicinal plant its local name of "Erba Tornabuona."[142] Captain Orazio Tornabuoni apparently pursued similar interests in healing and the natural world. Around 1610 he concocted a special plaster *(cerotto)* to relieve gout pain, which sparked great interest at court since Cosimo II suffered from this ailment. Tornabuoni vigorously touted his cure when writing to the ducal secretary Andrea Cioli in March 1613. The captain reported that he had tested his secret on two gout sufferers, "both of whom were immediately cured, that is, their pains ceased within just six hours" of application. Tornabuoni even included a small jar of the remedy with his letter. Although he was willing to send more on request, his posting in the Tuscan hinterland made it impossible to obtain a key ingredient, which he asked the court pharmacist to supply.[143] Subsequent letters suggest that the grand duke had become an avid user of the plaster, whose analgesic properties gave it wider use. A few weeks later, Tornabuoni advised Cioli that applying the plaster to his painful corns would bring immediate relief and even prevent their reappearance; the secretary would be freed from pain "in no time at all . . . as happened to me and many others who used the same remedy."[144] Apparently Cioli was delighted with the results. Soon after, he sent this remedy as a diplomatic gift to a French nobleman living in Vienna, this time to relieve the man's persistent toothache.[145]

When doubts about its efficacy surfaced five years later, the grand duke asked the court pharmacist Niccolò Sisti to verify the remedy's composition and production method with Tornabuoni's aunt, who retained a copy of the original recipe. Sisti reported the results of his excursion in August 1618. To his surprise, he had encountered unexpected resistance when visiting the convent. "First [the nun] replied that I ought to write to the captain," Sisti told the grand duke, in order "to release her from suspicion" of having revealed the secret. Caught between ducal authority and family loyalties, Tornabuoni's aunt tried to shift the burden of disclosure to the owner. Sisti continued: "then I showed her the recipe and asked that we compare it with her copy." The court apothecary expressed dismay when she replied "that she doesn't have the recipe for making the plaster." The nun admitted, however, that "the convent had arranged for the plaster's uses . . . to be printed . . . so that it could be distributed to those who wanted . . . to make the plaster themselves."[146]

The type of cheap print to which the nun referred was an integral part of marketing strategies after 1550. These forerunners of modern medical advertisements often lured customers with extravagant claims in order to boost sales. Local governments had great difficulty regulating the content of these handbills, which sometimes sparked vicious quarrels between itinerant healers.[147] Some cheap medical print was more instructional in nature, providing information about the timing, dosage, and administration of remedies. The Medici court itself printed hundreds of little flyers to accompany the medicines they dispensed as diplomatic gifts.[148] Convent pharmacies in seventeenth-century Bologna issued similar leaflets instructing clients how to take their medicines, including details about advance purging and dietary restrictions.[149] Most of this cheap medical print has been lost, but it laid the groundwork for modern advertising techniques and consumer protections that warrant further study.

Back at the convent, Sisti remained unpersuaded by the nun's evasions. "I believe without a doubt that she has the recipe for the plaster," he reported to Cosimo, "and that in listing its uses [the nuns] have credited the convent and not the captain." He admitted, however, that "I don't know if they sell the plaster or give it away." Whether or not the nuns stood to profit, this perceived violation of ownership rankled Sisti, a known experimenter with medicines, porcelain, and crystal. His patience exhausted, the court apothecary finally denounced the remedy itself: "to tell the truth . . . , I don't value this empirical secret one bit. Two years ago this same plaster was declared a miracle cure that relieved oozing, swelling and pain in the teeth, but it never had a good effect, or any effect at all."[150]

To determine its efficacy, Sisti persuaded Christine of Lorraine to stage a trial on gout patients the following year (1619). Testing remedies became increasingly commonplace in the late Renaissance, when sovereigns, physicians, and empirics regularly began trying out cures on human and animal subjects.[151] It is not clear how subjects for the trial were to be recruited. Regardless, close observation was paramount, since eye-witnessing by experts publicly certified both the authenticity of the testing process and its outcomes.[152] Tornabuoni himself requested that a surgeon and physician witness the proceedings; Christine agreed but reserved the right to appoint them. She further directed that the test be held in the ducal *fonderia* and that the remedy be manufactured by Sisti according to Tornabuoni's specifications.[153] Treading carefully here, the grand duchess demonstrated her political sense as well as her empirical interests. Since the plaster had been distributed as a diplomatic gift on several occasions, it would have been foolhardy to discredit it without convincing proof.

What was the outcome of this drug trial? Unfortunately, the documents leave us in suspense. Still, this episode reveals the extent to which religious women had become enmeshed with—and invested in—the commercial culture surrounding secrets. Asked to safeguard a valuable remedy, Tornabuoni's aunt and her convent walked a fine line between protecting its secrecy, on the one hand, and advertising its efficacy on the other. Whether they did so for commercial gain remains an open question. Despite its fragmentary documentation, the history of this intriguing cure illuminates one of the many ways in which female agents of health helped constitute the medical culture of late Renaissance Italy.

This chapter has shown that Renaissance nuns acquired and transferred medical skills and knowledge through apprenticeship, in much the same way as guildsmen. Seasoned practitioners trained their younger charges by combining verbal instruction with hands-on learning. These methods aimed at developing embodied skills, perceptual focus, and the ability to improvise in the moment. Although convent workshops paralleled the organization of guild-affiliated shops, they distributed technical know-how in ways that best suited house traditions. Most convents opted for expertise born of long experience, which propelled talented practitioners into leadership positions while giving rise to close working relationships. Nuns supplemented their learning through medical reading of a surprisingly wide nature, from canonical authorities to contemporary masters. Stashed in nearby cupboards for easy reference, these well-thumbed texts likely were subject to intensive reading and consultation.

To remain commercially viable, religious women also relied on new knowledge flowing into their communities via rich information networks. Kinship played a fundamental role in organizing these complex social webs. Thanks to these connections extending across court, guild, and academy, nuns accessed technologies that enabled product development in different market sectors. The use of the latest commercial-grade equipment in female institutional settings highlights both the widespread diffusion of innovative technologies in late Renaissance Italy and women's engagement with them. Moreover, Italian nuns made significant capital investments in storage rooms and maiolica vessels, which underscores both the materiality of pharmacy work and its importance for convent economies. At the same time, nuns astutely adapted the semiotics of trust underlying market exchange by emphasizing their role as religious intercessors. These multiple investments distinguished nuns as commercial producers from noblewomen who fabricated remedies for charitable purposes or gift-giving.

Finally, three case studies shed new light on Italian women's immersion in the culture of secrets, broadening the reach of their empirical interests in the process. Scholars customarily measure the production of medical knowledge by the yardstick of print authorship, but the cultural force of secrets extended well beyond these limits. Poggi never attempted to publish her recipe but shared it freely with Cardinal Borghese in hopes of building social credit. By trying out her Medici cousin's secrets, Caterina Eletta gained proficiency in household alchemy, which tightened links within court circles. Religious women could be drawn into contests over commercially valuable knowledge through their traditional custodial roles. All three episodes argue that scholars need to utilize diverse metrics to achieve a full understanding of women's engagement with the pursuit of health in early modern Europe.

CHAPTER FIVE

Restoring Health

Care and Cure in Renaissance Pox Hospitals

In this chapter I turn from the literate worlds of court and cloister to consider the relationship between women's care practices and knowledge production in the Renaissance hospital. At the beginning of the sixteenth century, Italian hospitals were renowned across Europe for their high standards of care.[1] Over the course of the Cinquecento, the web of charitable institutions blanketing the peninsula became more complex in response to new disease threats, growing poverty, and Catholic welfare initiatives. There were important local variants in the medical organization and activities of sixteenth-century Italian hospitals, but one of the most visible signs of change across the peninsula was the proliferation of specialized facilities and their integration into increasingly coordinated networks of care.[2] New healing sites cropping up in sixteenth-century Italy included pox hospitals devoted to treating syphilis; quarantine facilities that forcibly isolated sufferers during plague outbreaks; and convalescent hospitals offering extended respite. Italian hospitals also became "larger, more bureaucratic and more authoritarian" after 1550, in keeping with the aggressive, soul-saving nature of Counter-Reformation charity.[3] Complementing these facilities were myriad custodial institutions that sheltered foundlings, at-risk girls, and women in distress.[4] All of these institutions were populated overwhelmingly by the urban poor, who lacked the resources to manage a medical crisis or handle long-term adversity. Whatever their charitable goals, these facilities firmly cemented cultural associations between poverty and disease.

Renaissance hospitals have long been recognized as spaces of healing, but their role as epistemic spaces for the circulation of practice-based knowledge remains only partially understood.[5] By the sixteenth century, these institutions had acquired "an educational or proto-clinical function" but they were not formal teaching facilities, unlike medieval Islamic hospitals.[6] Nevertheless, as the institutional nexus between poor relief and public health initiatives, Renaissance hospitals stimulated and translated new medical thinking by putting new medical principles into practice on a broad scale. Within these complex institutions, female hospital nurses who carried out the bulk of day-to-day caregiving were central to this implementation. Their role in Renaissance hospital life, and in the broader process of medicalization, has not been well studied. In medieval and early modern Europe, hospital nursing provided a way for lower-class women to make a living while making a life.[7] Prior to the establishment of nursing orders like the Daughters of Charity, Renaissance nurses acquired expertise through observation, experimentation, and repetition of tasks over a lifetime. Their duties ranged from preparing and administering remedies to feeding patients and attending to their physical and spiritual needs. Hospital nurses also supported the broader institutional mission by preparing meals, doing laundry, and sometimes spinning linen in idle moments. Despite the centrality of nursing to the healing process, the corporeal nature of caregiving and its naturalization as "women's work" consigned this labor to the bottom rungs of the Renaissance medical hierarchy.[8]

In this chapter I focus on the medical activities of nurses working in the new specialized facilities dedicated to treating "the pox," which burst onto the European stage after 1495. Commonly associated with venereal syphilis, this disease ravaged the entire continent with lightning speed, thanks to frequent troop movements and the culture of prostitution.[9] Pox hospitals stood out as the most morally charged institutions within local hospital ecosystems, owing to the stigma attached to venereal disease. Blending therapeutic and redemptive goals, these cutting-edge institutions showcased simultaneously the coercive nature of sixteenth-century charity and innovative approaches to patient care. Responses to this new disease threat grafted new strands of medical thinking onto older institutional models, ranging from the use of novel wonder drugs to stringent management of the non-naturals. Taking a case study approach, I concentrate on the Florentine pox hospital of the Holy Trinity, founded in 1521 under confraternal auspices. Called the "Incurabili" because the pox initially was thought to be incurable, this institution was quickly integrated into the coordinated healthcare network emerging under the early Medici dukes.[10]

Several challenges loom large in recuperating both the medical agency and knowledge practices of Renaissance hospital nurses. Since most of these women were illiterate, they generated few letters or other narrative accounts, in contrast to the cultivated nuns and noblewomen examined thus far. Consequently, we see their work activities and subjective responses primarily through the eyes of others, such as well-born women and hospital administrators. Although extant sources make clear that hospital nurses developed a high degree of proficiency in the healing arts, especially in pharmacy, their perceived contributions to the overall medical economy were strongly conditioned by considerations of class as well as gender. Their low social status only reinforced the low value attached to their labor; in addition, their lack of corporate organization placed them at a distinct disadvantage compared to Renaissance physicians, surgeons, and apothecaries, who developed fine calibrations in professional standing in these same years. Moreover, the kind of emotional solace and bodywork nurses provided defied easy economic valuation, while the experiential nature of their medical knowledge makes it less visible in existing sources. Despite the paucity of firsthand accounts, however, we know from supporting evidence that these "invisible technicians" used medical work as a springboard for experimenting with remedies and patient care practices.[11] They too formed part of the larger social matrix linking the early modern medical economy to emerging empirical cultures.

With these caveats in mind, I first highlight common social concerns about pox across Europe, before developing a comparative social profile of Florentine pox nurses. To confront the many challenges posed by syphilis, the Florentine pox hospital drew on an unconventional pool of adolescent girls to serve as both medical and moral agents. In carrying out their work, these caregivers faced multiple health hazards recognized in contemporary medical thinking, such as contagion and poor ambient air. Noblewomen who patronized these facilities openly acknowledged these dangers while articulating the repugnance associated with hospital nursing. It is their voices that often serve as proxies for nurses. Next I advance earlier discussions of women's pharmaceutical work by considering differences in scope when performed in a hospital setting. Among the most accomplished pharmacists of their day, hospital nurses produced medicaments on an industrial scale unmatched by other female practitioners, even though we cannot put a price tag on their output. In the final section I examine how hospital nursing gained greater coherence as a body of knowledge and practice in late Renaissance Italy. This development stemmed in part from the renewed interest in the non-naturals as both preventive and curative principles. Diligent attention to these precepts strengthened the theoretical backbone of nursing, which heightened its medical standing over time. Vernacular treatises that insisted on the value of sensory-based knowledge for effective patient care further consolidated nursing as a healing art. Although their voices remain muted, Renaissance hospital nurses figured prominently among the female knowledge-makers who created and circulated medical know-how within networked communities of practice.

Caring for the Incurable: The Florentine Incurabili in Context

In the early sixteenth century, the origins and treatment of pox were hotly debated. Explanations for its appearance ran the gamut from corrupt air and poor health habits to divine punishment and infection by sexual contact.[12] Equally contentious was the very name of this scourge. Variously called the "French disease," the "Neapolitan sickness," and the "Spanish sickness," the pox consistently projected blame on others as the source of contamination.[13] The growing recognition by midcentury that syphilis was an endemic venereal disease gave rise to moralizing attitudes about the dangers of prostitution. Pox magnified long-standing associations between physical and moral corruption in contemporary understandings of disease, leading some municipal governments to target prostitutes.[14] Unlike many French and German towns, however, only a few Italian cities closed brothels as a preventive measure. The sex trade in Renaissance Italy was simply too vital to local commerce and the broader sexual economy to be shuttered completely.[15]

Urban centers across Europe responded to this disease threat by creating specialized pox hospitals. Italian cities establishing new purpose-built institutions between 1499 and 1526 included Genoa, Rome, Naples, Brescia, Florence, Venice, Ferrara, and Padua. At least twenty pox hospitals were founded in early sixteenth-century Germany, mainly in the southwest; London hospitals began treating poor pox patients from the 1550s.[16] Since the pox was a highly visible disease rife with sensory assaults, sufferers were evaluated in both medical and social terms, leading some sufferers to mask their symptoms.[17] A 1522 decree of the Venetian health board insisted that beggars suffering from pox be taken to the local Incurabili hospital to avoid infecting others by their poisonous stench. Indeed, it was the fetid odor associated with the condition that gave rise to its English nickname, the "foul disease."[18] The sheer number of sufferers lining Italian city streets in the early years of the epidemic overwhelmed available resources. The Florentine humanist Francesco Guicciardini observed around 1540 that many sufferers "became extremely deformed and rendered useless" by the ravages of disease and the damaging effects of remedies like mercury ointments.[19] Debilitated by illness and forced to move around the city on hand trestles, many of the afflicted died in the streets. Compounding these miseries were tricksters who reportedly applied "fake ointments" that mimicked pox sores in order to collect additional alms.[20] Hence Italian pox hospitals simultaneously aimed to assist needy sufferers and remove them from public view, by force if necessary.

The 1521 foundation of the Florentine Incurabili hospital reflected these complex social and sensorial reactions. The French disease first appeared in the city in May 1496, but it took two decades to organize a coherent institutional response, in part because of political instability.[21] Spearheading the establishment of a local pox hospital was the Medici pope Leo X, who bestowed numerous spiritual privileges on the institution; his nephew, Archbishop Giulio de' Medici (later Clement VII) followed suit. Records written by Medici sympathizers privilege their involvement, whereas hospital sources stress instead the central role played by Savonarolan activists. Among the latter were two men who helped establish the pharmacy at Santa Caterina discussed previously: the apothecary Bernardo Mini, who furnished the hospital with its initial stock of medicinals; and Simone Ginori, who bankrolled the purchase.[22] Strong Savonarolan backing meant that from its inception the Florentine Incurabili was woven into larger charitable networks ranging from girls' shelters to convent pharmacies. Despite the contentious political relationship between the Medici and Savonarolan sympathizers, they shared the common goal of creating a therapeutic space where pox sufferers would be "diligently nourished, tended and treated until healed."[23]

Running the hospital was the newly founded confraternity of the Holy Trinity, which attracted donors up and down the social spectrum. Prominent among them were well-born women with Savonarolan leanings, whose support echoed female activism in establishing syphilis hospitals in Venice, Genoa, Naples, and other Italian cities.[24] Women's financial backing was exceptionally robust in the early years of operation, even among poorer women without surnames. Throughout the 1520s, female members of the confraternity contributed 30 percent more than their male counterparts, despite paying only half the dues.[25] Supplementing this revenue stream were civic and private donations that permitted the purchase of property, medicines, furnishings, and other essentials. Begging boxes were set up in major churches across the city; female hospital nurses made the rounds of local shops every Saturday to beg alms.[26] In 1534, Duke Alessandro de' Medici directed one-third of the proceeds from a new citywide tax to the hospital, which stabilized finances. Ten years later, Paul III assigned the patrimonies of small, suppressed religious institutions to it as a diplomatic gesture. Beginning in 1541, Eleonora of Toledo gave the hospital ten ducats monthly before donating a lump sum of 500 ducats in 1560.[27] As private donations waned later in the century, ducal funding assumed greater importance, which integrated the Incurabili more thoroughly into the city's public health matrix.

Construction on the new hospital complex began circa 1522. Its location on Via San Gallo on the urban periphery was economically advantageous, since property values were lower outside the city center. Implicit in this choice, however, was the desire to contain both the moral and physical contagion of this scourge, echoing the placement of pox hospitals on the outskirts of many German towns.[28] Siting the hospital on the urban periphery addressed notions that harmful material could be transmitted through corrupt air or infected goods, notions commonplace even before Girolamo Fracastoro published his treatise on contagion in 1546.[29] Yet this site, on one of the most sacred Florentine streets, lined with convents, also protected the symbolic purity of the city. By 1520, pox had become irrevocably associated with prostitution and other forms of illicit sexuality. Moralistic tales and popular visual imagery throughout Italy depicted the pox hospital as "a symbol of disgrace and hopelessness," where diseased courtesans and other debauched women met a tragic end.[30] By surrounding carriers of moral contagion with the redemptive purity of female religious communities, Florentine authorities helped contain other anxieties about the pox. Italian cities frequently used these kinds of compensatory spatial strategies; the first Verona pox hospital, for example, was built on the site of a former brothel.[31]

Inside the facility, pox patients were divided into separate wards by sex, a typical practice at the time. Owing to site constraints, the Florentine Incurabili opted for a compact spatial arrangement, with the men's ward placed on the ground floor and the women's section on the floor above.[32] In its first half century of operation, this hospital remained a multipurpose facility catering to a diverse patient population. Persons suffering from other incurable ailments were permitted to lodge at the complex into the 1570s.[33] During these years, the Incurabili may have separated pox patients within the wards, much like the general hospital in Valencia, which isolated syphilitics from abandoned infants, the mad, and patients suffering from wounds and fevers.[34] Although the Florentine facility became more exclusionary after 1574, it apparently never created separate chambers for those suffering from late-stage syphilis, as did the Strasbourg pox hospital.[35]

Filling the beds of Italian Incurabili hospitals were artisans, foreigners, and the working poor. Their medical care was fully subsidized by the hospital; in Florence, patients paid no fees for treatment until 1645. However, after 1574 the hospital board tried to filter out nonresidents and other "undeserving" patients by requiring entrants to produce two certificates for admission: one from the hospital physician, attesting to their need for treatment; the other from their parish priest, confirming their inability to pay for it.[36] This fraud prevention measure wove the

Incurabili more tightly into the existing administrative structures of church and state. Poor sufferers seeking admission to the hospital thus had to navigate two levels of bureaucracy—ecclesiastical and medical—in order to obtain care. Hospital physicians may have screened entrants by performing a physical examination, as they did in Augsburg, although no clinical protocols from Florence have survived. Still, inmates apparently were not subjected to the harsh moral censure facing the London poor, who were vetted by a disapproving hospital board prior to admission.[37] As numbers mounted by the 1590s—a decade marked by severe famine throughout Tuscany—the Incurabili was forced to revise its admission policy once again, treating only the sick poor from the city itself.[38]

Despite the lack of patient registers, hospital account books offer some insight into patient identities.[39] The Incurabili was already treating hundreds of patients annually by 1540, when administrative records became more regularized. One of the most striking features of patient profiles is its skewed gender ratio. Renaissance hospitals typically admitted two or three times more men than women. Scholars have argued that this imbalance reflected women's more extensive social resources in the form of friends, family, and neighbors able to provide home-based care, as well as men's greater access to credit in the form of loans from employers, guild colleagues, or co-workers.[40] The Incurabili evidence is consonant with these trends. Between 1572 and 1576, sixty-five men and twenty-one women—a ratio of roughly three to one—paid some kind of voluntary alms to the Incurabili.[41] It should be emphasized that this number severely underrepresents the total number of inmates treated, since most patients were too poor to pay anything. Still, these gender disparities were less acute than those seen at the Roman Incurabili hospital of San Giacomo, the largest facility on the peninsula. There, women represented less than 20 percent of admissions in the second half of the sixteenth century. This exceptional figure can be explained by San Giacomo's open-door policy, as well as by the highly masculinized nature of Rome, where men constituted 65 percent of the population in the 1590s.[42]

Despite being anecdotal, the Florentine evidence sheds light on two important aspects of the local sexual economy. First, the bulk of named male patients in the mid-1570s were foreign migrants, mainly Spanish soldiers stationed at local garrisons, who had limited support networks. Other occupational identities listed in hospital registers included millers and wool workers from the city and environs. Second, virtually all of the men (sixty-four of sixty-five) captured in this snapshot were adults, whereas roughly half of the women (ten of twenty-one) were young girls. The latter were described as the unmarried daughters of tailors, mercers, masons, innkeepers, and other artisans, who may have contracted syphilis congenitally or through some form of coercive sex that permeated sixteenth-century urban life.[43] Their fathers probably sent them for treatment to improve their marital prospects. Adult women in this group included wives and widows of wool weavers, carders, and secondhand dealers, who may have contracted the disease from their husbands. Former prostitutes and domestic servants, whose occupations exposed them to inherent health risks, must have figured among patients as well.

Closing the circle of institutional care was the newly founded convalescent hospital of San Paolo, in operation by 1592. The establishment of this facility was bound up with local ecclesiastical politics as well as what Samuel Cohn has called "a new public health consciousness in medicine" emerging after the 1575 plague.[44] The juridical status of San Paolo had been mired in controversy for decades and would not be fully resolved until 1614, when it was decommissioned as a female religious house.[45] Meanwhile, the city lacked a dedicated site for short term transitional care for patients discharged from local hospitals. Syphilis treatment left manual laborers too debilitated to resume work immediately, while recurrent famines and a collapsing textile industry further undermined the health of the Tuscan labor force. Taking cues from the Roman convalescent hospital—the first of its kind in Europe—Grand Duke Ferdinando solved two problems with one stroke by transforming San Paolo into a convalescent facility between 1588 and 1592. In this new 120-bed hospital, "poor convalescents could be received with charity until they have recovered their former strength and health."[46] By giving the concept of convalescence an institutional base, the Medici both added a critical piece to the public health infrastructure and integrated its elements more tightly.

Building a Hospital Community

Renaissance hospitals borrowed from the dual conceptual models of household and monastery, which informed each other in both practical and symbolic ways. Many charitable facilities throughout early modern Europe utilized the notion of a "constructed family" to allocate responsibilities. By appropriating familial language and structures, these institutions created and legitimized important opportunities for women's medical work.[47] Contrasting with this model was a powerful monastic construct emphasizing hierarchy and penitential discipline. Many of the new charitable initiatives animated by Catholic reform—Magdalen houses for repentant prostitutes, shelters for vulnerable girls—were patterned along monastic lines or became more explicitly conventualized in the sixteenth century.[48] Although the Florentine Incurabili initially combined both models, the hospital increasingly resembled an enclosed convent in both structure and concept by 1600.

This trajectory can be seen in changing responsibilities for patient care among practitioners, including physicians and surgeons, as well as caregivers, pharmacists, and spiritual counselors. The first hospital statutes (1521) stressed active participation on the part of all confraternal members. Four "brothers" took turns scouring the streets for "those afflicted by the French disease or other incurable ailment." Before transporting them to hospital, however, these men filtered out persons suffering from leprosy or plague in order to avoid contagion.[49] Public health systems in premodern cities relied on similar surveillance teams comprising ordinary citizens, ranging from charitable volunteers to elderly female "searchers" employed by London parishes to inspect the sick and dead for signs of plague.[50] This community-based arrangement presumed considerable diagnostic skill on the part of ordinary folk, since pox was often conflated with leprosy in the early years of the outbreak.[51] Similar health assessments were performed by older women who vetted applicants at the Florentine girls' shelter called the Ceppo in the 1550s. Its statutes instructed the prioress and head female nurse to "carefully inspect the head, throat and the rest of [a girl's] body" for any signs of "incurable or contagious ailments" that would bar admission.[52] The fact that laypersons were called on to deploy their own interpretive resources when diagnosing infectious diseases speaks to an assumption that significant health literacy suffused both household and street.

Only in 1574 did the Incurabili codify arrangements for patient care by salaried staff. Statutes revised that year required a physician and surgeon to visit the sick daily. Renaissance physicians typically had limited physical contact with patients, normally taking an extended verbal history of symptoms in lieu of a detailed examination.[53] Less prestigious hospital surgeons treated external symptoms, which involved greater hands-on contact and the use of a sensory-based repertoire in clinical encounters. These surgeons earned a small monthly salary but did not live on-site; in the 1540 and 1550s they remained on staff for only one or two years.[54] Like Renaissance physicians, they probably used hospital service as an adjunct to private practice, since the salary was not especially lucrative and their purview relatively limited.[55]

Accompanying these practitioners on their rounds was the head male nurse. This live-in caregiver oversaw the men's ward and supervised a staff of about a dozen male servants, but exercised no authority in the women's sector. The small size and predominantly local origins of hospital staff in Florence contrasted sharply with the Roman pox hospital. Drawn to the epicenter of Catholic reform, at least a hundred men from Italy, Spain, Flanders, Provence, and France worked there as religious brothers and laymen in 1580.[56] Florentine hospital records indicate that, by contrast, male staff members were short-term employees paid a meager monthly salary in addition to room and board—an arrangement that led to high turnover.[57] Their persistent contact with dirt and excrement place them at the bottom of the Renaissance medical hierarchy. These working men may have found more attractive employment opportunities in the textile trades or in public works projects launched by the Medici dukes. Their more finite affiliation with the hospital tracks changing conceptions of service in charitable institutions influenced by early modern capitalism. Unlike their medieval predecessors, these hospital servants were seen less as household members than as short-term contract labor.[58]

In contrast to this transient staff, the eighteen to twenty-four women who constituted the female nursing community at the Incurabili made a lifetime commitment to the institution. Longevity of service enabled them to develop deep expertise as both caregivers and administrators. In this sense, the Incurabili nurses resembled the larger group of one hundred women serving the general hospital of Santa Maria Nuova.[59] Not all Italian pox hospitals adopted this model. Local labor markets as well as diverse institutional mandates conditioned staff arrangements. Even in Italy, where custodial institutions were highly conventualized, hospital nursing formed part of an economy of makeshift that allowed working women to cobble together resources as circumstances permitted. At the Roman Incurabili hospital, the prioress and her three salaried assistants, mainly migrants from other parts of Italy, served only a few months before moving on to other work. The entire female staff turned over every one to two years.[60]

In recruiting female staff, the Florentine Incurabili drew from an unconventional pool. Traditionally, hospital nurses in late medieval Europe were mature women, often widows, who had accumulated considerable experience in illness management and already had fulfilled various domestic and reproductive responsibilities.[61] Indeed, the Incurabili hospital in Pavia was staffed by twelve widowed "matrons" responsible for female patient care. Other Italian and English hospitals required nurses to be between thirty-six and fifty years old when entering service.[62] In choosing mature women to tend the sick, Renaissance hospitals balanced the physically taxing demands of nursing against the need to maintain hospital discipline. Contemporaries commonly assumed—not always correctly—that older women would refrain from youthful indiscretions such as brawling, public drunkenness, and sexual flirtations that could plunge the community into disarray.[63] Although these wives, widows, and single women entered service from different points in their life cycle, they nevertheless capitalized on practical experience that could be put to good use in a hospital setting. Scholars probably have underestimated the extent to which experiential know-how flowed into a hospital through its nurses, who circulated their expertise via daily chores and bedside encounters.

The Florentine pox hospital radically departed from these norms by recruiting single girls between the ages of sixteen and eighteen—the customary age at which local girls first married. The 1574 hospital statutes stipulated that these sexually pure girls *(oneste fanciulle)* had to be at least sixteen years old before joining the staff, which echoed Tridentine norms for entering a convent.[64] The nurses' quasi-monastic lifestyle further showcased the Incurabili's redemptive functions. These women pledged permanent celibacy, brought a small dowry, followed highly regulated routines, shared common living spaces, and could not leave the premises without permission. Instead of being paid directly for their work, they received room and board, clothing, and other necessities. This setup differed from staffing arrangements at Elizabethan hospitals, where nurses earned a small annual income supplemented by fees for additional services, from which they bought their own food. Unlike their English counterparts, Florentine nurses could not marry or establish their own households.[65]

By adhering to a monastic lifestyle, these recruits played a compensatory role in the urban sexual economy. Pox nurses helped purify the city as moral agents who offset the effects of sexual vice, much like the nuns surrounding them on Via San Gallo. Florentines had long deployed civic strategies of ritual cleansing, such as directing fines levied on sex crimes to houses for repentant prostitutes *(convertite)*.[66] In fact, Incurabili hospitals often sustained close ties to *convertite* communities because of their common sexual associations. In Venice, repentant prostitutes formed a distinct subgroup among hospital patients, whereas in Verona they occupied part of the same complex. Women hoping to enter the Convertite house in Naples first worked in the pox hospital as a kind of pre-novitiate; the two communities operated as sister institutions after 1538.[67]

Choosing young girls to nurse pox sufferers back to health embodied other aspects of the "new philanthropy" associated with Catholic reform. Part of the hospital's mission was to prevent vulnerable girls from falling into prostitution by recruiting them into nursing service instead. Applicants were brought to the attention of the hospital board by confraternal members, who often paid the required entrance fee of thirty to eighty ducats and provided necessary clothing and linens. The case of sixteen-year-old Francesca provides an illuminating example of these practices. Hailing from the nearby town of Prato, Francesca was described by the hospital board as "extremely poor . . . without father, mother or other close relations . . . and good-natured, with good habits." She lacked the means to become a serving nun in a local convent and was already too old to be placed in one of the new conservatories springing up around the city after 1540.[68] Hearing from numerous people that Francesca wanted to serve God and the sick poor, and "considering the dangers awaiting such a girl," the hospital board voted unanimously in 1570 to admit her as a nurse, thereby permitting her "to conserve her chastity and enable her to live a celibate life."[69] This recruitment strategy thus met twin objectives: securing an able caregiver while rescuing a poor girl from prostitution. Physical strength and good health obviously were important in meeting the rigors of hospital service. Several Incurabili recruits were described as being "healthy, obedient, and willing to work hard" or "broad-chested and robust." Many of them were country girls who were considered to be stronger and healthier than urban residents.[70]

Changing recruitment patterns for nurses reveal that the hospital had evolved into a patronage site for local elites by the 1580s, much like other charitable institutions in Medici Florence. In recommending candidates for nursing positions, sponsors created chains of clientage that reached into the lower social tiers; in this way they appropriated institutions for their own use. Hospital wardens and confraternal officials still sought to rescue girls like Francesca, but they increasingly secured placements for their own relatives, tenants, and clients. In 1572, the hospital prior Marcello Acciaiuoli won admittance for two young nurses—one a local draper's daughter, the other a country girl—based on his personal recommendation. Fifteen years later, the Incurabili accepted the daughter of a rural tenant working for the former hospital prior Antonio Salviati, again based on his recommendation.[71] At times, satisfying patron demands led to poor outcomes. After the hospital of San Paolo accepted a young country girl in 1592 at the grand duke's request, the warden soon discovered that "her slight build and gentle complexion" made it unlikely "that she possessed the strength to perform the many tasks" nursing required. Seeing that she was a skilled lace-maker, the official suggested that the grand duke find a more suitable placement in his wife's retinue instead.[72]

A clear example of how patronage reshaped the Incurabili nursing community by 1600 involves the hospital warden Niccolò Zerbinelli, whose esteemed medical career was capped by election to the College of Physicians. A relative newcomer to hospital service in 1597, Zerbinelli obtained permission to board his orphaned nephew at the Incurabili for an extended period free of charge. The following year, he proposed that his twelve-year-old niece Solomea be admitted as a nurse, despite being under the statutory age. In his petition to the governing board, Zerbinelli explained that his niece found herself "without a father and abandoned by her mother, who has remarried." If she proved capable of performing the required tasks, he promised to provide the necessary entrance fees and other goods. Judged to be "healthy and apt for service," Solomea cleared this hurdle and later matured into a distinguished administrator, as discussed below.[73]

Once the hospital community became firmly established, the Incurabili nurses themselves exercised some control over personnel, bringing personal experience to bear on their decisions. Each trainee spent a probationary period of one to nine months rotating through the women's ward, hospital pharmacy, kitchen, and laundry to test her aptitude. This trial period combined a trade apprenticeship with a monastic novitiate. It enabled current staff to gauge a candidate's mental quickness, manual dexterity, emotional temperament, and physical stamina—all of which had a bearing on her suitability for lifelong service. At the end of the probationary period, the governing board consulted the nursing staff about her performance. These peer assessments seem to have been taken seriously through the end of the sixteenth century. Hospital administrators rejected two applicants in 1574, for instance, "because they were unfit for the work of this hospital and unsuited to its purpose, according to the information provided by the majority of all the other nurses, who were diligently consulted."[74]

The nursing staff also had a voice in selecting the hospital prioress from among their ranks. The governing board interviewed each of the nurses annually to "hear their opinions" regarding which candidates showed sufficient aptitude to supervise staff and patients.[75] By statute, the prioress was required to be at least forty years old; she could serve two consecutive annual terms before stepping down, but it was not uncommon for talented administrators to serve repeated rotations, much like convent officers. Between 1571 and 1582, three experienced nurses—Antonia, Lucia, and Alessandra, the first two from the Tuscan town of Castelfranco—alternated in this leadership position every two years.[76] This balance between distributing authority, on the one hand, and concentrating experience, on the other, was a hallmark of Savonarolan charitable institutions in the first century of their existence. Italian orphanages and conservatories relied on similar systems of internal promotion for top administrators. By tapping accomplished women who had spent their entire adult lives within these enclaves, female charitable institutions ensured the continuity of personnel and house traditions.

The Incurabili prioress supervised both female staff and patients, who were obliged to obey her directives. She also appointed a treasurer to handle earnings from textile work and to make everyday purchases for staff members.[77] Overseeing her was the male warden and governing board; they could chastise her behavior, audit her accounts, and overturn her decisions if warranted. Even though the prioress acted as a liaison among hospital personnel, she exercised no formal decision-making power over the institution as a whole. As the Incurabili moved away from its founding ideals by 1600, the nurses lost much of their autonomy in selecting their own officers. Instead of consulting female staff as a matter of course, the hospital board relied more heavily on recommendations by the warden when choosing the prioress.[78]

These new leadership patterns are exemplified by the medical career of Solomea Zerbinelli, who entered service as her uncle's protégé in 1598. Known in hospital records by the religious name Sister Barbera, Zerbinelli was elected prioress in 1634 at the age of forty-eight, after spending thirty-six years on the job. She dominated the position of prioress until her death in 1657 at age seventy-one, serving without interruption for the last fifteen years.[79] In that role, she exerted tremendous influence over the practical training and moral formation of incoming nurses, setting disciplinary standards for the nursing community as a whole. During her tenure as prioress, Zerbinelli also helped coordinate the hospital's transition from a charitable to a fee-based institution. Tuscany experienced a sharp economic downturn in the mid-seventeenth century, while the hospital itself was plagued by massive financial problems and waning lay support. The Incurabili enacted sweeping reforms in 1645 in response to these problems. After steep budget cuts proved insufficient, the governing board for the first time began charging patients an admission fee of three and a half lire to offset the cost of treatment. This nominal sum represented about 18 percent of monthly earnings for an unskilled construction worker.[80] To generate additional revenue, Zerbinelli also began to market in-house remedies more vigorously. Under her supervision, the hospital pharmacy expanded its commercial dealings with local convents, hospitals, and charitable groups in the 1650s. These transactions centered primarily on ready-made sarsaparilla remedies to treat syphilis and malarial fevers—two endemic health problems that undercut the productivity of the local labor force.[81]

Despite the obvious influence of conventual models on hospital organization, the Incurabili nurses never took religious vows that would have rendered them legally "dead to the world." Instead, these women straddled two worlds. When the orphaned teenager Francesca officially donned the russet-colored uniform of Incurabili nurses—the public marker of her affiliation—she entered an ambiguous legal category that was neither fully religious nor fully secular. This in-between status was marked by a unique occupational label. As Margaret Pelling has observed, female caregivers were known by a wide range of titles, the linguistic profusion of which reflected the fluidity of occupational labeling throughout Renaissance Europe.[82] The term "nurse" *(infermiera)* was used almost exclusively in convents to identify the nun charged with tending sick peers. Nurses working at Santa Maria Nuova were prosaically called "women of the hospital" *(donne dell'ospedale);* a few practitioners showing exceptional expertise in both nursing and pharmacy were dubbed "house physician" *(medica di casa).*[83] The term *pinzochere* had a capacious reach. It was applied loosely to tertiaries who had taken simple, rather than solemn, religious vows; it also described hospital nurses whose pious motives for entering service were understood, despite their lack of formal religious affiliation. Some women who engaged in hospital work were simply called "Mona," the respectful form of address for mature women, while others were described as annuitants *(commesse).* The latter contractually exchanged property in return for an annual stipend and sometimes residential accommodations. Their duties varied considerably and might preclude work altogether; becoming an annuitant at a monastic or charitable institution was a common way of securing long-term care in old age.[84]

Despite this linguistic exuberance of the time, the Incurabili nurses were given a new, unique name. Hospital records called them "servant girls" *(fanciulle serventi)*—a term that placed them in a perpetual state of service while diminishing their expertise as practitioners. The same designation was later used at San Paolo following its reorganization by the Medici grand dukes as a convalescent hospital circa 1590.[85] Service was a model for human relationships in the early modern period; the terms used to define and mediate these relationships carried great cultural weight and merit exploration beyond the present scope. Calling Incurabili nurses "servant girls" rather than tertiaries or "hospital women" showcased household duties tied to the materiality of bodywork, such as changing bed linens and emptying chamber pots.[86] Instead of highlighting their pharmacy skills or religious service, this term devalued the role played by these care workers in a complex healing process, making them less visible to others. This new occupational label also tied nurses at these new facilities directly to secular oversight, which helped the early Medici dukes avoid jurisdictional problems with the Church. Even though the Incurabili nursing community became more conventualized, hospital records insisted that these women never took vows. Divested of ecclesiastical affiliations as well as professional respect, Italian hospital nurses unwittingly advanced the conceptual transformation of hospital service into a form of gendered labor.

Once accepted into service, the Incurabili nurses confronted strenuous working conditions and regular exposure to health hazards. One of the most obvious health risks they faced was "bad air," a root cause of illness in the Galenic system. Medical thinking about the relationship between air and human health had its roots in Hippocrates' *Air, Water and Places,* which appeared in an Italian print edition in 1526. In the late Middle Ages, miasma—the malignant vapors released from sewers, stagnant water, and other organic waste—was commonly blamed for causing plague and other contagions. Concerns about miasma formed the backbone of public health initiatives throughout the early modern period.[87] In contrast to these macro-level concerns, late Renaissance health manuals increasingly paid attention to the health effects of indoor ambient air.[88] Influential guides to healthy living discussed factors such as temperature, humidity, and correct ventilation that contributed to the soundness of air within homes and other interior spaces. Physicians agreed that air quality in enclosed spaces could easily become adulterated through stagnation and bodily exhalations. The dangers of self-created pollution were exacerbated by summer heat as well as the fumes emanating from stoves and fireplaces, all of which could do grievous harm by suffocating the brain. Since air quality also affected mood, it affected the delicate humoral balance underlying a sense of well-being.

Although Italian health manuals targeted primarily domestic environments, the medical principles they articulated applied equally well to the institutional spaces of hospital wards. There was both a conceptual and practical fluidity between domestic and institutional interiors, since many early modern people shuttled between these residential spaces over their lifetimes.[89] Echoing vernacular medical advice, Florentine hospital accounts demonstrate this new attentiveness to the role of ambient air in the healing process. In 1569 the hospital warden at San Paolo lamented the "continuous putrid stink" emitted by the tombs located inside the facility. These old tombs corrupted the air throughout the hospital, harming both patients and staff. The odor reportedly was so pungent that it penetrated the walls of the storage vault, spoiling the wine there. To create a more salubrious environment, the warden emptied the tombs and installed new windows that opened and closed easily, thereby admitting better light and ventilation.[90]

All Renaissance hospitals confronted the problem of corrupt air, but pox hospitals faced special challenges due to both the nature of the disease and its highly regimented cure. As noted earlier, the French disease was notorious for its stench. Hospital visitors in late sixteenth-century Valencia reportedly tapped their noses as they passed the pox ward, often refusing to enter because of the foul odor and fear of contagion.[91] Administering the wood cure created additional problems. Braziers were used to heat pox wards to a high temperature in order to induce copious sweating, which removed peccant matter from patients' bodies, but their simultaneous emission of heat and smoke polluted these microenvironments. Fetid odors issuing from bodily excretions as well as thick exhalations of patients jammed into wards further degraded ambient air. Nurses were instructed to keep windows closed during sweat sessions to avoid patients becoming chilled. Presumably these conditions pertained in the men's ward as well, although the exposure of male staff remained limited, considering high turnover. Hospital nurses signaled their awareness of these health issues. Finding little respite outside the ward, they complained to the governing board in 1600 that crowded conditions in their living quarters created an unhealthy environment. In response, the board decided to expand their rooms a few years later and to enlarge the women's wing at the same time to better serve the sick.[92]

In carrying out their duties, caregivers also coped with anxieties about becoming ill themselves. The lack of firsthand accounts puts their subjective responses out of reach, but the fictive dialogue written in 1629 by the Barnabite priest Biagio Palma (1577–1635) offers a window into culturally pervasive attitudes toward hospital nursing; it also provides rare glimpses of patients' experiences of hospitalization in this period. Palma served as rector to a group of Roman noblewomen who aimed to perfect their spiritual lives by "visiting poor sick women in hospital and assisting them in whatever way possible."[93] Members took turns visiting every hospital in the city on a weekly basis to feed and console the sick poor. This consorority was the Roman branch of the Congregation of the Umiltà di San Carlo; it was headed by Costanza Barberini, sister-in-law of Urban VIII. Inspired by Carlo Borromeo's activities during the 1577 plague in Milan, the group embodied both the lofty spiritual aspirations and profound social biases characteristic of Counter-Reformation welfare.

To stimulate devotion among his charges, Palma scripted an imaginary dialogue between himself and a lay sister, in which he cast hospital visitation as a kind of spiritual heroics. Visiting the sick had a long tradition in Christianity as one of the seven acts of mercy, but this practice took on more performative, class-based dimensions in the sixteenth and seventeenth centuries. Illness presented sufferers with an opportunity for spiritual regeneration; similarly, visiting the sick in the hospital invited self-reflection on one's own shortcomings. Hearing, seeing, and smelling sickness in others could only reinforce and amplify the lessons preached in sermons or devotional literature to restrain one's appetites or correct other flaws; these encounters probably carried extra punch for wealthy aristocratic women, who normally could control their domestic ambience and daily activities to a considerable degree. As one important element in the economy of salvation, these voluntary acts of charity were meant to elicit profound or even transformative emotional responses in pious noblewomen that aided their spiritual development.

In Palma's eyes, what stimulated these advances was the range of sensory assaults inflicted by the hospital environment. Palma depicts the late Renaissance hospital as a sensorium replete with distressing stimuli. Indeed, he argues that it is the visceral disgust provoked by hospital visitations that makes them more spiritually meritorious than fasting or penance. By conquering their disgust, charitable visitors not only gained self-discipline but also earned the prayers and moral indebtedness of the poor.[94] This priest represents the distinctive hospital soundscape as a cacophony of ordinary sounds like bell-ringing and murmured prayers punctuated by more objectionable, disturbing noises—involuntary sighs, cries of pain, loud quarrels among patients—that unsettled hospital visitors and inmates alike. In keeping with miasmatic theories, smell occupied a unique place among these sensory threats. Speaking in the voice of the lay sister, Palma candidly declares that visiting hospital patients is "repugnant," "disgusting," and "very difficult work." The priest responds in his own voice that the "stench and stink" encountered on the wards is "nauseating, especially for those women who are naturally delicate." Here the hospital ward is figured as a micro-contact zone in which two sensory orders—one represented by poor patients and staff, the other by upper-class hospital visitors—have to be negotiated. Palma advises his charges to recall the strategies used by earlier pious women, such as Angela of Foligno, Elizabeth of Hungary, and Empress Placida, who confronted similar sensory challenges when nursing the sick poor. The medieval Italian penitent Angela and her followers, for instance, reportedly imagined the vile, stinking sores they washed and kissed to be "fragrant red roses." Fortified by these female exemplars, the lay sister realizes at some length that it is the very repulsion of hospital encounters that cleanses sin. Finally, she confesses her greatest fear, of contracting an illness from a patient. Marshaling abundant pastoral examples, Palma replies that the noblewoman must perform these good deeds without regard for her own health and place her trust in God.[95]

Given their limited options, the Incurabili nurses presumably accepted the risks inherent in their work, which provided both a livelihood and a road to salvation. Palma seems to imply that the lower-class origins of hospital nurses would have inured them to the sensory threats that noblewomen found so distressing. However meritorious it might be from a religious standpoint, the daily grind of hospital nursing lacked the dazzling performative aspects of charity work done by noblewomen. In the 1620s, Costanza Barberini ceremonially abased herself by serving poor women at table in the Roman convalescent hospital, where these theatricalized performances carried little risk of contagion; nor did they subject her to the intense physicality of suffering.[96] The continual health hazards to which hospital nurses were exposed, coupled with the corporeal nature of caregiving, explain why nursing carried great weight as a charitable activity in Counter-Reformation Italy, yet remained an occupation chiefly reserved for the lower classes.

"Invisible Technicians": Medical Know-How and Patient Care

The 1574 hospital statutes tasked female staff at the Incurabili with three main areas of patient care: preparing the wood cure for all inmates; tending the women in the female ward "with charity and diligence"; and supporting the general needs of patients and staff by preparing meals and doing laundry.[97] In addition to these formal work assignments, nurses performed a number of unwritten duties, from ensuring compliance with a grueling regimen to defusing tensions between patients or visitors. They managed complex gender interactions implicit in clinical encounters by offering a social buffer at bedside examinations by male physicians. In the winter months, the Incurabili nurses spun linen used in-house or sold on the local market, with proceeds filling the common purse, or sewed caps worn by inmates.[98] By vesting multiple responsibilities in female staff, the Florentine Incurabili avoided the two-tiered system emerging in some late medieval English hospitals, where routine care of the sick was handled by female servants and less demanding chores by professed nuns.[99]

The extensive pharmaceutical work female staff carried out at charitable institutions should not be surprising, since this allocation of responsibilities capitalized on a cheap labor pool and eliminated the need for a salaried apothecary. Indeed, hospital dispensaries provided the backbone of local healthcare across Europe.[100] Their long service and encounters with many clinical variations placed hospital nurses among the most accomplished pharmacists of their day. Several recipes for ointments and salves devised by female staff were inscribed in the 1515 *ricettario* of Santa Maria Nuova; presumably these remedies had been tried and tested on hospital patients to determine their efficacy. The hospital of San Paolo also ran a commercial pharmacy staffed by religious women in the 1560s and 1570s, which retailed many of the same products as convents and guild apothecaries.[101]

Even large custodial facilities like the Innocenti foundling hospital (est. 1445) allocated medical labor and resources in similar ways, since they worked on the same gendered models of charity and institutional self-help. Plagued by recurrent financial problems, this civic orphanage housed over 700 women and girls in 1581. By this time, the Innocenti provided some of its own medical personnel in the form of four female "pharmacists and physicians" *(spetiale e mediche)* living on-site, who treated residents "according to their art." Three nurses rounded out the internal medical staff; their role was to administer medications and care for the sick "with charity and diligence."[102] These women were not hired from outside but had been foundlings themselves, trained in-house via the same apprenticeship system that saw other orphans working at silk looms within the complex itself. The orphanage also was preparing for the future by training fourteen young girls "in the art of pharmacy," which included instruction in reading and writing.[103] The latter skill set was an important requirement for orderly administration—an area in which the Innocenti often fell short. As part of earlier reform efforts, the hospital prior, Vincenzo Borghini, had recommended that the pharmacy keep two separate ledgers: one to record external "sales and purchases" of various medicinal items; the other an account of "the medicines, syrups and everything else consumed in the orphanage and hospital" that clarified internal distribution patterns.[104]

The pharmacy workshop offered an important proving ground for future administrators at the Innocenti, with capable apothecaries moving from the dispensary to the head office as they did in Italian convents. Replacing the prioress Mona Maria upon her death in 1574 was fifty-year-old Mona Fiammetta, who had grown up in the orphanage and had served in the pharmacy for the previous ten years. These administrative officers, who had spent their entire lives within the institution, earned praise and affection from peers and supervisors.[105] In its hands-on training methods and complex management systems, this foundling home resembled a monastic community writ large—one that contributed to social welfare by retailing medicines to the public as well as housing orphans. By neglecting healthcare practices at institutional sites such as hospitals and custodial institutions, scholars have underestimated both the prevalence of women's pharmaceutical expertise and the medical resources dispersed throughout Italian cities.

Staffed by skilled women, the pharmacy at the Florentine pox hospital was among the most important civic institutions braided into a revitalized public health apparatus. By the mid-sixteenth century, the Incurabili nurses were making virtually all of the medicaments consumed by hospital patients, whether for their internal or external use. These remedies ranged from unguents applied to pox sores and other skin conditions to assorted distillates, syrups, and electuaries taken internally. Indeed, one of the hospital's first purchases in the 1540s was "a small hatchet for chopping herbs in the pharmacy." In the same years, the hospital bought bulk quantities of betony, roses, violets, almonds, and other medicinal herbs from the pharmacist at the nearby hospital of Bonifazio.[106]

In performing this large-scale pharmacy work, the nurses operated as designated agents in public health structures being reimagined by the early Medici dukes. In 1560, Cosimo singled out the hospital as the only institution in the dominion that could administer the wood cure. This move formed part of a larger ducal initiative to bring local charitable institutions under state purview.[107] Consumers could still purchase ready-made guaiac cures from local pharmacies, but only female nurses could prepare the remedy used at the pox hospital. This measure ensured some standardization and quality control in addressing a massive public health problem; at the same time, it enmeshed these medical artisans directly in public health strategies for containing pox. Cosimo's son Ferdinando further harnessed the nurses' medical expertise to the political goals of the Medici state. When establishing the convalescent facility of San Paolo in 1592, the grand duke chose one of the experienced Incurabili nurses to oversee the new community as governess and instruct new recruits in practical nursing knowledge and hospital organization.[108] These shared practices irrevocably linked the two facilities until Pietro Leopoldo's sweeping institutional reforms circa 1800.

The nurses' primary task at the Incurabili pharmacy was to prepare the guaiac remedy, administered seasonally as part of a comprehensive regimen. A New World wonder drug, guaiac enjoyed great popularity as a pox treatment throughout the sixteenth century. Recipes for making this remedy circulated widely in Latin treatises and various vernaculars in both print and manuscript form after 1525.[109] By 1533, the hospital substituted guaiac decoctions for earlier topical ointments made from rose oil mixed with gum resin. Consonant with pox treatment elsewhere on the Italian peninsula, the Florentine Incurabili offered the wood cure annually for a few months between mid-February and the end of October, when mild temperatures assured that patients would not undermine treatment by becoming chilled upon leaving the well-heated wards.[110] In contrast to German pox hospitals, however, the use of mercury in any form was strictly prohibited. This different therapeutic approach reflected the greater penetration of Paracelsian ideas into German medicine compared with Italy, where Galenism continued to reign supreme.[111] "Taking the wood" aimed at reducing pain, healing lesions, and improving functionality, rather than curing the disease completely. It was not uncommon for sufferers to undergo several rounds of the wood cure from one year to the next, which imposed steep burdens of illness on local societies in the form of lost productivity.

As the centerpiece of treatment, guaiac figured among the hospital's biggest expenditures. Around 1525, "holy wood" reportedly retailed on the Florentine market for the enormous sum of eleven scudi (77 lire) per pound—an inflated price that probably reflected both the novelty and scarcity of this commodity.[112] Around 1541, the Florentine Incurabili began purchasing most of its guaiac in bulk allotments of 200 to 300 pounds from the Pinadori apothecary firm, which also held the ducal contract on sugar imports. With its ability to handle trans-Atlantic commodity chains, this firm became the hospital's principal supplier for the next three decades. The increasing volume of imports by the 1560s dramatically dropped the price of holy wood to roughly two and a half to three and a half lire per pound, although it could be purchased through monastic brokers for less, depending on quality and preparation.[113] For instance, in 1567 the nuns of Santa Lucia bought two pounds of guaiac "cut into large pieces" to distill for one of their sisters, paying sixteen soldi per pound (less than one lira) to the guild apothecary "at the sign of the Moor." When their confessor showed the bill to the pharmacist at San Marco, he claimed they had been overcharged; according to the friar, the fair price for that grade of guaiac was only twelve soldi.[114] Prices continued to fall in the ensuing decades. When Grand Duke Ferdinand established an official price list for common medicinals in 1593, a pound of guaiac was pegged at roughly one and a half lire. Ready-made distillates and oils prepared from the wood were also widely available, attesting to the spread of this cure.[115]

It is difficult to establish the per capita cost of taking the cure with any precision, since both the price and quality of guaiac fluctuated. Individual patients also responded to the regimen differently. German pox hospitals figured that two to ten pounds of guaiac per person were required for a three-to-six week course of treatment.[116] Hence one course of guaiac therapy in late sixteenth-century Florence would have cost anywhere from six to thirty lire—the equivalent of 2 to 13 percent of annual earnings by an unskilled construction worker in the 1580s.[117] Even though guaiac became known as a poor man's remedy by the end of the century, paying for the wood cure obviously remained out of reach for thousands of sufferers.[118]

Preparing this remedy in volume required a combination of strength, patience, and attention to detail. First the tough wood bark had to be filed or shaved and the resulting product steeped in twelve parts water for one day. Although shaved guaiac could be purchased on the retail market by the late sixteenth century, hospital records show that nurses continued to prepare the wood from scratch well into the 1630s as a cost-saving measure.[119] The steeped liquid then was boiled in large vats until reduced to a third or half its original volume. Nurses tended this steamy brew carefully, using various visual cues to achieve the desired result; several of the Incurabili nurses wore eyeglasses to improve their visual acuity.[120] "Wood maids" in Augsburg pox hospitals were forbidden from leaving the "kitchen" during preparation to ensure that the mixture did not burn or boil over. This concentrated liquid, adjusted for seasonal and humoral variations, formed the core of treatment.[121] Sufferers drank the bitter decoction as a hot liquid twice a day for thirty to forty days in order to purge the body of peccant matter through copious sweating and evacuation. During that time, they remained largely bedridden. A second, weaker batch was brewed to be drunk with meals, while the foam produced during decoction was skimmed, dried, and applied topically to pox sores. Because decoctions were judged to be most effective when fresh, the Incurabili nurses produced huge batches every two or three days during the cure months. The industrial scale on which they worked is indicated by the hospital's consumption of one to three tons of guaiac annually between 1542 and 1588.[122]

When preparing this remedy, the Incurabili nurses worked in pairs for six months at a time before switching to other duties. This regular rotation ensured that the skills needed to manufacture guaiac on this scale were distributed broadly throughout the staff. One nurse served as the senior member of the team, the other assisted her; they swapped positions weekly during their rotation since the work itself was both intense and rather tedious.[123] This arrangement also equalized distinctions among staff members and prevented the formation of internal hierarchies that might challenge the authority of the head nurse. From a long-term institutional perspective, this system allowed the Florentine Incurabili to replicate training and transmit technical know-how across cohorts at no additional cost. By circulating skills in this manner, staff nurses mirrored the modes of knowledge transfer already documented at convent pharmacies.[124]

Administering the wood cure on such a large scale required considerable internal coordination among hospital staff. Each of the three decoctions, which possessed specific therapeutic properties, was labeled and stored separately so that they could be administered at the appropriate time in treating individual patients. Drinking vessels had to be warmed to the right temperature before being filled. Nurses also had to track the intervals between intake of food, drink, and medicament to maximize the cure's efficacy. These detailed protocols must have given rise to a record-keeping system of some type, most likely in the form of small daybooks kept by the prioress, which were commonplace in female-run communities.[125] Tending patients for thirty to forty days also required a kind of nimble watchfulness. Some patients responded more quickly to treatment than others, leading to continual adjustments. Although the hospital physician undoubtedly supervised the timing of major transitions in diet and dosage, good nursing care still hinged on a continuous reading of the body and the ability to spot changes quickly. Caring for a ward of pox patients over many months was rife with surprises, whether in the form of noncompliance or forbidden gifts of food smuggled into wards. Once the physician had finished rounds and left the premises, nurses had to improvise on the spot, exercising their own judgment or consulting more experienced staff.

Caregivers relied on the five senses as primary "instruments of knowing" when gauging a patient's progress.[126] Sensory evaluation had played a significant role in medical diagnosis and treatment since antiquity. Whether taking a pulse or examining a urine flask, both physicians and empirics relied on their powers of sensory observation to identify underlying pathologies, develop differential diagnoses, and determine appropriate therapies.[127] Yet the senses were not considered equal. In traditional sensory hierarchies, sight stood at the top, with touch and smell frequently consigned to the lower rungs. As noted in Chapter 1, offensive smells could have important diagnostic value, since ailments like intestinal worms and plague were considered to emit characteristic odors. Touch may have been the most ambiguous and problematic of all the senses on which Renaissance nurses relied. Depicted as both the "king of senses" and the basest sense, touch was the most somatically diffused among sensorial instruments. In contrast to seeing and hearing, which could function despite a distance between bodies, touching and being touched relied on physical proximity for its operations.[128] Saints and kings possessed miraculous powers of healing through touch; in fact, belief in the curative properties of the royal touch reached its peak in seventeenth-century England under Charles II.[129] At the same time, the corporeal nature of touch embodied both a "legacy of erotic prohibition" and the potential for contagion that made it suspect at a pivotal moment in the rise of empiricism.[130] Haptic knowledge acquired through years of experience figured among the principal therapeutic tools used by Incurabili nurses, who treated hundreds of patients in a single season. The development of mindful or "knowing" hands through cleaning and dressing pox sores, feeding and cleansing patients, among other tasks, contributed to broader inter-sensorial care techniques and assessments of recovery.

Although Incurabili nurses accrued a wealth of sensory-based insights in the course of service, the status value of this knowledge was circumscribed by class, gender, and professional standing. The reconsideration of empirical knowledge in the early modern period largely mirrored the traditional division of medical labor and ultimately consolidated physicians' standing atop a recognized hierarchy. Celebrated for innovative bedside teaching, the Paduan lecturer Giovanni Battista da Monte nevertheless took great pains to avoid any tinge of empiricism when acknowledging the value of sensory evidence in clinical encounters.[131] Similarly, his colleague Andreas Vesalius distanced himself from "ignorant" barber-surgeons even while insisting that medical students needed to feel the structure of internal organs with their own hands. This appreciation for experiential learning within an academic context enhanced the growing stature of what has been termed "learned empiricism" among university and court physicians.[132]

Without the legitimation provided by guilds and universities, however, the practice-based insights nurses brought to patient care were often denigrated within professional circles.[133] These rhetorical take-downs underscore the extent to which the value attached to particular kinds of medical work depended on the practitioner's social identity. Both their humble origins and their subordinate gender undercut the perceived value of nurses' sensory-based knowledge in the eyes of medical men eager to assert superior professional competencies. As Mary Fissell has argued, the "hierarchies of value" first created in the early modern period were reproduced by later generations.[134] Despite their vital role in healing, Renaissance hospital nurses remained "invisible technicians" in contemporary hospital records and most subsequent historical narratives.

Caregiving and the Non-Naturals

Although care and cure were tightly interwoven in Renaissance hospitals, the conceptual dimensions of care have been overshadowed by an emphasis on more spectacular curative practices. The embodied knowledge implicit in seemingly simple caregiving techniques such as dressing chronic sores or feeding patients who were unable to sit up either has eluded historical analysis or has been emptied of specific medical meanings.[135] Much of this know-how could be transferred experientially between cohorts, but some circumstances required quick improvisation. While there is still much to be learned about hospital care practices in the sixteenth and seventeenth centuries, examining daily routines at the Florentine Incurabili reveals that a renewed focus on the non-naturals contributed to a larger shift in medical thinking then underway. These six principles—diet, air, sleep, evacuation, exercise, emotional equilibrium—had been fundamental to conceptions of health and healing since the days of Galen. The proliferation of vernacular health manuals after 1550 sparked renewed interest in their role in healthy living. The result was a rich culture of prevention that penetrated Italian everyday life, marked by the creation of health-related objects such as wash basins, wine coolers, and foot warmers.[136]

What has not yet been recognized, however, is the extent to which this discourse also gave institutional nursing greater coherence as a body of knowledge and practice around 1600. Renewed attention to the non-naturals both shaped daily hospital routines and gave nursing activities more explicit theoretical foundations that underpinned medicine as a profession. This influence can be seen in the publication of the first European manual devoted to the theory and practice of nursing, titled *La prattica dell'infermiero.* Written by the Italian Capuchin friar Francesco Dal Bosco (1564–1640), this key work in the history of nursing placed the non-naturals at the very heart of caregiving. Indeed, it defined the essence of skilled nursing as the art of helping the sick to "manage the six non-naturals well."[137] The author spoke from years of firsthand experience. Dal Bosco served as nurse-apothecary at the Venetian monastery of the Redentore from 1629 to 1640 and had launched a commercial pharmacy there. Intended as a memory aid and instructional manual for his apprentices, the work comprises 116 "observations" explaining how to recognize and treat common ailments such as fevers, epilepsy, skin conditions, joint ailments, and venereal diseases. A local printer spotted its commercial potential and published it posthumously in 1664. Subsequent reprints made this manual a standard reference work in Italian homes, hospitals, and other institutions until the nineteenth century.[138]

Throughout the book, Dal Bosco emphasized the value of practical knowledge—that is, everything he had "seen, learned and done" over many years of tending the sick. Although he never challenged the superior position of physicians, he nevertheless carved out an autonomous space for nursing by integrating copious practical advice into his manual. For instance, he advised nurses that their hands should not be too hot or too cold when taking a pulse, so that any "feverish heat" could be detected more easily. He described the best ways for drawing blood, giving enemas, treating ulcerations, and bathing sore limbs, based on years of trial and error.[139] Noting that sick bodies are in a constant state of flux, Dal Bosco allowed nurses considerable leeway in medical decision-making. A physician might prescribe medication appropriate for a patient's condition at the time of his visit; since the body inevitably changed after his departure, however, nurses judged when, whether, and how much medication to administer, without interfering with the natural healing process. This friar made clear that the gravest area of nurses' responsibility lay in proper management of the non-naturals. Good caregivers skillfully monitored the balance between rest and exercise, between consumption and evacuation, between emotional highs and lows. Invoking firsthand experience, Dal Bosco impressed upon readers that "even the slightest error committed by a nurse" in regulating these fundamentals could be detrimental or even fatal.[140]

Although Dal Bosco's instructions applied to general nursing care, they had particular relevance for nurses working at Italian pox hospitals. Treating the pox involved a carefully orchestrated regimen centered on the non-naturals, whose performance could make or break the cure. Successful treatment required nurses to enforce a stringent diet, control the ambient air, maintain proper patient hygiene, and monitor other behaviors. A crucial part of pox treatment revolved around diet, long considered "the first instrument of medicine" in the Galenic tradition.[141] There was a decidedly penitential aspect to the guaiac regimen. Sufferers could drink little besides watered-down wine while undergoing treatment, even though the decoctions induced tremendous thirst made more acute by sweating and purging. Patients' food intake was equally limited, consisting at first solely of almond biscuits, which the nurses baked in enormous quantities—some 4,500 pounds over a ten-month period alone in 1542.[142] Portions were carefully rationed; patients consumed a few ounces, along with a spoonful of pureed grapes twice a day in tandem with the decoction.[143] "One must learn to put up with hunger" was an oft-quoted maxim that accompanied the cure, and the patient who best endured starvation supposedly made the speediest recovery. Not until the twentieth day of treatment was a full diet permissible.[144]

Equally crucial to success was the regulation of ambient air. Hospital staff had to strike a balance between the quality and temperature of air within this microenvironment. As noted earlier, noxious odors were considered detrimental to both health and recovery. Stench was a common problem in all Renaissance hospitals, but it was exacerbated in pox wards by the need to keep them unusually warm to promote sweating.[145] While opening windows might have improved ventilation, it could chill the air and wreak havoc on sweaty bodies. Protocols used at the Augsburg pox hospital instructed nurses to open the ward windows for a few minutes several times a day, during which time patients remained well-covered in their beds.[146] It is not clear exactly how the Florentine Incurabili nurses managed this balance. Ward windows may have been covered with thick curtains to avoid draughts and sudden changes in temperature. It is also possible that pox nurses utilized the corrective power of scent, since poor ambient air affected patients' emotional equilibrium and even their spiritual well-being. Certainly other institutions introduced corrective measures to redress problems with indoor air. The 1625 statutes of the largest girls' shelter in Verona directed the head nurse to bring flowers, leafy boughs, or fragrant cedar into the infirmary in order to freshen the atmosphere and cheer the young patients.[147]

Intertwined with this regimen was a restricted set of hygienic practices. Washing by immersion in water was controversial in the early modern period because it opened the pores to the atmosphere, making the body more vulnerable to the entry of pernicious substances. Judging from the disappearance of bath chambers in sixteenth-century Italian palaces, the practice of full immersion apparently declined in late Renaissance Italy, with considerable micro-variation.[148] The fragrant soaps produced in such abundance by apothecaries were meant for washing the hands and other visible parts of the body. The rest of the body was kept "clean" by frequent changes of linen undergarments and bed sheets, as well as the use of scent. Since the wood cure already made the body vulnerable through copious sweating, bathing was not practiced at Italian Incurabili hospitals. Instead, nurses used "frictions" to eliminate superfluities from the pores by rubbing patients' hands and faces with rough cloths. Frictions were described as particularly suitable for the elderly, serving as an alternative to exercise. Patients' faces were washed only at rare intervals, and their hands never permitted to come into contact with cold water.[149]

New understandings of contagion after 1540 heightened the medical implications of these activities.[150] A primary concern in Italian hospitals and households was laundering bed linens to a high standard. Vernacular health manuals often included detailed instructions about the correct performance of household chores that contributed to a salutary environment.[151] At the Incurabili, laundering bed linens until they were spotless was far more consequential, since contact with bodily fluids excreted by pox sufferers presented clear dangers. Contaminated matter could linger in these items for long periods, leading to infection of healthy bodies by the "French poison" if they were not cleansed properly. In his popular 1548 pox treatise, the Nuremberg barber-surgeon

Franz Renner warned readers never to exchange clothes or share bed linens with a pox sufferer, at the risk of becoming ill themselves.[152]

Hence the Incurabili and other Italian hospitals paid close attention to washing sheets thoroughly on a daily basis. Doing laundry on an institutional scale was physically demanding work that exposed nurses to further health hazards. These women soaked and scalded sheets, bed shirts, and other linens before boiling them with strong soap in enormous vats holding eight to eleven barrels of water. Troughs located inside the hospital complex drained the water away from buildings; the sheets were then hung out to dry on open-air terraces situated on the upper floors where they could catch the breeze.[153] Even though sixteenth-century Italians recognized the health value of clean sheets and clothes, the actual work of removing filth placed washerwomen on the lowest rung of the Renaissance literary imagination.[154]

Because of its labor-intensive nature and massive consumption of materials, laundering linens on this scale stimulated innovative technologies in the late Renaissance. Local engineers and inventors designed myriad new devices that would save time, money, and natural resources when performing everyday tasks. If they wanted to patent these devices, however, they needed to demonstrate their efficiency by testing them at institutional sites like hospitals and convents. One example among many must suffice. In 1604 the Florentine engineer Cosimo Bottegari patented his energy-efficient boiler that consumed only half the normal fuel. Bottegari demonstrated its value to local patent officials by conducting a "proof test" at three institutions: the hospital of Bonifazio and the convents of Montedomini and the Murate. Writing the affidavit in her own hand, the hospital prioress testified that it normally took two hours to bring their laundry vats to the boil and another ten to complete the entire cleansing process. A single wash, she stated, consumed 300 large pieces of firewood plus a dozen bundles of kindling. Similar autograph affidavits were submitted by the two abbesses on behalf of their communities, each of which housed close to 200 women. Using these figures as a baseline, Bottegari documented the savings achieved by his boiler, which earned him a ten-year patent on its manufacture. Growing institutional populations in hospitals, convents, and various custodial facilities prompted other technological advances to handle large-scale tasks such as food preparation.[155]

When an emerging health consciousness penetrated custodial settings, however, it acquired a more disciplinary cast. Cleanliness possessed social as well as medical meanings that signaled moral order. Across Italy, the hygienic practices implemented at shelters, asylums, and orphanages were intended to instill moral discipline in residents as well as maintain their health. The 1590 statutes of Santa Caterina, a Florentine civic shelter for abandoned girls, devoted two full chapters to describing routines for doing the laundry and cleaning the dormitory. Since "so many young girls live in such a confined space and in such poverty," read one chapter, "one should strive for great cleanliness throughout the facility for the benefit of their health." House officials were instructed to keep workrooms clean and swept daily, followed by regular inspections. Two girls were appointed to open dormitory windows every day, sweep the premises, check that beds were made correctly, and report any "dirty" girls to the prioress. Similar provisions obtained at one of the largest girls' shelters in late Renaissance Verona.[156] Surveillance of this kind, whether undertaken at pox hospitals, orphanages, or asylums, formed a constituent element of the coercive charity marking late Renaissance Italy.

Perhaps the most elusive task assigned to Incurabili hospital nurses concerned the "passions of the soul." Maintaining a healthy equilibrium between mind and body was the most complicated of the six non-naturals. Physicians, philosophers, and theologians discussed what we would call the emotions using different conceptual vocabularies and frames of reference. Despite these inflections, the link between illness and emotional turmoil was widely accepted among Renaissance Italians.[157] The need for some form of consolation must have been particularly acute in pox hospitals, where prolonged stays, grueling regimens, and physical pain might induce vastly different responses among patients. Common reactions such as despondence and boredom left a much smaller footprint in historical sources than offensive behaviors and outright defiance of hospital rules; hospital nurses who spent a lifetime in medicine had to manage the full spectrum of emotional and behavioral responses. Internal ordinances from the Roman pox hospital of San Giacomo cited patient transgressions that ran the gamut from playing cards, quarreling, and blaspheming to "immodest acts"—a euphemism for illicit sexual activities of various kinds. All of these behaviors resulted in expulsion from the hospital.[158]

Indeed, the very rationale for visiting poor hospital patients in Counter-Reformation Italy was to provide the emotional support and spiritual consolation needed to heal body and soul. These bedside encounters have left few documentary traces, but the 1629 fictive dialogue written by Biagio Palma discussed earlier offers insight into the possible scope and tenor of these conversations among patients, visitors, and staff. Palma was one of many religious writers who promoted the therapeutic value of consolation as part of an emerging health consciousness. In so doing, he assumed a considerable level of familiarity between female nurses and their patients. This priest advised charitable visitors, for instance, to confer with the hospital prioress or other female staff before entering the wards in order to grasp the patients' state of mind and emotional needs. Anticipating multiple scenarios, Palma scripted sample conversations between noble visitors and poor women, complete with cues to make these cross-class exchanges flow more easily.[159] Presumably the noblewomen boosted their confidence by rehearsing portions of the dialogue, which then could be adapted on the spot. In crafting responses to dozens of difficult questions about illness, Palma provided a training manual that authorized pious noblewomen to act as spiritual guides to the poor. Their scripted replies reinforced the sacramental piety of the Counter-Reformation church, explaining illness as a divine instrument that tested faith and expunged sins. In turn, poor women were assured that they too could become models for other inmates by displaying patience in the face of suffering. Throughout the dialogue, Palma depicted the hospital as a spiritual proving ground for women from opposite ends of the social spectrum.

Palma likely drew on his personal experience as a priest when voicing patient anxieties, whose range and specificity give them a strong sense of credibility. One sick woman on the brink of despair tells her noble visitor: "I'm lying in this bed like an animal, with no feeling of God"; another complains that "the physician still doesn't know what's wrong with me." Other sufferers articulate deeply human concerns. "When will my illness end? Will it return?" asks one woman. Yet another poses the age-old question: "Why me?" Still others confess their fear of dying or resolve to lead better lives if cured. Similar "sickbed promises" made by early modern Protestants speak to a common Christian understanding of illness that transcended confessional differences.[160] In an exchange that seems especially apt for pox sufferers, one exasperated patient exclaims that "I've been lying in this bed for more than two months, completely beaten and broken, and the remedies haven't worked." An equally frustrated woman laments bitterly that, although she has offered votive images and prayed for the return of good health, she remains "nailed to this bed." Once again Palma arms noble visitors with ample advice for spiritual sustenance, urging steadfast patience and faith in God that could even heal "the ulcers of our souls." In another scenario, a humble woman worries aloud that she has become a nuisance to caregivers. To this the visitor replies that if physicians or nurses should ever speak to her sharply, it may be due to fatigue or because they have many other duties to perform.[161] Most likely Incurabili hospital nurses drew on the same rich pool of spiritual advice when going about their work. Even though we know little about how these care workers actually managed this therapeutic role, their healing skills and monastic lifestyle made them ideal agents for helping patients construct a meaningful framework for suffering.

In confronting the challenges of syphilis, Renaissance pox hospitals braided innovative medical thinking with the "new philanthropy" associated with Catholic reform. Italian urban magistrates and charitable confraternities deployed commonplace spatial strategies for containing infection and purifying the city by siting these facilities on the urban periphery or by appropriating former brothels for hospital use. Florentines took this symbolic containment one step farther than citizens of other locales by drawing female hospital nurses from the ranks of young, at-risk girls. These care workers played a compensatory role in the urban sexual economy by acting as the mirror images of prostitutes blamed for spreading syphilis. In tapping this pool of nurses to combat a significant public health threat, the Florentine Incurabili simultaneously set out to save souls and provide low-cost medical services in a conventualized environment. Entering service at the Incurabili hospital offered low-born women a respectable career path, while validating them as unheralded agents of public health. They occupied a liminal category between secular and religious life—one that emphasized their subordinate, lifelong institutional service. Many of the demographic features, organizational arrangements, and governance practices seen at the Florentine Incurabili became normative for female nursing staff at European hospitals well into the modern era.

Like their convent counterparts, hospital nurses at the Florentine Incurabili were accomplished pharmacists who transferred technical knowledge and skills through apprenticeship. This hands-on training gave practitioners room to improvise in ways that refreshed the production process between cohorts. Nurses and nun apothecaries shared similar skills in the stillroom, but the complexity of administering the wood cure, as well as the sheer volume of guaiac produced on a regular basis, placed Incurabili nurses in an artisanal category of their own. The industrial scale on which they worked, whether making remedies or doing laundry, both encouraged and capitalized on the development of new technologies such as fuel-efficient boilers and stoves, whose merits they certified as knowledgeable witnesses. These mutually productive exchanges figured among the many small acts that integrated nuns and nurses into the matrix of "new sciences" emerging in early modern Europe.

The fictive dialogue penned by Biagio Palma highlights some of the sensory aspects of Renaissance hospital life that enveloped nurses, patients, and visitors alike. Chief among these was a cacophony of disturbing noises and fetid odors that exposed nurses, who spent long hours on the wards, to various health risks. Palma depicts the hospital ward as a micro-contact zone where two sensory orders collided along class lines. Because hospital nursing was perceived to be both hazardous and repugnant, it carried great redemptive weight in Catholic reform schemes, while remaining within the occupational purview of the lower classes. Pious Roman noblewomen capitalized on these cultural and religious attitudes when performing hospital visitations. The solace they offered the sick poor as a means of perfecting their own spirituality drew women from different ends of the social spectrum not only into a shared economy of salvation but also into relationships of interdependence. Although the ongoing financial support of well-born women was critical, these theatricalized visitations had little in common with the daily grind of hospital nursing. Indeed, the intensely corporeal nature of hospital caregiving placed it at the bottom of emerging hierarchies of value that accompanied the rise of empiricism.

Nevertheless, hospital nursing began to acquire greater coherence as a body of knowledge and practice in late Renaissance Italy, thanks to growing interest in the non-naturals. These Galenic precepts traditionally loomed large in hospital care, influencing decisions about architecture, lighting, and the disposition of space. The renewed focus on foundational principles in the late Renaissance, however, also gave nursing a stronger theoretical backbone that heightened its medical standing over time. Francesco Dal Bosco's pioneering manual further validated nursing as an autonomous skill set by incorporating practical knowledge derived from long experience. At the same time, this emerging health consciousness assumed a highly disciplinary bent when translated to institutional settings. In this sense, Renaissance hospitals and charitable facilities embodied the distinctive moral concerns of their day.

Still, we can recognize important continuities between the holistic care offered by Renaissance hospitals and modern healthcare practices. The dialogues of Biagio Palma remind us that skilled caregiving involved not only the technique of taking a pulse, administering clysters, or bathing sore limbs. Renaissance hospital nurses also tended to complex mind-body connections by helping patients grapple with the emotional and spiritual fallout of suffering. Although humoral frameworks have long been supplanted by biomedical thinking and patient support has been placed on a secular footing, holistic and compassionate notions of healing still condition the quality and experience of hospital care.

Conclusion

TAKING AN INTEGRATIVE, GENDERED APPROACH to healthcare practices and knowledge production in late Renaissance Italy offers a powerful set of analytical tools for understanding how people actually treated illness, maintained good health, and developed experiential knowledge about the body. By placing gender squarely at the center of analysis, we apprehend more clearly the multiple registers in which health and healing were understood and how these different ways of knowing helped to generate the dynamism of Renaissance medicine. Using the prism of class as an organizing tool, I have argued that Italian women from different social strata—queens and consorts, aristocratic nuns, artisan women, orphaned teenagers—played significant roles in health promotion, illness management, and medical experimentation in the sixteenth and early seventeenth centuries. The vastly different spaces in which they worked argues for the multiplicity of sites for healing and knowledge-making beyond the traditional settings of guilds, universities, and academies.

These medically informed women greatly enlarged the medical resources available in Italian society, on the one hand, and produced and circulated a wide spectrum of skills and knowledge, on the other.

Bridging micro and macro levels of analysis, I have shown that household caregivers, hospital nurses, and nun apothecaries helped anchor a diversified medical economy that ballooned in response to increased demand for healthcare services in the early modern period. Throughout the long sixteenth century, many innovative forms of medical charity—convent pharmacies, pox hospitals, girls' shelters—capitalized on women's cheap medical labor and extensive practical knowledge. The glaring need for poor relief resulting from warfare and social dislocation, economic decline, recurrent famines, and new disease threats prompted Italian nuns among others to simultaneously serve the public good and boost their income by commercializing their pharmaceutical expertise. Some types of bodywork, such as hospital nursing, exposed caregivers to recognized health risks that carried great redemptive weight even while sitting at the bottom of the medical hierarchy.

Adding impetus to this expanding healthcare sector was a preventive health paradigm that both augmented the health literacy of vernacular reading publics and made female householders important guardians of healthy living. The surging interest in preventive health throughout the long sixteenth century can be seen in new hygienic practices that spurred the production of soaps and perfumes, which opened new opportunities for small-scale female vendors to enter local markets. Similarly, the enthusiasm for fabricating medicines and wellness products in household and convent kitchens speaks not only to financial considerations but also to the rapid circulation of vernacular health precepts enabled by Italian print culture. Attentiveness to the quality of ambient air in custodial institutions, hospital wards, and nurses' living quarters provides yet other evidence for the burgeoning culture of prevention whose implementation involved a wide range of historical actors.

The performance of this varied medical work was deeply intertwined with the production and circulation of practical knowledge about the body and the natural world. Portions of this study have highlighted the epistemic and commercial contributions of Renaissance religious women, whose substantial numbers, extensive social connections, and recognized charitable role gave them greater visibility and access to unusual resources. Working at the nexus of market and laboratory, convent communities operated as important knowledge spaces in both pharmacy and botany. Nuns as well as female practitioners of different stripes operated as hybrid experts who worked along a differentiated continuum of knowledge. Much of this knowledge was acquired and transferred experientially through hands-on instruction by seasoned practitioners across cohorts. Like their counterparts in early modern craft guilds, female medical artisans learned by seeing and doing in the collaborative epistemic spaces of the workshop. These training methods aimed to develop embodied skills, perceptual focus, and the ability to improvise in the moment. The result was a maker's knowledge rife with insights into materiality and techniques, but one that remained result-oriented rather than geared toward general theoretical principles. Collaborative labor created a rich variety of female alliances—between master and apprentice, between partners in the production process, between institutional leaders—that merit further exploration. But medical collaborations also gave rise to new forms of internal stratification within large convent communities that allocated work assignments on the basis of social standing.

Household and institutional practitioners relied heavily on sensory-based repertoires, whether diagnosing the health conditions of newborns or judging the readiness of a medicinal preparation. Indeed, one of the subtle yet consistent elements running throughout this study is the extent to which the senses signified in a medical context. Pharmacy and nursing work in particular owed much to inter-sensorial experiences with materials, methods, production processes, microenvironments, the natural world, and the human body itself. Caregivers used the characteristic smell of intestinal worms to diagnose the ailment afflicting a little Medici princess; Incurabili nurses visually gauged when huge vats of guaiac remedy had boiled down sufficiently; hospital visitors reimagined fetid sores as fragrant flowers in order to perform their charitable work. These habits of sensing give scholars unique insight into Renaissance healthcare practices and the lived experiences of health and sickness.

In the case of literate women, these hands-on competencies were augmented by vernacular medical reading of surprisingly wide scope, although the role of both manuscript books and printed books in circulating medical knowledge among Italian women warrants further study. In other cases, the pursuit of health opened the door to more advanced experiments in pharmacy and household alchemy. In virtually all situations, however, practitioners both benefited from and contributed to the rapid flow of information through social networks. The evidence presented here helps conceptualize and map the capillary nature of these networks, in which knowledge flowed sideways and upward, not merely from the top down. Even within the limitations of current evidence, it is already clear that Italian women, both lay and religious, played a significant role as producers and brokers of medical knowledge. The forgotten healers discussed here facilitated the development of early modern medicine as a body of knowledge and practice in myriad ways.

In exploring these diverse activities, this study has lifted up aspects of healing that are commonly overlooked because they did not rise to the level of visible economic exchange. Redefining what counts as medical work showcases the contributions made by Medici caregivers, for instance, in their roles as narrative agents as well as skilled practitioners. Maria Salviati provided a trusted voice in conveying medical information to the ducal couple through her detailed health bulletins; similarly, matrons and nurses organized their observational insights into an intelligible diagnosis that affected the next steps in medical decision-making. These communicative acts paralleled and intersected with the physical dimensions of bodywork as part of an integrated process. Emotional solace also played a significant role in healing, whether proffered by hospital nurses, charitable visitors, or friends and relatives. The case study of Orsola Fontebuoni shows in considerable detail how the priceless gift of consolation enabled her to win the trust of Medici consorts, which indirectly influenced the political fortunes of the Tuscan and Mantuan states. This vaunted healer also trafficked in the unseen powers of magico-religious remedies, now repackaged in ways that capitalized on print technologies and new religious currents. The circulation of such time-honored remedies at the highest levels of Italian society testifies both to their plasticity in form and to their integral place within the Renaissance healing armament. Similarly, Roman noblewomen routinely consoled female hospital patients as part of their larger charitable program. Moved by the sights, sounds, and smells of sickness, they learned how to empathize and dispense succor in line with Counter-Reformation teachings.

Several important conclusions emerge from this examination of women's pursuit of health in late Renaissance Italy. First, my analysis confirms the proficiency of early modern women in pharmacy noted by recent scholars, but also shows that a wide spectrum of practitioners made cures on different scales for different purposes. Good household managers like Maria Salviati and Eleonora of Toledo fabricated everyday remedies for members of their princely households, despite having ready access to a range of medicaments commonly sold or exchanged on local and international markets. Their activities place Italian urban and aristocratic women squarely in the frame of domestic medical practices studied in other parts of Europe; at the same time, they demonstrate how vernacular health literacy played a constitutive role in court medicine. At the other end of the social spectrum, poor Incurabili hospital nurses engaged in industrial-scale production of guaiac and other remedies as unheralded agents of an expanding public health apparatus. Monastic women showed similar breadth of scale and purpose when making medicines. Nuns' activities in the stillroom ranged from commercial ventures motivated by charity, like the Santa Caterina pharmacy run by Giovanna Ginori, to the crafting of secrets for personal gain by self-confident noblewomen like Semidea Poggi. Still other evidence reveals that Medici court women such as Christine of Lorraine, her daughter Caterina Medici Gonzaga, and her daughter-in-law Maria Maddalena of Austria used components of their pharmaceutical dowries to facilitate cultural exchanges between courts. These diverse scenarios highlight not only gradations in skill level, but also the extensive networks, suppliers, and synergies involved in making medicines on disparate scales. The sheer volume of medicaments produced by both hospital and convent women exposed them to innovative technologies such as furnaces, boilers, and distilling devices. In short, pharmacy provided a nexus for women from all walks of life to facilitate productive exchanges between and within the "new sciences" that were transforming early modern Europe.

My study also highlights the extent to which the pursuit of health was a social activity—one that both relied on and stimulated a high degree of connectivity. In seeking effective remedies or proffering medical advice, Italian urban women mobilized local and wide-ranging networks of exchange. As the case of Orsola Fontebuoni and the Hapsburg archduchess reveals, the circulation of fertility aids and sacred healing objects mediated cross-class interactions within local female circles. Yet the economy of medical gift exchange not only provided early modern women with an important vehicle for exercising agency and enhancing sociability; in some instances, these exchanges incorporated commercial dimensions as well. In creating religious and commercial ties with Italian nuns across the Mediterranean, Queen Leonor of Portugal simultaneously extended her charitable reach and opened new markets for maritime goods. In turn, convents mediated the performance of aristocratic largesse, transforming them into important distribution points for medical goods and services. Similarly, Eleonora of Toledo became both a magnet for exotic materia medica flowing into Tuscany from the Levant and a conduit for shipping prized Medici court remedies back to Iberia. This young duchess put medicine to different uses than the Portuguese queen, however; dispensing medical goods and favors offered a means of calibrating rank within the court and building clientage networks outside it. Regardless of their geographical scope, the diplomatic and commercial transactions negotiated by both nuns and noblewomen highlight the unstudied role of women as secondary distributors within both local and global trade mechanisms.

In addition, this study showcases the centrality of familial relationships to the flow of new medical thinking into and out of religious communities. Kinship was foundational to Italian life, but the evidence presented here affords fresh insight into its role in knowledge transfers integral to the pursuit of health. As nodes in larger webs of information exchange, convents built extensive networks on ties of family, political alliance, and mutual obligation. Acknowledging the familial and social nature of their networks is crucial for understanding the ways in which information moved in the early modern world. The gift of secrets from Don Antonio de' Medici to his cousin, the reluctant nun Caterina Eletta, provides one of the best examples of capillary circulation, but similar circuits of exchange operated outside court circles as well. Sister Orsola Del Riccio shared insights about pharmacy practices with her brother, the master gardener Agostino Del Riccio, which allowed convent women to profit from and perhaps contribute to some of the innovations emerging from the Medici ducal laboratories. Vested family interests allowed expert practitioners like Stefano Rosselli and members of the Bettini clan to maintain close business ties and informational relationships with particular religious houses. Far from being "dead to the world," religious women facilitated further exchanges among family, friends, patrons, and clients that amplified health literacy throughout urban society.

One of the most striking features to emerge from this examination is the extent to which Renaissance women became immersed in the medical marketplace as commercial agents and innovators. Convent pharmacies stand out prominently in this regard. By commercializing their knowledge, Renaissance nuns both generated crucial revenue for their communities and developed an important public presence as healers. These female-run businesses shed new light on the production of specialized knowledge outside traditional occupational settings. Capitalizing on growing demand for ready-made wares after 1550, Italian nuns created new products and pursued aggressive marketing strategies such as branding. The significant capital investment these institutions made in commercial-grade equipment, storage rooms, and maiolica vessels underscores both the materiality of pharmacy work and its importance for convent economies. To be successful in a competitive marketplace, nuns adapted the semiotics of trust underlying commercial exchange by highlighting their core religious identity—marking vessels with images of patron saints, dispensing medicines in view of miracle-working icons, reiterating charitable motives when challenged by authorities or competitors. Moreover, Renaissance religious women engaged in larger cultural contests over the ownership of medical knowledge when they traded in "secrets" and specialty items. These multiple investments distinguished nuns as commercial producers from early modern noblewomen who fabricated remedies for charitable purposes or gift-giving.

The commercialized nature of Italian medicine also meant that urban women shaped local health markets by exercising consumer preferences for particular retail products. The ability of convent pharmacies to accommodate restrictive gender norms made them especially attractive to female customers—a fact that gave these businesses a competitive advantage over lay apothecary shops. Female pharmacy patrons probably seized on the sociability afforded by convent dispensaries to extend their social networks while exchanging therapeutic advice. Convent pharmacies emerge from available evidence as lively retail outlets, hubs of sociability, and significant transfer points for medical knowledge. Although we know less about the consumer choices female householders made when selecting vendors, the example of Maria Salviati is instructive here. The business relationships and patterns of medical provisioning she established with religious communities in the early sixteenth century shaped Medici court practice for years to come.

Another thread running through this analysis is the receptivity that female practitioners showed to new medical thinking, especially the renewed emphasis on the non-naturals. This engagement can be seen in both household medicine and hospital care. Maria Salviati's practical expertise in the realm of pediatrics both reflected and contributed to its emergence as a distinct body of knowledge and practice—a trend fostered by an intense concern with family life throughout the Renaissance. As household healers working within a court setting, both Salviati and Eleonora of Toledo displayed a pronounced interest in preventive health measures, particularly when the well-being of the Medici heirs was at stake. Yet they did not always interpret current health advice in identical ways. Even within the same household, medically informed women might advocate different points of view about remedies and regimens.

Similarly, hospital nursing integrated both new and traditional medical thinking into pox treatment. This disease threat, which sparked a massive public health crisis with lightning speed, forced Italians to graft innovative measures for containment onto treasured Galenic principles. The renewed focus on hygiene, ambient air, and contagion surfacing in late Renaissance medical discourse intersected with moral concerns about discipline, now implemented on an institutional scale. The result was a constellation of caregiving facilities that emphasized charity and coercion in equal measure. Taking a holistic approach to patient care, Renaissance hospitals—and the agents of health who staffed them—advanced the use of new therapies like guaiac while constructing all-encompassing care routines showcasing the importance of the non-naturals. Lifting up the significance of these fundamentals to the healing process gave hospital nursing a stronger theoretical backbone, which gradually heightened its medical standing. In addition, the incorporation of valuable practical knowledge into nursing manuals, like the pioneering work of Francesco Dal Bosco, validated nursing as an autonomous skill set. Because skilled caregiving required attention to the whole person, nurses worked on the front lines in helping patients grapple with the emotional and spiritual fallout of illness.

Finally, my analysis offers new perspectives on the relationship between professional and practice-based medicine in an era of increased licensing and medical provisioning by state actors. As shown in Chapter 1, household medicine, as a set of practices and a body of knowledge, played an integral role in health and healing at the early Medici court. In this most politicized version of the early modern household, domestic and learned medicine could still remain complementary as ways of understanding human health, with many points of intersection. Social relationships at court were always mediated by considerations of power, but the emergent Medici court made full use of noblewomen's medical knowhow. Beyond the household, I have indicated that current notions of medical professionalization can obscure the ways that Italian states both capitalized on women's medical labor and constrained it. The early Medici dukes, for instance, clearly recognized the value of convent pharmacies within the local healthcare infrastructure. The inability to shutter monastic pharmacies over the years speaks not only to the limited reach of early modern regulatory mechanisms, but also to the deep reliance on religious and charitable institutions for sheer survival. Without the low-cost medical services women provided as convent apothecaries, hospital nurses, orphanage staff, and charitable administrators, the social welfare structure of the Medici state simply could not have functioned.

Overall, in this study I have argued for the value of taking an integrative approach to health and society, for denaturalizing forms of medical labor, and for thinking about medical knowledge-making as a networked enterprise involving a rich assemblage of actors. In so doing, I have attempted to paint a more inclusive picture of Renaissance healthcare that recognizes the contributions of forgotten healers and situates the pursuit of health as a driving force in historical development.

ABBREVIATIONS

AAB	Archivio Arcivescovile, Bologna
AAF	Archivio Arcivescovile, Florence
AOI	Archivio storico dell'Ospedale degli Innocenti, Florence
ASF	Archivio di Stato, Florence
Auditore	Auditore dei Benefici Ecclesiastici poi Segreteria del Regio Diritto
Aud. Rif.	Auditore delle Riformagioni
CRSGF	Corporazioni Religiose Soppresse dal Governo Francese
Incurabili	Ospedale della SS. Trinita detto degli Incurabili
LCF	Libri di Commercio e di Famiglia
MSS	Manoscritti
MAP	Mediceo avanti il Principato
MdP	Mediceo del Principato
OSMN	Ospedale di S. Maria Nuova
OSP	Ospedale di S. Paolo
S. Agata	S. Agata di Bibbiena
S. Jacopo	S. Jacopo di Ripoli
ASR	Archivio di Stato, Rome
BLF	Biblioteca Laurenziana, Florence
BMF	Biblioteca Marucelliana, Florence
BNCF	Biblioteca Nazionale Centrale, Florence
BRF	Biblioteca Riccardiana, Florence
Wellcome	Wellcome Library, London

FREQUENTLY CITED JOURNALS

AMAISF	*Atti e memorie dell'Accademia italiana di storia della farmacia*
BHM	*Bulletin of the History of Medicine*
RQ	*Renaissance Quarterly*

NOTES

INTRODUCTION

1. Useful overviews of recent historiography include Monica Green, "Gendering the History of Women's Healthcare," *Gender & History* 20 (2008): 487–518; Mary E. Fissell, "Introduction: Women, Health, and Healing in Early Modern Europe," *Bulletin of the History of Medicine* (hereafter *BHM*) 82 (2008): 1–17; and Sharon T. Strocchia, "Introduction: Women and Healthcare in Early Modern Europe," *Renaissance Studies* 28 (2014): 496–514.
2. Margaret Pelling, "'Thoroughly Resented?' Older Women and the Medical Role in Early Modern London," in *Women, Science and Medicine 1500–1700,* ed. Lynette Hunter and Sarah Hutton (Stroud: Sutton, England, 1997), 63–88; Seth Stein LeJacq, "The Bounds of Domestic Healing: Medical Recipes, Storytelling, and Surgery in Early Modern England," *Social History of Medicine* 26 (2013): 451–468.
3. Susan Broomhall, *Women's Medical Work in Early Modern France* (Manchester, UK: Manchester University Press, 2004); Deborah E. Harkness, "A View from the Streets: Women and Medical Work in Elizabethan London," *BHM* 82 (2008): 52–85; Annemarie Kinzelbach, "Women and Healthcare in Early Modern German Towns," *Renaissance Studies* 28 (2014): 619–638.
4. Cathy McClive, "Blood and Expertise: The Trials of the Female Medical Expert in the Ancien-Régime Courtroom," *BHM* 82 (2008): 86–108; Debra Blumenthal, "Domestic Medicine: Slaves, Servants and Female Medical Expertise in Late Medieval Valencia," *Renaissance Studies* 28 (2014): 515–532; Elizabeth Cohen, "Miscarriages of Apothecary Justice: Un-separate Spaces of Work and Family in Early Modern Rome," *Renaissance Studies* 21 (2007): 480–504.
5. Alisha Rankin, *Panaceia's Daughters: Noblewomen as Healers in Early Modern Germany* (Chicago: University of Chicago Press, 2013); Sara Pennell, "Perfecting Practice? Women, Manuscript Recipes, and Knowledge in Early

Modern England," in *Early Modern Women's Manuscript Writing*, ed. Victoria E. Burke and Jonathan Gibson (Aldershot, UK: Ashgate, 2004), 237–258.

6. Richelle Munkhoff, "Searchers of the Dead: Authority, Marginality and the Interpretation of Plague in England, 1574–1665," *Gender & History* 11 (1999): 1–29; Richelle Munkhoff, "Poor Women and Parish Public Health in Sixteenth-Century London," *Renaissance Studies* 28 (2014): 579–596; Jane Stevens Crawshaw, "Families, Medical Secrets and Public Health in Early Modern Venice," *Renaissance Studies* 28 (2014): 597–618.
7. Fissell, "Introduction," 15.
8. Margaret Pelling, "Trade or Profession? Medical Practice in Early Modern England," in *The Common Lot: Sickness, Medical Occupations and the Urban Poor in Early Modern England* (London: Longman, 1998), 230–258. Monica Green, "Women's Medical Practice and Health Care in Medieval Europe," *Signs* 14 (1989): 434–473, at 445–446, defines female medical practitioners as "women who at some point in their lives would have either identified themselves in terms of their medical practice or been so identified by their communities."
9. Monica Green, "Bodies, Gender, Health, Disease: Recent Work on Medieval Women's Medicine," *Studies in Medieval and Renaissance History* 4 (2005): 1–46; Montserrat Cabré, "Women or Healers? Household Practices and the Categories of Health Care in Late Medieval Iberia," *BHM* 82 (2008): 18–51; Fissell, "Introduction."
10. I have adapted the notion regarding a continuum of skills and knowledge among early modern artisans from Ursula Klein and E. C. Spary, "Introduction: Why Materials?," in *Materials and Expertise in Early Modern Europe: Between Market and Laboratory*, ed. Ursula Klein and E. C. Spary (Chicago: University of Chicago Press, 2010), 1–23.
11. Monica Green has called for taking an "integrative" approach to medieval history that would incorporate the history of health, medicine, and body maintenance into other subfields such as social, cultural, and political history; Monica Green, "Integrative Medicine: Incorporating Medicine and Health into the Canon of Medieval European History," *History Compass* 7 (2009): 1218–1245.
12. Alisha Rankin, "Becoming an Expert Practitioner: Court Experimentalism and the Medical Skills of Anna of Saxony (1532–85)," *Isis* 98 (2007): 23–53; Elaine Leong, "Making Medicines in the Early Modern Household," *BHM* 82 (2008): 145–168; Meredith K. Ray, *Daughters of Alchemy: Women and Scientific Culture in Early Modern Italy* (Cambridge, MA: Harvard University Press, 2015).
13. Elaine Leong and Sara Pennell, "Recipe Collections and the Currency of Medical Knowledge in the Early Modern "Medical Marketplace," in *Medicine and the Market in England and Its Colonies, c. 1450–c. 1850*, ed. Mark S.

Jenner and Patrick Wallis (New York: Palgrave Macmillan, 2007), 133–152; Rebecca Laroche, *Medical Authority and Englishwomen's Herbal Texts, 1550–1650* (Aldershot, UK: Ashgate, 2009); Alisha Rankin, "Exotic Materials and Treasured Knowledge: The Valuable Legacy of Noblewomen's Remedies in Early Modern Germany," *Renaissance Studies* 28 (2014): 533–555.

14. Elaine Leong, "'Herbals she peruseth': Reading Medicine in Early Modern England," *Renaissance Studies* 28 (2014): 556–578.
15. Tessa Storey, "Face Waters, Oils, Love Magic and Poison: Making and Selling Secrets in Early Modern Rome," in *Secrets and Knowledge in Medicine and Science, 1500–1800,* ed. Elaine Leong and Alisha Rankin (Aldershot, UK: Ashgate, 2011), 143–163.
16. Studies by Nicholas Terpstra and Brian Pullan have been fundamental for understanding both the charitable services offered to women and women's work within these institutions. Nicholas Terpstra, *Abandoned Children of the Italian Renaissance: Orphan Care in Florence and Bologna* (Baltimore, MD: Johns Hopkins University Press, 2005); *Lost Girls: Sex and Death in Renaissance Florence* (Baltimore, MD: Johns Hopkins University Press, 2010); and *Cultures of Charity: Women, Politics and the Reform of Poor Relief in Renaissance Italy* (Cambridge, MA: Harvard University Press, 2013). Brian Pullan provides a magisterial synthesis of earlier scholarship in *Tolerance, Regulation and Rescue: Dishonoured Women and Abandoned Children in Italy, 1300–1800* (Manchester, UK: Manchester University Press, 2016). Despite their outstanding contributions, these studies have not focused expressly on issues of health.
17. Sandra Cavallo and David Gentilcore, "Spaces, Objects and Identities in Early Modern Italian Medicine," *Renaissance Studies* 21 (2007): 473–479.
18. Sandra Cavallo, "Secrets to Healthy Living: The Revival of the Preventive Paradigm in Late Renaissance Italy," in *Secrets and Knowledge,* 191–212; and Sandra Cavallo and Tessa Storey, *Healthy Living in Late Renaissance Italy* (Oxford: Oxford University Press, 2013).
19. Jon Arrizabalaga, John Henderson, and Roger French, *The Great Pox: The French Disease in Renaissance Europe* (New Haven, CT: Yale University Press, 1997); Daniela Lombardi, "Poveri a Firenze. Programmi e realizzazioni della politica assistenziale dei Medici tra Cinque e Seicento," in *Timore e Carità. I Poveri nell'Italia moderna,* ed. Giorgio Politi, Maria Rosa and Franco della Peruta (Cremona: Libreria del convegno, 1982), 165–184.
20. Samuel K. Cohn Jr., *Cultures of Plague: Medical Thinking at the End of the Renaissance* (Oxford: Oxford University Press, 2010).
21. Terpstra, *Cultures of Charity;* Pullan; *Tolerance, Regulation and Rescue;* Philip Gavitt, *Gender, Honor, and Charity in Late Renaissance Florence* (Cambridge, UK: Cambridge University Press, 2011).
22. Key works include Pamela H. Smith, *The Body of the Artisan: Art and Experience in the Scientific Revolution* (Chicago: University of Chicago Press, 2004);

Making Knowledge in Early Modern Europe: Practices, Objects and Texts, 1400–1800, ed. Pamela H. Smith and Benjamin Schmidt (Chicago: University of Chicago Press, 2007); *The Mindful Hand: Inquiry and Invention from the Late Renaissance to Early Industrialization,* ed. Lissa Roberts, Simon Schaffer, and Peter Dear (Amsterdam: Edita, 2007); and *Ways of Making and Knowing: The Material Culture of Empirical Knowledge,* ed. Pamela H. Smith, Amy R. W. Meyers, and Harold J. Cook (Ann Arbor: University of Michigan Press, 2014).

23. Pamela O. Long, *Artisan/Practitioners and the Rise of the New Sciences, 1400–1600* (Corvallis: Oregon State University Press, 2011).
24. One in every eight Florentine women lived in a convent in 1552, and more than half of Venetian patrician women were nuns by the 1580s. Their numbers continued to rise over the next half century; Sharon T. Strocchia, *Nuns and Nunneries in Renaissance Florence* (Baltimore, MD: Johns Hopkins University Press, 2009), 1–2.
25. Gianna Pomata, "Medicina delle monache. Pratiche terapeutiche nei monasteri femminili di Bologna in età moderna," in *I monasteri femminili come centri di cultura fra Rinascimento e Barocco,* ed. Gianna Pomata and Gabriella Zarri (Rome: Edizioni di storia e letteratura, 2005), 331–363; Gianna Pomata, "Practicing between Earth and Heaven: Women Healers in Seventeenth-Century Bologna," *Dynamis* 19 (1999): 119–143; and Sharon T. Strocchia, "The Nun Apothecaries of Renaissance Florence: Marketing Medicines in the Convent," *Renaissance Studies* 25 (2011): 627–647.
26. Margaret Pelling, "Nurses and Nursekeepers: Problems of Identification in the Early Modern Period," in *The Common Lot,* 179–202; Carol Rawcliffe, "Hospital Nurses and Their Work," in *Daily Life in the Late Middle Ages,* ed. Richard Britnell (Gloucestershire, UK: Sutton, 1998), 43–64; John Henderson, *The Renaissance Hospital: Healing the Body and Healing the Soul* (New Haven, CT: Yale University Press, 2006), 186–221.

CHAPTER ONE ~ THE POLITICS OF HEALTH AT THE EARLY MEDICI COURT

1. Margaret Pelling, "'Thoroughly Resented?' Older Women and the Medical Role in Early Modern London," in *Women, Science and Medicine 1500–1700: Mothers and Sisters of the Royal Society,* ed. Lynette Hunter and Sarah Hutton (Stroud: Sutton, 1997), 63–88, at 70.
2. See for instance Montserrat Cabré, "Women or Healers? Household Practices and the Categories of Health Care in Late Medieval Iberia," *BHM* 82 (2008): 18–51; Elaine Leong, "Making Medicines in the Early Modern Household," *BHM* 82 (2008): 145–168; Debra Blumenthal, "Domestic Medicine: Slaves, Servants and Female Medical Expertise in Late Medieval Valencia," *Renaissance Studies* 28 (2014): 515–532; Richelle Munkhoff,

"Poor Women and Parish Public Health in Sixteenth-Century London," *Renaissance Studies* 28 (2014): 579–596; Seth Stein LeJacq, "The Bounds of Domestic Healing: Medical Recipes, Storytelling, and Surgery in Early Modern England," *Social History of Medicine* 26 (2013): 451–468.

3. Katharine Park, *Secrets of Women: Gender, Generation, and the Origins of Human Dissection* (New York: Zone Books, 2006), 82.
4. Sandra Cavallo, "Secrets to Healthy Living: The Revival of the Preventive Paradigm in Late Renaissance Italy," in *Secrets and Knowledge in Medicine and Science, 1500–1800,* ed. Elaine Leong and Alisha Rankin (Aldershot, UK: Ashgate, 2011), 191–212.
5. Mar Rey Bueno, "The Health of Philip II, A Matter of State. Medicines and Medical Institutions in the Spanish Court (1556–1598)," in *Être médicin à la cour (Italie, France, Espagne, XIIIe–XVIIIe siècle),* ed. Elisa Andretta and Marilyn Nicoud (Florence: SISMEL, 2013), 149–159; Valeria Finucci, *The Prince's Body: Vincenzo Gonzaga and Renaissance Medicine* (Cambridge, MA: Harvard University Press, 2015), 4.
6. Alisha Rankin, "Becoming an Expert Practitioner: Court Experimentalism and the Medical Skills of Anna of Saxony (1532–1585)," *Isis* 98 (2007): 23–53; Alisha Rankin, *Panaceia's Daughters: Noblewomen as Healers in Early Modern Germany* (Chicago: University of Chicago Press, 2013); Meredith K. Ray, *Daughters of Alchemy: Women and Scientific Culture in Early Modern Italy* (Cambridge, MA: Harvard University Press, 2015); Sheila Barker, "Christine of Lorraine and Medicine at the Medici Court," in *Medici Women: The Making of a Dynasty in Grand Ducal Tuscany,* ed. Giovanna Benadusi and Judith C. Brown (Toronto: University of Toronto Press, 2015), 155–181.
7. Lorenzo Cantini, *Vita di Cosimo de' Medici primo granduca di Toscana* (Florence: Albizziniana, 1805), 131; Mario Battistini, "Il medico Andrea Pasquali," *Rivista di storia delle scienze mediche e naturali,* anno 17, vol. 8 (1926): 231–233.
8. For instance, the House of Savoy retained a physician to care for noble consorts by the late fourteenth century and employed yet another to tend the princely children in the mid-fifteenth; Irma Naso, "I Savoia e la cura del corpo. Medici a corte nel tardo Medioevo," in *Être médicin à la cour,* 51–85. Cosimo also employed the astrologer Giuliano Ristori da Prato, whom he consulted about the impending births of his children; Carolina Acerboni, "L'infanzia dei principi di Casa Medici: Saggio storico sulla vita privata fiorentina nel Cinquecento," ser. 2, anno 38, vol. 5 (1916): 108–125. After 1547 the court kept the surgeon Carlo Cortesi on retainer; MSS. vol. 321, p. 31.
9. Gabrielle Langdon, *Medici Women: Portraits of Power, Love and Betrayal from the Court of Duke Cosimo I* (Toronto: University of Toronto Press, 2006), 23–35; Natalie Tomas, "Commemorating a Mortal Goddess: Maria Salviati

de' Medici and the Cultural Politics of Duke Cosimo I," in *Practices of Gender in Late Medieval and Early Modern Europe,* ed. Megan Cassidy-Welch and Peter Sherlock (Turnhout, Belgium: Brepols, 2008), 261–278.

10. Salviati nevertheless drew a monthly stipend of fifty scudi from court coffers; Maria Pia Paoli, "Di madre in figlio: per una storia dell'educazione alla corte dei Medici," *Annali di storia di Firenze* 3 (2008): 65–145, at 75. Natalie Tomas examines the productive political partnership between Salviati and Cosimo in "'With His Authority She Used to Manage Much Business': The Career of Signora Maria Salviati and Duke Cosimo I de' Medici," in *Studies on Florence and the Italian Renaissance in Honour of F.W. Kent,* ed. Peter Howard and Cecilia Hewlett (Turnhout, Belgium: Brepols, 2016), 133–148.
11. Natalie Tomas, "Eleonora di Toledo, Regency, and State Formation in Tuscany," in *Medici Women: The Making of a Dynasty in Grand Ducal Tuscany,* ed. Giovanna Benadusi and Judith C. Brown (Toronto: University of Toronto Press, 2015), 59–89.
12. Sheila Barker, "The Contributions of Medici Women to Medicine in Grand Ducal Tuscany and Beyond," in *The Grand Ducal Medici and Their Archive (1537–1743),* ed. Alessio Assonitis and Brian Sandberg (Turnhout, Belgium: Brepols, 2016), 101–116. Eleonora's sizeable entourage is analyzed by Chiara Franceschini, "*Los scholares son cosa de su excelentia, como lo es toda la Compañia:* Eleonora di Toledo and the Jesuits," in *The Cultural World of Eleonora di Toledo, Duchess of Florence and Siena,* ed. Konrad Eisenbichler (Aldershot, UK: Ashgate, 2004), 181–206.
13. See for instance MdP. vol. 4, fol. 167, dated 13 Dec. 1541.
14. Bruce Edelstein, "Eleonora di Toledo e la gestione dei beni familiari: una strategia economica?," in *Donne di potere nel Rinascimento,* ed. Letizia Arcangeli and Susanna Peyronel (Rome: Viella, 2008), 733–764.
15. The Italian physician Gaetano Pieraccini tapped this correspondence in the 1920s, but his analysis was skewed by beliefs in inborn personality traits and an antipathy to female power. Gaetano Pieraccini, *La stirpe de' Medici di Cafaggiolo: Saggio di ricerche sulla trasmissione ereditaria dei caratteri biologici,* 3 vols. (Florence: Vallecchi, 1924; reprinted 1986). Judith Brown analyzes Pieraccini's views in her "Introduction," in *Medici Women,* at 21–24.
16. Rankin, *Panaceia's Daughters;* Sandra Cavallo and Tessa Storey, *Healthy Living in Late Renaissance Italy* (Oxford: Oxford University Press, 2013), 55–61. The phrase "trusted community of knowers" is borrowed from Elaine Leong, "'Herbals she peruseth': Reading Medicine in Early Modern England," *Renaissance Studies* 28 (2014): 556–578, at 568.
17. *Il Tesoro dei poveri. Ricettario medico del XIII secolo,* ed. Luca Pesante (Arezzo: Aboca, 2007).
18. BNCF. Ms. Landau Finaly. 268 is a fifteenth-century copy of the bloodletting regimen authored by the Paduan physician Pietro d'Abano (d. 1316); the 1515 hospital *ricettario* is catalogued as BNCF. Magliabechiana. XV. 92.

19. Cavallo and Storey, *Healthy Living,* 113; see also Mary E. Fissell, "The Marketplace of Print," in *Medicine and the Market in England and Its Colonies, c. 1450–c. 1850,* ed. Mark S. R. Jenner and Patrick Wallis (New York: Palgrave Macmillan, 2007), 108–132.
20. The literature on early modern recipe books is extensive. Basic bibliography includes Cabré, "Women or Healers?"; Leong, "Making Medicines"; Elaine Leong and Sara Pennell, "Recipe Collections and the Currency of Medical Knowledge in the Early Modern 'Medical Marketplace,'" in *Medicine and the Market,* 133–152; Rebecca Laroche, *Medical Authority and Englishwomen's Herbal Texts, 1550–1650* (Aldershot, UK: Ashgate, 2009).
21. Francesca Trivellato, "Guilds, Technology and Economic Change in Early Modern Venice," in *Guilds, Innovation and the European Economy, 1400–1800,* ed. Steven R. Epstein and Maarten Prak (Cambridge, UK: Cambridge University Press, 2010), 199–231; Alisha Rankin, "Exotic Materials and Treasured Knowledge: The Valuable Legacy of Noblewomen's Remedies in Early Modern Germany," *Renaissance Studies* 28 (2014): 533–555; Leong, "'Herbals she peruseth.'"
22. *Experimenti de la Ex.ma S.r Caterina da Furlj Matre de lo Illux.mo S.r Giouanni de Medici,* ed. Pier Desiderio Pasolini (Imola: Ignazio Galeati e Figlio, 1894). This volume has been studied in the context of Sforza's scientific activities by Ray, *Daughters of Alchemy,* 14–45; Fabrizia Fiumi and Giovanna Tempesta, "Gli 'experimenti' di Caterina Sforza," in *Caterina Sforza: Una Donna del Cinquecento* (Imola: La mandragora, 2000), 139–146; Marco Viroli, *Caterina Sforza, Leonessa di Romagna* (Cesena: Il ponte vecchio, 2008), 207–211; Natale Graziani and Gabriella Venturelli, *Caterina Sforza* (Milan: Mondadori, 2001), 150–159.
23. "Lettere di Giovanni de' Medici detto delle Bande Nere," *Archivio Storico Italiano* n.s. Vol. 9, part 2a (1859): 109–147, at 127, dated 29 Dec. 1525. "Ci trovamo manco nelli forzeri a Roma uno libro scritto a mano di ricette di più et varie cose operate: che sensa falla nisuno lo ritrovamo, chè in ogni modo lo volemo."
24. This discovery was first noted by Barker, "Contributions of Medici Women," 103. These volumes are catalogued respectively as BNCF. Palatino. 1021, fols. 50r–54v; and BNCF. Magliabechiana. XV. 14.
25. Sheila Barker and Sharon Strocchia, "Household Medicine for a Renaissance Court: Caterina Sforza's *Ricettario* Reconsidered," in *Gender, Health, and Healing, 1250–1550,* ed. Sara Ritchey and Sharon Strocchia (Amsterdam University Press, forthcoming).
26. MAP. filza 85, no. 397, dated 8 April 1527; filza 106, no. 54, dated 17 Oct. 1525, and no. 56, dated 15 July 1523.
27. Barker, "Contributions of Medici Women," 103; Alfredo Perifano, *L'Alchimie à la cour de Côme Ier de Médicis: saviors, culture et politique* (Paris: Honoré Champion, 1997).

28. Juan Luis Vives, *The Education of a Christian Woman: A Sixteenth-Century Manual,* ed. and trans. Charles Fantazzi (Chicago: University of Chicago Press, 2000), 263.
29. Archivio della Quiete, Florence. A VII, n. 46, fols. 3v–4r (Eleonora Ramirez di Montalvo, *Istruzione alle maestre,* circa 1656–1657). My thanks to Jennifer Hariguchi for sharing her transcription with me. The community established a commercial pharmacy in 1674. See Jennifer Hariguchi, "*Istruzione alle maestre* (Instruction for Teachers): A Model Text for Women's Lay Conservatories in Seventeenth and Eighteenth-Century Tuscany," *Early Modern Women* 10 (2016): 3–21.
30. Rankin, *Panaceia's Daughters;* Leong, "Making Medicines."
31. Pieraccini, *Stirpe de' Medici,* 1: 471.
32. George Frederic Still, *The History of Paediatrics* (Oxford: Oxford University Press, 1931), 159–162. In his domestic accounts, the merchant Francesco Castellani recorded three recipes for ridding children of worms between 1459 and 1485; James Shaw and Evelyn Welch, *Making and Marketing Medicines in Renaissance Florence* (Amsterdam: Rodopi, 2011), 236.
33. MdP. vol. 1171, ins. 6, fol. 285, dated 9 Jan. 1544. "Secondo l'ordine et solito di quella benedecta anima della Sig.ra Maria [Salviati], havevono fatti certi remedii allo stomacho." Caroline P. Murphy, *Murder of a Medici Princess* (Oxford: Oxford University Press, 2009), 20, translates portions of this letter.
34. Barker, "Contributions of Medici Women," 103.
35. Tessa Storey, "Face Waters, Oils, Love Magic and Poison: Making and Selling Secrets in Early Modern Rome," in *Secrets and Knowledge,* 143–163.
36. Rankin, *Panaceia's Daughters,* 10–14.
37. Shaw and Welch, *Making and Marketing Medicines,* 81–156.
38. MdP. vol. 1170, fol. 392; Pieraccini, *Stirpe de' Medici,* 1: 467. On the convent pharmacy, see Sharon T. Strocchia, "The Nun Apothecaries of Renaissance Florence: Marketing Medicines in the Convent," *Renaissance Studies* 25 (2011): 627–647.
39. Dionisio Pulinari da Firenze, *Cronache dei Frati Minori della Provincia di Toscana,* ed. Saturnino Mencherini (Arezzo: Cooperativa tipografica, 1913), 258–264, at 262: "il Duca Cosimo e la Duchessa, sua consorte, tutto quello che s'aveva a fare d'importanza per le persone loro o dei loro figliuoli, volevano che si facesse per le mani sue." See also Enrica Viviani della Robbia, *Nei monasteri fiorentini* (Florence: Sansoni, 1946), 21. In addition, the convent supplied the nourishing tonic distilled from capons and plantago prescribed for Cosimo during an extended illness. MdP. vol. 1170, ins. 6, fol. 342, dated 25 Oct. 1543.
40. Danielle Jacquart, "Naissance d'une pédiatrie en milieu de cour," *Micrologus* 16 (2008): 271–294.

41. Romana Martorelli Vico, "Madri, levatrici, balie e padre: Michele Savonarola, l'embriologia e la cura dei piccoli," in *Michele Savonarola. Medicina e cultura di corte,* ed. Chiara Crisciani and Gabriella Zuccolin (Florence: SISMEL, 2011), 127–135; Gabriella Zuccolin, "Nascere in latino e in volgare. Tra la *Pratica Maior* e il *De regimine pregnantium,*" in *Michele Savonarola. Medicina e cultura di corte,* 137–209. Savonarola's treatise has been edited by Luigi Belloni, *Il trattato ginecologico-pediatrico in volgare (Ad mulieres Ferrarienses De regimine pregnantium et noviter natorum usque ad septennium)* (Milan: Stucchi, 1952).
42. Jacquart, "Naissance d'une pédiatrie," 272; Still, *History of Paediatrics,* 58–66.
43. Simon de Vallambert, *Cinq Livres, de la maniere de nourrir et gouverner les enfans dès leur naissance,* ed. Colette H. Winn (Geneva: Droz, 2005), 9–15.
44. Bruce Edelstein, "*La fecundissima Signora Duchessa:* The Courtly Persona of Eleonora di Toledo and the Iconography of Abundance," in *The Cultural World of Eleonora di Toledo, Duchess of Florence and Siena,* ed. Konrad Eisenbichler (Aldershot, UK: Ashgate, 2004), 71–97.
45. Quoted in Langdon, *Medici Women,* 41; Carolina Acerboni, "L'infanzia dei principi di Casa Medici. Saggio storico sulla vita privata fiorentina nel Cinquecento. Vestiti ed appartamenti dei Principi Medicei," *Rassegna nazionale,* ser. 2, anno 39, vol. 8 (1917): 202–211.
46. Langdon, *Medici Women,* 44–46. Barker, "Contribution of Medici Women," 106, locates Salviati's "comprehensive pediatric medicine" within the broader context of Renaissance prince-practitioning.
47. Carolina Acerboni, "L'infanzia dei principi di Casa Medici. Saggio storico sulla vita privata fiorentina nel Cinquecento. Vita di famiglia e vita pubblica," *Rassegna nazionale,* ser. 2, anno 39, vol. 7 (1917): 301–316.
48. Caroline Castiglione, "Peasants at the Palace: Wet Nurses and Aristocratic Mothers in Early Modern Rome," in *Medieval and Renaissance Lactations: Images, Rhetorics, Practices,* ed. Jutta Gisela Sperling (Burlington, VT: Ashgate, 2013), 79–99.
49. Still, *History of Paediatrics,* 48–50, 162–163.
50. Christiane Klapisch-Zuber, "Blood Parents and Milk Parents: Wet Nursing in Florence, 1300–1530," in Christiane Klapisch-Zuber, *Women, Family, and Ritual in Renaissance Italy,* trans. Lydia G. Cochrane (Chicago: University of Chicago Press, 1985), 132–164.
51. Leon Battista Alberti, *The Family in Renaissance Florence,* trans. Renée Neu Watkins (Columbia: University of South Carolina Press, 1969).
52. Julia L. Hairston, "The Economics of Milk and Blood in Alberti's *Libri della famiglia:* Maternal versus Wet-Nursing," in *Medieval and Renaissance Lactations,* 187–212.
53. Diana Bullen Presciutti, "Picturing Institutional Wet-Nursing in Medicean Siena," in *Medieval and Renaissance Lactations,* 129–146, at 132;

MdP. vol. 361, fol. 34, dated 13 June 1543: "È grassa e fresca et pulita come una rosa."

54. MdP. vol. 5926, fol. 31, dated 5 Nov. 1542: "La Signora Donna Maria ha granato tutti li canini et tagliatone uno di mo' che imparrebbe da svezarla et non penso che la patisse tanto, che la non patischa più a usare il latte della balia, giudicandolo focoso."
55. Valentina Giuffra and Gino Fornaciari, "Breastfeeding and Weaning in Renaissance Italy: The Medici Children," *Breastfeeding Medicine* 8 (2013): 1–6.
56. MdP. vol. 358, fol. 626, dated 15 Nov. 1542: "Donna Maria è grassellina al possibile et la nocte sta quietissima della perdita della balia, ma qualche volte al giorno glie ne vien la ricordanza, et intenerisce et subito leva un pocho di pianto, il che a poco a poco cesserà."
57. For instance, Giovanna of Austria endorsed a property request made by the wet nurse Emilia, who had nurtured young Princess Anna, "because in nourishing this baby girl, [she] behaved with the utmost vigilance, diligence and affection, leaving us very satisfied." ["perche in questa nurritura di questa figliuolina s'è portata con quella vigilanza, diligenza, et amorevolezza possibile, da restarne molto satisfatta."]. MdP. vol. 5926, fol. 117, dated 25 Nov. 1572.
58. Jacqueline Musacchio, *The Art and Ritual of Childbirth in Renaissance Italy* (New Haven, CT: Yale University Press, 1999), 21.
59. MdP. vol. 269, fol. 45, dated 9 May 1586; vol. 1175, ins. 5, fol. 2, dated 11 July 1549.
60. The full report by unnamed physicians is published in Pieraccini, *Stirpe de' Medici,* 2: 131. See also Musacchio, *Art and Ritual of Childbirth,* 25–26.
61. Carolina Acerboni, "L'infanzia dei principi di Casa Medici. Saggio storico sulla vita privata fiorentina nel Cinquento. Balocchi, giuochi, feste tradizionali di famiglia, onoranze funebri," *Rassegna nazionale,* ser. 2, anno 39, vol. 9 (1917): 34–43.
62. MdP. vol. 1170, ins. 2, fol. 77, dated 17 Nov. 1542: "dicessi che si come lei teneva non piccolo dispiacere che S.S. et detti Signori figliuoli li stessino presente in Fiorenza, dove conoscie essere cattivissime aere per loro, così ha preso contentezza non piccolo della resolutione fatta per la stanza loro alla Badia di Fiesole, et si rallegra assai che se sia ritrovato un luogo salubre per la sanità loro."
63. Sandra Cavallo, "Health, Air and Material Culture in the Early Modern Italian Domestic Environment," *Social History of Medicine* 29 (2016): 695–716. For Salviati's remarks, see Murphy, *Murder of a Medici Princess,* 22, and Acerboni, "Vestiti ed appartamenti," 209.
64. "Lettere di Giovanni de' Medici," 21, dated 24 Sept. 1520.
65. MdP. vol. 345, fol. 12, dated 4 July 1540; fol. 364, dated 21 July 1540.

66. Gianna Pomata, *Contracting a Cure: Patients, Healers, and the Law in Early Modern Bologna* (Baltimore, MD: Johns Hopkins University Press, 1998), 122–123.
67. Michael Stolberg, *Experiencing Illness and the Sick Body in Early Modern Europe* (New York: Palgrave, 2011), 95–100; Ulinka Rublack, "Fluxes: The Early Modern Body and the Emotions," *History Workshop Journal* 53 (2002): 1–16.
68. Rankin, *Panaceia's Daughters,* 194–195.
69. MdP. vol. 1170, ins. 6, fol. 320, dated 26 Oct. 1543; fol. 336, dated 23 Oct 1543.
70. MdP. vol. 1171, ins. 3, fol. 121, dated 11 Sept. 1544: "Fatelo più presto mezo dito maggiore che min[o]re."
71. MdP. vol. 1173, ins. 1, fol. 13, dated 9 Jan. 1547; vol. 1171, ins. 9, fol. 425, dated 6 July 1545; Finucci, *Prince's Body,* 64.
72. Carolina Acerboni, "L'infanzia dei principi di Casa Medici. Saggio storico sulla vita privata fiorentina nel Cinquento," *Rassegna nazionale,* ser. 2, anno 38, vol. 5 (1916): 108–125, at 112.
73. MdP. vol. 5926, fol. 30, dated 10 Oct. 1542: "Et vi si disse quanto sino a quell'hora la S.ra Donna Isabella haveva patito delli singhiozi et dell'apparir negra et del batter della fontanella che piu per l'adrieto non s'era veduto et se ne incolpava la luna; ma questa notte di poi si'è molto observata et non ha posato quie tormente come e solita l'havendo havuti tali singhiozi piu volte. Et questa mattina qualcuno nel resto la sta assai bene l'inceso va purgando al solito"; fol. 31, dated 5 Nov. 1542: "La Signora Donna Isabella sta assai bene non se le vede per hora segnio alcuno del suo accidente, bisogna pensare che la luna l'habbi a dare qualche alteratione." See also Murphy, *Murder of a Medici Princess,* 20–21.
74. MdP. vol. 5926, fol. 30, dated 10 Oct. 1542: "demonstrandosi per l'interiore che è quello che sempre ci ha fatto più temerare, perché l'exteriore si p[u]o più facilmente cognoscere et farli delli remedii che a quello di dentro."
75. Bruce T. Moran, "Prince-Practitioning and the Direction of Medical Roles at the German Court: Maurice of Hesse-Kassel and His Physicians," in *Medicine at the Courts of Europe, 1500–1837,* ed. Vivian Nutton (London: Routledge, 1990), 95–116.
76. Gabriella Zuccolin, "Medici a corte e formazione del signore," in *Costumi educative nelle corti europee (XIV–XVIII secoli),* ed. Monica Ferrari (Pavia: Pavia University Press, 2010), 77–102; Gabriella Zuccolin, "Sapere medico e istruzioni etico-politiche: Michele Savonarola alla corte estense," *Micrologus* 16 (2008): 313–326.
77. Chiara Crisciani, "Cura e educazione a corte: note su medici e giovani principi a Milano (sec. XV)," in *I bambini di una volta,* ed. Monica Ferrari (Milan: Franco Angeli, 2006), 41–50; Irma Naso, "I Savoia e la cura del

corpo. Medici a corte nel tardo medioevo," in *Être médicin à la cour,* 51–85; Marilyn Nicoud, "Diététique et alimentation des élites princières dans l'Italie medievale," in *Pratiques et discours alimentaires en Méditerranée de l'Antiquité à la Renaissance,* ed. Jean Leclant, André Vauchez, and Michel Sartre (Paris: PERSEE, 2008), 317–336.

78. Allen J. Grieco, "Medieval and Renaissance Wines: Taste, Dietary Theory, and How to Choose the 'Right' Wine (14th–16th centuries)," *Medievalia* 30 (2009): 15–42.

79. Monica Ferrari, "*Ordini da servare nella vita* ed *Emploi du temps.* Il ruolo pedagogico del medico in due corti europee tra '400 e '600," *Micrologus* 16 (2008): 295–313.

80. Jacquart, "Naissance d'un pédiatrie," 275, notes that the age of six was seen as a watershed in learned medical literature, based on Avicenna's *Canon.*

81. Guglielmo Enrico Saltini, "Di una visita che fece in Genova nel 1548 il fanciullo Don Francesco di Cosimo I de' Medici al Principe Don Filippo di Spagna," *Archivio storico italiano,* ser. 4, vol. 4 (1879): 19–34, at 32: "Perchè maestro Andrea, quando non voleva che S.S. Illma. mangiasse di qualche cosa che li fusse posta innanzi, li diceva, o l'è troppo dura, o la non è ben cotta, o la è di mala digestione."

82. Quoted in Saltini, "Di una visita," 32: "Et credo che al ritorno suo, madonna Giulia harà assai difficultà et fatica di governarlo e tenerlo sotto la custodia solita, per che ha cominciato [d]a gustare la grandezza, et la libertà et largheza che ha di presente."

83. Élodie Lequain, "Le bon usage du corps dans l'education des princesses à la fin du Moyen Âge," in *Cultures de cour, cultures du corps (XIVe–XVIIIe siècle),* ed. Catherine Lanoë, Mathieu da Vinha, and Bruno Laurioux (Paris: Publications de l'université Paris-Sorbonne, 2011), 115–125.

84. Symphorien Champier, *La nef des dames vertueuses,* ed. Judy Kem (Paris: Honoré Champion, 2007), 123–173.

85. Champier, *La nef des dames vertueuses,* 135–137; Lequain, "Bon usage du corps," 117.

86. Anne of France, *Les enseignements d'Anne de France a sa fille Susanne de Bourbon* (Marseille: Lafitte, 1978), 24–25, 27–28; Lequain, "Bon usage du corps," 120–121.

87. Cantini, *Vita di Cosimo de' Medici;* Battistini, "Il medico Andrea Pasquali." For information about the Pasquali family, see Stefano Calonaci, "Gli angeli del testamento. Donne fedecommissarie e fedecommittenti nella Toscana moderna," in *Nobildonne, monache e cavaliere dell'ordine di Santo Stefano. Modelli e strategie femminili nella vita pubblica della Toscana granducale,* ed. Marcella Aglietti (Pisa: ETS, 2009), 79–96, at 94–95.

88. Several remedies developed by female house staff were included in the hospital's 1515 recipe book: BNCF. Magliabechiana. XV. 92.

89. Comparative salary data is provided by Paoli, "Di madre in figlio," 76, 80. Apparently Pasquali authored several medical writings that have not survived; Cantini, *Vita di Cosimo de' Medici,* 131.
90. Pieraccini, *Stirpe de' Medici,* 2: 88: "L'ortetica della balia di donna Maria è tornato in dentro, et tiene della febre. Però la Signora Maria ha fatto consultare il caso suo dal Ripa: et lui ha concluso di dare a detta balia un pocho di cassia che pensa li abbia a giovare, et perchè donna Maria non vol altra poppa, ch'altra poppa la chiama cacca, et non vol pappine, però il predetto physico consente che la sua balia ordinaria li dia la poppa per un mancho male et così seguirà di dargliela, et la puttina sta bene."
91. MdP. vol. 1171, ins. 6, fol. 285: "parendo loro che la bocchina sentisse di bachi." Murphy, *Murder of a Medici Princess,* 20, translates "sentisse" as "felt," but the context suggests a reliance on smell rather than touch.
92. Mark S. R. Jenner, "Follow Your Nose? Smell, Smelling, and Their Histories," *American Historical Review* 116 (2011): 335–351, at 348.
93. Murphy, *Murder of a Medici Princess,* 20.
94. Annemarie Kinzelbach, "Women and Healthcare in Early Modern German Towns," *Renaissance Studies* 28 (2014): 619–638, at 633.
95. Michael Stolberg, *Experiencing Illness and the Sick Body in Early Modern Europe* (New York: Palgrave, 2011), 69–72; Rankin, *Panaceia's Daughters,* 168–203.
96. MdP. vol. 5, fol. 321, dated 20 Sept. 1543.
97. MdP. vol. 1169, fol. 383, dated 26 Oct. 1543: "Et quel che mi pesa et preme sino all'anima è che sua excellentia non vuol patire se li faccino i servitiali che sono la medicina vera di questi mali. Et quel che peggio è si vuol governare d'ogni cosa a suo modo, laonde il medico si trova disperato et la illustrissima signora sua madre desperatissima." See also vol. 1170, ins. 7, fol. 375, dated 2 Nov. 1543.
98. Gianna Pomata, *Contracting a Cure: Patients, Healers, and the Law in Early Modern Bologna* (Baltimore, MD: Johns Hopkins University Press, 1998), 42.
99. MdP. vol. 346, fols. 289 and 292, dated 13 and 14 Sept. 1540, respectively, quote at latter: "benchè dica non volere cosa alcuna per intrinseco ma assai sara usar li remedii extrinseci in li quali non pocho confido . . . e spero si habia concluder un bagno che rinfreschi e astringa quella parte."
100. Olivia Weisser, *Ill Composed: Sickness, Gender, and Belief in Early Modern England* (New Haven, CT: Yale University Press, 2015), quote at 116; Silvia De Renzi, "Tales from Cardinals' Deathbeds: Medical Hierarchy, Courtly Etiquette and Authority in the Counter Reformation," in *Être médicin à la cour,* 235–258.
101. MdP. vol. 5926, fol. 14, dated 11 May 1542: "sendo cessate le doglie et li dolori."

102. MdP. vol. 5926, fol. 55, dated 4 Dec. 1543; vol. 1170a, ins. 3, fol. 276, dated 6 Dec. 1543.
103. Weisser, *Ill Composed,* 104–128.
104. Pieraccini, *Stirpe de' Medici,* I: 483–484.
105. MdP. vol. 1170, ins. 8, fol. 414, dated 12 Dec. 1543.
106. Tomas, "Commemorating a Mortal Goddess," 271–272.
107. Barker, "Contributions of Medici Women," 107.
108. MdP. vol. 1170, ins. 6, fol. 352, dated 11 Oct. 1543: "L'Ill.ma Sig.ra Duchessa vole che la S. V. [. . .] facci chiamar a se lo spetiale et li comandi che stilli otto dieci fiaschi d'aqua de' cocomeri in quel modo che sta scritto nella aggiunta poliza che si ha da dare a lui et dirl che tutto exeguischa con buona deligentia." MdP. vol. 1171, ins. 1, fol. 4, dated 3 Jan. 1544. See also Barker, "Contributions of Medici Women," 107.
109. MdP. vol. 1171, ins. 1, fol. 25, dated 21 Feb. 1544.
110. Kathy Stuart, *Defiled Trades and Social Outcasts: Honor and Ritual Pollution in Early Modern Germany* (Cambridge, UK: Cambridge University Press, 1999), 156–158, 183. Rankin, "Exotic Materials," notes the presence of human ingredients in the apothecaries of two German noblewomen known for their healing skills.
111. Beth A. Conklin, *Consuming Grief: Compassionate Cannibalism in an Amazonian Society* (Austin: University of Texas Press, 2001), 10–11.
112. Mark S. R. Jenner and Patrick Wallis, "The Medical Marketplace," in *Medicine and the Market,* 1–23.
113. *El ricettario dell'arte, et universita de medici, et spetiali della città di Firenze. Riveduto dal collegio de medici per ordine dello illustrissimo et eccellentissimo signore duca di Firenze,* (Florence: Lorenzo Torrentino, 1550), 3: "il tempo mostrato nuove sorti di medicina." On the evolution of Florentine pharmacopeia, see Alfonso Corradi, *Le prime farmacopee italiane, ed in particolare: Dei Ricettari Fiorentini. Memoria* (Milan: Rechiedei, 1887), 50. Some newly approved remedies were developed by "modern" masters like Gabriele Falloppio. Teresa Huguet Termes, "Standardising Drug Therapy in Renaissance Europe? The Florence (1499) and Nuremberg Pharmacopoeia (1546)," *Medicina e storia* 8 (2008): 77–101.
114. MdP. vol. 413a, fol. 1261, dated 20 March 1553.
115. Montserrat Cabré, "Keeping Beauty Secrets in Early Modern Iberia," in *Secrets and Knowledge,* 167–190, quote at 171; Montserrat Cabré, "Beautiful Bodies," in *A Cultural History of the Human Body in the Medieval Age,* ed. Linda Kalof (Oxford: Berg, 2010), 121–139.
116. Ray, *Daughters of Alchemy,* 16; Daniela Pizzagalli, *La signora del Rinascimento, vita e splendori di Isabella d'Este alla corte di Mantova* (Milan: BUR Biblioteca Univ. Rizzoli, 2013), 360, 369, 401, 405, 430.
117. Langdon, *Medici Women,* 90–91.

118. Craig Koslofsky, "Knowing Skin in Early Modern Europe, c. 1450–1750," *History Compass* 12 (2014): 794–806; Kevin Siena and Jonathan Reinarz, "Scratching the Surface: An Introduction," in *A Medical History of Skin: Scratching the Surface,* ed. Jonathan Reinarz and Kevin Siena (London: Pickering & Chatto, 2013), 1–15.
119. Three manuscript versions of Rosselli's secrets have survived. The earliest one, cited here, is conserved as BLF. Ms. Antinori 151, recipe at fol. 99r. A second recipe book is conserved as BMF. Cod. C. CXLV. The third, begun 10 August 1593, records the provenance of recipes with much greater frequency. That version was intended as a legacy for Rosselli's sons Francesco and Vincenzo, who followed him into the business. Housed in a private collection in Seville, it has been published as Stefano Francesco di Romolo Rosselli, *Mes secrets à Florence au temps des Médicis 1593,* ed. Rodrigo de Zayas (Paris: J.-M. Place, 1996). Rosselli's shop at the Canto del Giglio is discussed by Shaw and Welch, *Making and Marketing Medicines,* 304–307.
120. The Florentine apothecary Natale Martini, for instance, boasted that his proprietary remedy for gout "had been used by Marchese Antonmaria Malespina" and its efficacy "confirmed many times." LCF. vol. 3374, fols. 34v–35r, dated 1584: "ricetta contra alla podagra, la quale l'usava il marchese Antonmaria Malespina . . . confermato piu volte."
121. MdP. vol. 401, fol. 402, dated 14 February 1551: "li mandi il suo medico et un buono speciale, et maximamente Pegna, perché si trova malato di terzana et non si fida de' Senesi"; vol. 638, fol. 106, dated 17 June 1545: "un poco della nostra manna et reubarbaro della meglio et più perfetta che ci sia"; vol. 1171, ins. 9, fol. 425, dated 6 July 1545.
122. Naso, "I Savoia," 54. Elborg Forster, "From the Patient's Point of View: Illness and Health in the Letters of Liselotte von der Pfalz (1652–1722)," *BHM* 60 (1986): 297–320, notes that it was considered a polite gesture at the French royal court to dispatch one's personal physician to ailing friends and relatives.
123. Gregory Lubkin, *A Renaissance Court: Milan under Galeazzo Maria Sforza* (Berkeley: University of California Press, 1994), 130–131; AOI. vols. 12720, 12721, 12722, 12954, 12955, 12965.
124. Anatole Tchikine, "Horticultural Differences: The Florentine Garden of Don Luis de Toledo and the Nuns of S. Domenico del Maglio," *Studies in the History of Gardens and Designed Landscapes* 30 (2010): 224–240.
125. MdP. vol. 1170a, ins. 3, fol. 156, dated 22 Oct. 1545: "mandi uno sachetto o qualche unzione per lo stomacho del Signor don Luigi e subito." This episode is also discussed by Barker, "Contributions of Medici Women," 107.
126. MdP. vol. 1175, ins. 1, fol. 50, dated 11 Dec. 1549.

127. Anna Foa, "The New and the Old: The Spread of Syphilis (1494–1530)," in *Sex and Gender in Historical Perspective,* ed. Edward Muir and Guido Ruggiero (Baltimore: Johns Hopkins University Press, 1990), 26–45.
128. Robert S. Munger, "Guaiacum, the Holy Wood from the New World," *BHM* 4 (1949): 196–229; John Parascandola, "From Mercury to Miracle Drugs: Syphilis Therapy over the Centuries," *Pharmacy in History* 51 (2009): 14–23.
129. Claudia Stein, *Negotiating the French Pox in Early Modern Germany* (Aldershot, UK: Ashgate, 2009), 60–62.
130. MdP. vol. 1171, ins. 9, fol. 425, dated 6 July 1545: "non si deva più differire."
131. Incurabili. vol. 58, fol. 131v.
132. Munger, "Guaiacum," 202.
133. Stein, *Negotiating the French Pox,* 101–103.
134. MdP. vol. 1173, ins. 3, fol. 123r.
135. MdP. vol. 1173, ins. 3, fol. 145, dated 5 April 1547; fol. 148, dated 7 April 1547.
136. Eleonora summoned Pasquali to attend some of her ladies-in-waiting "either that same evening or early the next morning"; MdP. vol. 1170, ins. 5, fol. 253, dated 9 June 1543.
137. MdP. vol. 1171, ins. 9, fol. 423, dated 5 July 1545; vol. 1171, ins. 3, fol. 107, dated 29 Aug. 1544.
138. MdP. vol. 1170a, ins. 3, fols. 144 and 187, dated 28 Oct. 1545; fol. 203, dated 31 Oct. 1545; ins. 1, fol. 79, dated 14 Nov. 1545.
139. Edelstein, *"La fecundissima Signora Duchessa."*
140. MdP. vol. 1170a, ins. 3, fol. 92, dated 21 Aug. 1545: "Et dicie che la duchessa comanda che io scriva a V. S. che li sia dato ciò che l'à di bisognio." MdP. vol. 1172, ins. 6, fol. 43, dated 5 Dec. 1546.
141. In 1608 the flamboyant Duke of Mantua Vincenzo Gonzaga distributed a famed epilepsy ointment to the public; Finucci, *Prince's Body,* 17.
142. MdP. vol. 1170, fol. 344, dated 25 April 1543: "Queste son le parole formali."

CHAPTER TWO ~ GIFTS OF HEALTH

1. Edward L. Goldberg, "Artistic Relations between the Medici and the Spanish Courts, 1587–1621: Part 1," *Burlington Magazine* 138 (1996): 105–114, at 107.
2. Marcello Fantoni, "Feticci di prestigio: il dono alla corte medicea," in *Rituale cerimoniale etichetta,* ed. Sergio Bertelli and Giuliano Crifò (Milan: Bompiani, 1985), 141–161.
3. Jean-Francois Dubost, "Liberalità calcolate: politiche del dono tra corte di Francia e corti italiane al tempo di Maria de' Medici," in *Medici Women*

as Cultural Mediators (1533–1743). Le donne di casa Medici e il loro ruolo di mediatrici culturali fra le corti d'Europa, ed. Christina Strunck (Milan: Silvana, 2011), 207–225.

4. Meredith K. Ray, *Daughters of Alchemy: Women and Scientific Culture in Early Modern Italy* (Cambridge, MA: Harvard University Press, 2015), 14–45; Sheila Barker, "The Contributions of Medici Women to Medicine in Grand Ducal Tuscany and Beyond," in *The Grand Ducal Medici and Their Archive (1537–1743),* ed. Alessio Assonitis and Brian Sandberg (Turnhout, Belgium: Brepols, 2016), 101–116; Sheila Barker, "Christine of Lorraine and Medicine at the Medici Court," in *Medici Women: The Making of a Dynasty in Grand Ducal Tuscany,* ed. Giovanna Benadusi and Judith C. Brown (Toronto: University of Toronto Press, 2015), 155–181; Evelyn Welch, *Shopping in the Renaissance: Consumer Cultures in Italy, 1400–1600* (New Haven, CT: Yale University Press, 2005), 245–273.
5. Deanna Shemek, "In Continuous Expectation: Isabella d'Este's Epistolary Desire," in *Phaethon's Children: The Este Court and Its Culture in Early Modern Ferrara,* ed. Dennis Looney and Deanna Shemek (Tempe: Arizona Center for Medieval and Renaissance Studies, 2005), 269–300.
6. Sara Ritchey, "Affective Medicine: Later Medieval Healing Communities and the Feminization of Health Care Practices in the Thirteenth-Century Low Countries," *Journal of Medieval Religious Cultures* 40 (2014): 113–143.
7. MdP. vol. 4959, fol. 373, dated 7 Aug. 1632: "un vasetto d'olio da stomaco per il signor Conte suo marito."
8. Barker, "Contributions of Medici Women," 101–105.
9. MdP. vol. 6107, fol. 339, dated 25 Feb. 1617: "uno studiuoletto pieno di vasetti di argento et altre cosettine attenente il tutto a spezieria et chirugia stimata scudi dugento."
10. MdP. vol. 6110, fol. 315, dated 13 Nov. 1627. Barker, "Contributions of Medici Women," 104–105, discusses the exchange of French and Hapsburg remedies between Christine and her daughter-in-law Maria Maddalena.
11. Gabriella Zarri, "Matronage/maternage. Tipologie di rapporti tra corti femminili e istituzioni religiose," in *Le donne medici nel sistema europeo delle corti (XVI–XVIII secolo),* ed. Giulia Calvi and Riccardo Spinelli (Florence: Polistampa, 2008), 67–74.
12. R. Burr Litchfield, "Demographic Characteristics of Florentine Patrician Families, Sixteenth to Nineteenth Centuries," *Journal of Economic History* 29 (1969): 191–205; Jutta Gisela Sperling, *Convents and the Body Politic in Late Renaissance Venice* (Chicago: University of Chicago Press, 1999), 18; Gabriella Zarri, "I monasteri femminili a Bologna tra il XIII e il XVII secolo," *Atti e memorie: Deputazione di storia patria per le provincie di Romagna,* n.s. 24 (1973): 133–224.

13. Molly Bourne, "From Court to Cloister and Back Again: Margherita Gonzaga, Caterina de' Medici and Lucrina Fetti at the Convent of Sant'Orsola in Mantua," in *Domestic Institutional Interiors in Early Modern Europe,* ed. Sandra Cavallo and Silvia Evangelisti (Burlington, VT: Ashgate, 2009), 153–179; see also the essays in *I monasteri femminili come centri di cultura tra Rinascimento e Barocco,* ed. Gianna Pomata and Gabriella Zarri (Rome: Edizioni di storia e letteratura, 2005).
14. Friendships between early modern Catholic women have not been studied to the same extent as alliances between Protestant women; on the latter, see Amanda E. Herbert, *Female Alliances: Gender, Identity and Friendship in Early Modern Britain* (New Haven, CT: Yale University Press, 2014).
15. Meredith K. Ray, "Letters and Lace: Arcangela Tarabotti and Convent Culture in *Seicento* Venice," in *Early Modern Women and Transnational Communities of Letters,* ed. Julie D. Campbell and Anne R. Larsen (Aldershot, UK: Ashgate, 2009), 45–73; Lynn Westwater, "A Rediscovered Friendship in the Republic of Letters: The Unpublished Correspondence of Arcangela Tarabotti and Ismaël Bouilliau," *RQ* 65 (2012): 67–134.
16. Natalie Z. Davis, *The Gift in Sixteenth-Century France* (Madison: University of Wisconsin Press, 2000), 26, 30.
17. For the early history of the convent, see Sharon T. Strocchia, *Nuns and Nunneries in Renaissance Florence* (Baltimore, MD: Johns Hopkins University Press, 2009), 20–21, 102–104; and Kate Lowe, "Rainha D. Leonor of Portugal's Patronage in Renaissance Florence and Cultural Exchange," in *Cultural Links between Portugal and Italy in the Renaissance,* ed. Kate Lowe (Cambridge, UK: Cambridge University Press, 2000), 225–248.
18. Strocchia, *Nuns and Nunneries,* 169–170.
19. CRSGF. 81. vol. 100, #326, dated 25 Sept. 1499: "le pomegrante et altri frutti del orto vostro ne sono state gratissime per molti rispetti." Pomegranate juice was a common ingredient in Mediterranean eye remedies, including the ones recorded in her recipe book. See Efraim Lev, "Mediators between Theoretical and Practical Medieval Knowledge: Medical Notebooks from the Cairo Genizah and Their Significance," *Medical History* 57 (2013): 487–515.
20. CRSGF. 81. vol. 100, # 190, 196.
21. Isabel dos Guimarães Sá, "Between Spiritual and Material Culture: Male and Female Objects at the Portuguese Court, 1480–1580," in *Domestic Institutional Interiors,* 181–199, at 185.
22. Sheila ffolliott, "Catherine de' Medici as Artemisia: Figuring the Powerful Widow," in *Rewriting the Renaissance: The Discourses of Sexual Difference in Early Modern Europe,* ed. Margaret W. Ferguson, Maureen Quilligan, and Nancy Vickers (Chicago: University of Chicago Press, 1986), 227–241.
23. Ivo Carneiro de Sousa, *A Rainha D. Leonor (1458–1525). Poder, Misericórdia, Religiosidade e Espiritualidade no Portugal do Renascimento* (Lisbon: Fundaçao

Calouste Gulbenkian, 2002), 173–179; Fernando da Silva Correia, "O Julgamento da Rainha D. Leonor, seguido de tres relatorios medicos (I: A historia clinica d'El Rei D. Joao II; II: A historia clinica da Infanta Santa Joana; III: Rainha D. Leonor, pelo dr. Julio Dantas)," sep. da *Revista Ocidente,* Lisbon, 1943.

24. Carneiro de Sousa, *A Raihna D. Leonor,* 889–914; Isabel Vilares Cepeda, "Os Livros da Rainha D. Leonor, segundo o inventario de 1537 do Convento da Madre de Deus," *Revista da Biblioteca Nacional,* ser. 2, vol. 2 (1987): 51–81. The *Fasciculus de medicina* ("little bundle of medicine") was assembled by a Venetian publisher in 1494 from various minor tracts and attributed to Ketham in order to give it greater authority; see Tiziana Pesenti, *Fasiculo de medicina in volgare: Venezia, Giovanni e Gregorio De Gregori, 1494,* 2 vols. (Treviso: Antilia, 2001).
25. Sá, "Between Spiritual and Material Culture," 185; Lisbeth de Oliveira Rodrigues and Isabel dos Guimarães Sá, "Sugar and Spices in Portuguese Renaissance Medicine," *Journal of Medieval Iberian Studies* 7 (2015): 176–196.
26. Lowe, "Rainha D. Leonor," 231; Sá, "Between Spiritual and Material Culture," analyzes the contents of Leonor's oratory and chapel. On the role of charity in Portuguese state formation, see Isabel dos Guimarães Sá, "Catholic Charity in Perspective: The Social Life of Devotion in Portugal and Its Empire (1450–1700)," *e-Journal of Portuguese History* 2 (2004): 1–20.
27. Rodrigues and Sá, "Sugar and Spices," 178.
28. Carneiro de Sousa, *A Rainha D. Leonor,* 81–82; Lowe, "Rainha D. Leonor," 230; Katharine Park and John Henderson, "'The First Hospital among Christians': The Ospedale di Santa Maria Nuova in Early Sixteenth-Century Florence," *Medical History* 35 (1991): 164–188; Rodrigues and Sá, "Sugar and Spices," 180.
29. Rodrigues and Sá, "Sugar and Spices," 181, 183; Sá, "Catholic Charity in Perspective."
30. This paragraph is drawn from BNCF. II. II. 509, fol. 53r. I use the following edition unless my own translation differs significantly: Sister Giustina Niccolini, *The Chronicle of Le Murate,* ed. and trans. Saundra Weddle (Toronto: University of Toronto Press, 2011), 128–138; Lowe, "Rainha D. Leonor," 225–228. See also Denise Stocchetti, "La fondazione del monastero fiorentino delle Murate e la pellegrina Eugenia," *Archivio italiano per la storia della pietà* 18 (2005): 177–247.
31. CRSGF. 81. vol. 100, at # 234, dated 10 August 1497. The verso of this letter bears Leonor's royal seal.
32. *Chronicle of Le Murate,* 135; Lowe, "Rainha D. Leonor," 228, stresses the convent's connections to Savonarolan reform initiatives.
33. These letters are conserved as CRSGF. 81. vol. 100; letters to King Manuel at # 236, 237, and 240, among others. A full inventory of the correspondence

is published by Ivo Carneiro de Sousa, "A Rainha D. Leonor e as Murate de Florença (Notas de Investigação), *Revista da Faculdade de Letras do Porto,* serie historia, ser. 2, 4 (1987): 119–133.

34. James Shaw and Evelyn Welch, *Making and Marketing Medicine in Renaissance Florence* (Amsterdam: Rodopi, 2011), 207–209.
35. Vittorio Rotolo, "La storia medica dello zucchero," *Rivista di storia della medicina,* ser. 2, vol. 8 (1998): 15–25; Alisha Rankin, *Panaceia's Daughters: Noblewomen as Healers in Early Modern Germany* (Chicago: University of Chicago Press, 2013), 73.
36. William J. Bernstein, *A Splendid Exchange: How Trade Shaped the World* (New York: Grove / Atlantic, 2008), 205.
37. Francesco Guidi Bruscoli, *Bartolomeo Marchionni, 'homem de grossa fazenda,' (ca. 1450–1530). Un mercante fiorentino a Lisbona e l'impero portoghese* (Florence: L.S. Olschki, 2014), 94–105.
38. Shaw and Welch, *Making and Marketing Medicine,* 31, 193, 213. This shop was located at the Canto al Giglio near the Old Market, the heart of Florentine commercial life.
39. Shaw and Welch, *Making and Marketing Medicine,* 193; Rodrigues and Sá, "Sugar and Spices," 183–187; Rotolo, "Storia medicina dello zucchero," 20.
40. CRSGF. 81. vol. 100, # 242, undated: "solevare la extrema necessita nostra."
41. Guidi Bruscoli, *Bartolomeo Marchionni.*
42. I am using Portuguese weights and measures provided by Guidi Bruscoli, *Bartolomeo Marchionni,* xiv.
43. By my calculations, the convent received a total of 18,656 pounds of sugar. BNCF. II. II. 509, fol. 53r; *Chronicle of Le Murate,* 136–137.
44. Guidi Bruscoli, *Bartolomeo Marchionni,* 24. CRSGF. 81. vol. 100, # 301r, noted that the Portuguese crown kept a "libro della limosina delle monache delle Murate."
45. Shaw and Welch, *Making and Marketing Medicine,* 192.
46. Guidi Bruscoli, *Bartolomeo Marchionni,* 24; Rodrigues and Sá, "Sugar and Spices," 186.
47. Sharon T. Strocchia, "The Nun Apothecaries of Renaissance Florence: Marketing Medicines in the Convent," *Renaissance Studies* 25 (2011): 627–647.
48. *Chronicle of Le Murate,* 135–136; CRSGF. 81. vol. 100, # 252r, 253r, dated 6 Sept. 1510 and 14 April 1515. Both letters from Leonor to the abbess itemized goods consigned to the convent. Pennets were "twisted sticks of pulled sugar, mixed with starch and oil of sweet almonds"; Shaw and Welch, *Making and Marketing Medicine,* 210.
49. CRSGF. 81. vol. 100, # 244, dated 24 May 1505, to Francesco Corbinelli.
50. Shaw and Welch, *Making and Marketing Medicine,* 208.

51. Jo Wheeler, http://renaissancesecrets.blogspot.com/2014/09/stefano-rosselli-secrets-and-medici.html.
52. Stefano Rosselli, *Mes Secrets: A Florence au temps des Medici, 1593: patisserie, parfumerie, medecine,* ed. Rodrigo de Zayas (Paris: J.-M. Place, 1996), 186–187.
53. *Tariffa et prezzi dei medicinali et mercanzie attenente alli speziali della città di Firenze* (Florence: Giorgio Marescotti, 1593) [LCF. vol. 164.], 13.
54. CRSGF. 81. vol. 100, # 465, dated March 1510: "per aiutare la vostra infermeria."
55. CRSGF. 81. vol. 100, # 468r, dated 10 Feb 1504/5: "gran conforto e recreatione al numero delle nostre inferme che sono molte e in continuosa necessitate."
56. Shaw and Welch, *Making and Marketing Medicine,* 209–210.
57. Rodrigues and Sá, "Sugar and Spices," 186.
58. Kate Lowe, "Understanding Cultural Exchange between Portugal and Italy in the Renaissance: Social and Institutional Relations," in *Cultural Links between Portugal and Italy,* 1–18; Guidi Bruscoli, *Bartolomeo Marchionni,* 3–36.
59. CRSGF. 81. vol. 100, # 213–216, dated 1543–1547.
60. Four of Marchionni's letters to the Murate dating from 1510 to 1520 are printed in Guidi Bruscoli, *Bartolomeo Marchionni,* 219–221.
61. CRSGF. 81. vol. 100, # 246r, dated Oct. 1499, from Abbess Elena Bini to Queen Leonor: "certe cosette per segno d'amore in picchola parte dimostro."
62. Dubost, "Liberalità calcolate," 213.
63. Kate Lowe, "Women's Work at the Benedictine Convent of Le Murate in Florence: Suor Battista Carducci's Roman Missal of 1509," in *Women and the Book: Assessing the Visual Evidence,* ed. Lesley Smith and Jane H. M. Taylor (London: British Library, 1997), 133–146.
64. CRSGF. 81. vol. 100, # 241r, 254r. In 1515 the nuns sent another prayer book made in their scriptorium as a gift for the queen's nephew. The full inventory of Leonor's private library is published by Carneiro de Sousa, *Rainha D. Leonor,* 889–914.
65. Sara Pennell examines the relationship between scale and value in "Mundane Materiality, or, Should Small Things Still Be Forgotten? Material Culture, Micro-Histories and the Problem of Scale," in *History and Material Culture: A Student's Guide to Approaching Alternative Sources,* ed. Karen Harvey (London: Routledge, 2009), 173–191.
66. CRSGF. 81. vol. 100, # 198 for the Agnus Dei, called "una cosa devota e bella . . . Et cussi subito cum devotione ge l'havemo facto mettere al colo"; gifts of floral garlands recorded at # 241r, 243r.
67. Sandra Cavallo and Tessa Storey, *Healthy Living in Late Renaissance Italy* (Oxford: Oxford University Press, 2013), 69.

68. Elissa Weaver, *Convent Theatre in Early Modern Italy: Spiritual Fun and Learning for Women* (Cambridge, UK: Cambridge University Press, 2002), 39–40.
69. Annette B. Weiner, *Inalienable Possessions: The Paradox of Keeping-While-Giving* (Berkeley: University of California Press, 1992).
70. Lowe, "Rainha D. Leonor," 245.
71. Guidi Bruscoli, *Bartolomeo Marchionni,* 187.
72. CRSGF. 81. vol. 100, # 206r, dated 7 April 1548; # 200r, dated 11 Sept. 1559.
73. CRSGF. 81. vol. 100, # 204r, dated 31 Oct 1564.
74. *Chronicle of Le Murate,* 197–203.
75. These letters are conserved as MdP. vol. 6081.
76. Ray, *Daughters of Alchemy,* 79; Mary G. Steegman, *Bianca Cappello* (London: Constable and Company, 1913), 218, 226. Writing to Cappello on 30 September 1581, Ricci requested some steel distillate from the Medici court pharmacy "in the form usually taken [to relieve] a blockage" *(oppilazione).* This term frequently referenced the cessation of menses in young women due to poor diet, overwork, excessive penitence, or pregnancy. Ricci explained that "there is a nun here, a niece of mine, who is in need of taking [the medicine], and the doctors say that there's some risk in giving it to her in the ordinary way, and that the distillate is much safer." Anticipating a positive reply, she asked Cappello to include instructions "about how [the remedy] should be administered, as well as the dosage, timing, and routines that one should follow in taking [it]." *Le lettere spirituali e familiari di S. Caterina de' Ricci fiorentina,* ed. Cesare Guasti (Prato: R. Guasti, 1861), 103–104. Other nuns suffering from blockages were prescribed similar purgatives; Maria Celeste Galilei, *Letters to Father: Suor Maria Celeste to Galileo (1623–1633),* trans. Dava Sobel (New York: Penguin, 2001), 47, 63.
77. Manuela Belardini, "'Piace molto a Giesù la nostra confidanza': Suor Orsola Fontebuoni a Maria Maddalena d'Austria," in *Per lettera. La scrittura epistolare femminile tra archivio e tipografia secoli XV–XVII,* ed. Gabriella Zarri (Rome: Viella, 1999), 359–383.
78. Estella Galasso Calderara, *La Granduchessa Maria Maddalena D'Austria: Un amazzone tedesca nella Firenze medicea del '600* (Genoa: SAGEP, 1985), 71–75, 83, 91; MdP. vol. 6081, dated 15 April 1616.
79. Judith C. Brown, "Introduction," in *Medici Women,* 17–57.
80. Maria Galli Stampino, "Maria Maddalena, Archduchess of Austria and Grand Duchess of Florence: Negotiating Performance, Traditions, and Taste," in *Early Modern Hapsburg Women: Transnational Contexts, Cultural Conflicts, Dynastic Continuities,* ed. Anne J. Cruz and Maria Galli Stampino (Burlington, VT: Ashgate, 2013), 41–56; Alice E. Sanger, "Maria Maddalena d'Austria's Pilgrimage to Loreto: Visuality, Liminality and Exchange," in *Medici Women as Cultural Mediators,* 253–265, esp. 253, 257; Kelly Harness,

Echoes of Women's Voices: Music, Art, and Female Patronage in Early Modern Florence (Chicago: University of Chicago Press, 2006).

81. Belardini, "'Piace molto a Giesù,'" 361–362.
82. Daniela Lamberini, "Il monastero di San Mercuriale a Pistoia. Lineamenti di storia dei secoli X–XIX dalla documentazione d'archivio," in *L'architettura del San Mercuriale a Pistoia. Un frammento di città,* ed. Francesco Gurrieri (Florence: Alinea, 1989), 15–25.
83. Nicholas Terpstra, *Lost Girls: Sex and Death in Renaissance Florence* (Baltimore, MD: Johns Hopkins University Press, 2010), 66–84.
84. Lamberini, "Monastero di San Mercuriale," 24, 48h; Stefania Capecchi, "Il ricettario della molto reverenda Madre Donna Maddalena Favilla, spetiale nel monastero di San Mercuriale della città di Pistoia nell'anno 1750," *AMAISF* 18 (2001): 55–60.
85. Lamberini, "Monastero di San Mercuriale," 25.
86. MdP. vol. 6081, dated 16 Jan. 1620, 28 June 1622, and 31 May 1624; see also Belardini, "'Piace molto a Giesù,'" 369.
87. Galilei, *Letters to Father,* 57.
88. Silvia Evangelisti, "To Find God in Work? Female Social Stratification in Early Modern Italian Convents," *European History Quarterly* 38 (2008): 398–416.
89. Belardini, "'Piace molto a Giesù,'" 363, 380; on Galantini, see Gilberto Aranci, *Formazione religiosa e santità laicale a Firenze tra Cinque e Seicento. Ippolito Galantini fondatore della Congregazione di San Francesco (1526–1620)* (Florence: Pagnini, 1997).
90. Adelisa Malena, *L'eresia dei perfetti: Inquisizione romana ed esperienze mistiche nel Seicento italiano* (Rome: Edizioni di storia e letteratura, 2003), 60; Schutte, *Aspiring Saints,* 212.
91. Spanning the years 1609 to 1625, Bartolomeo's letters indicate that correspondence to and from Orsola was sent but never received; BRF. Ms. 2712, cc. 131–143. The Japanese print ("un imprese del Giapone molto bello per alla granduchessa") is noted at c. 136.
92. MdP. vol. 6081, dated 12 Feb. 1620, 5 March 1620, and 23 Sept. 1620.
93. MdP. vol. 6113, fol. 250, dated 22 Feb. 1621. Fontebuoni simply called the herb "cina," which at the time could refer either to china root *(Smilax china)* or false chinaroot *(Smilax pseudochina).* Sheila Barker examines the linguistic confusion and overlap between these two herbs, and between china root and cinchona, in "Malaria and the Search for Its Cure in Granducal Tuscany," *Medicea* 5 (2010): 54–59.
94. MdP. vol. 6081, dated 2 Nov. 1620: "io in cambio gli mando due ampolle di liquore di San Niccolò di Bari che l'adoperi al suo ginochio." See also Barker, "Contributions of Medici Women," 106. Galasso Calderara, *Granduchessa Maria Maddalena,* 102, comments on Maria Maddalena's mobility problems.

95. Gianna Pomata, "Practicing between Earth and Heaven: Women Healers in Early Modern Bologna," *Dynamis* 19 (1999): 119–143.
96. Jacqueline Musacchio, *The Art and Ritual of Childbirth in Renaissance Italy* (New Haven, CT: Yale University Press, 1999).
97. Lea T. Olsan, "Charms and Prayers in Medieval Medical Practice," *Social History of Medicine* 16 (2003): 343–366.
98. Musacchio, *Art and Ritual of Childbirth*, 141–143.
99. MdP. vol. 6081, dated 3 June 1617. "Non sapendo se l'Altezza Sua l'arebbe a gradito ma assicurata dalla Ill.ma Signora Maddalena Nobili gliel'ho fatto e glielo mando per la medesima . . . Nel panno da stomaco non n'è cosa che gli possa nuocere et è della polvere d'ascentio et menta garofani e canella noce moschada et muschio. V'è del pan di S. Nicholaio da Tolentino e altre cose benedette che metto ne brevi " See also Barker, "Contributions of Medici Women," 106.
100. MdP. vol. 6101, fol. 583, dated 9 June 1617.
101. MdP. vol. 6081, dated 18 June 1617. "Questi panni durano e son buoni 3 o 4 mesi, ma si riscaldano ogni tanto in questo modo si piglia un testo caldo bene, poi vi si getta su della malvagia e si tien questo panno a scaldare a quel fumo e quando e caldo si mette allo stomaco, così da più conforto e giovamento." After use, the archduchess was told to "buttassero sul fuoco per amor di quelle cose santé che vi son dentro."
102. MdP. vol. 6081, dated 18 June 1617. "Amantissima mia signora m'è venuto in pensiero che gli mettessi in sullo stomaco la pezzetta della Madre Passitea che fossi il Signore gli potrebbe dare il suo aiuto. Questa pezzetta la tiene la signora Salviati spidalinga di San Gregorio."
103. Vincenzo Criscuolo, "Documenti vaticani su Passitea Crogi clarissa cappuccina senese (1564–1615)," *Collectanea franciscana* 62 (1992): 651–683; Alfonso Casini, *Passitea Crogi, donna senese* (Siena: Edizioni Cantagalli, 1991).
104. MdP. vol. 6081, dated 26 Nov 1617. "Gli mando una scatola con poca cosa et un cuoricino di reliquie per mettere al collo del nuovo principino . . . di piu gli dico che quella crocetta di reliquie, la quale gli detti quando erano al crocifisso che ha gratia particulare dal Signore sopra le donne di parto."
105. Quoted in Barker, "Christine of Lorraine," 166–167. I have altered the translation slightly.
106. Molly Bourne, "Medici Women at the Gonzaga Court, 1584–1627," in *Italian Art, Society, and Politics: A Festschrift for Rab Hatfield*, ed. Barbara Deimling, Jonathan K. Nelson, and Gary M. Radke (Syracuse, NY: Syracuse University Press, 2007), 223–243.
107. Christine of Lorraine, *Lettere alla figlia Caterina de' Medici Gonzaga duchessa di Mantova (1617–1629)*, ed. Beatrice Biagioli and Elisabetta Stumpo (Flor-

ence: Firenze University Press, 2015), 31–32. Barker, "Christine of Lorraine," 166–167, provides the original text plus a translated excerpt.

108. Sandra Cavallo, "Pregnant Stones as Wonders of Nature," in *Reproduction: From Antiquity to the Present Day,* ed. Nick Hopwood, Rebecca Flemming, and Lauren Kassell (Cambridge, UK: Cambridge University Press, 2018), E17.
109. Don C. Skemer, *Binding Words: Textual Amulets in the Middle Ages* (University Park: Penn State University Press, 2006), 18, 241–242.
110. Franco Cardini, "Il 'breve' (secoli XIV–XV): tipologia e funzione," *La ricerca folklorica* 5 (1982): 63–73.
111. Augusto Calderara, *Abraxas: Glossario dei termini di sostanze, formule e oggetti usati in pratiche magiche o terapeutiche, citati nei documenti di ABRATASSA* (Lucca: M. Pacini Fazzi, 1989), 38.
112. MdP. vol. 6113, fol. 309, dated 10 July 1624; vol. 6081, dated 5 March 1621.
113. Patrick Geary discusses the "priceless" nature of relics in "Sacred Commodities: The Circulation of Medieval Relics," in *The Social Life of Things: Commodities in Cultural Perspective,* ed. Arjun Appadurai (Cambridge, UK: Cambridge University Press, 1986), 169–194.
114. Musacchio, *Art and Ritual of Childbirth,* 126–147.
115. MdP. vol. 6113, fol. 102, dated 20 June 1618: "gli mando alcuni le nostri brevini sapendo che per la sua molta devotione gli son' grati et una Madonnina da tenersela nella corona. Non manco di tener' memoria delle sue intentioni."
116. MdP. vol. 6113, fol. 165, dated 24 July 1619: "Non manco ogni giorno insieme con tutte di mandare particular' preghiere a S. D. Maestà per lei per il Signor Duca e per cotesto stato desidero che Giesù sodisfaccia a i desiderii loro."
117. MdP. vol. 6081, dated 15 April 1616: "Giesù è il nostro medico e la nostra medicina."
118. George McClure, "Healing Eloquence: Petrarch, Salutati, and the Physicians," *Journal of Medieval and Renaissance Studies* 15 (1985): 317–346.
119. MdP. vol. 6113, fol. 231, dated 7 May 1620: "Non sappiamo quello che Giesù voglia far' di noi. [. . .] L'oratione non è mai persa ma sempre opera qualche cosa."
120. Christine of Lorraine, *Lettere alla figlia,* 18, 32.
121. Willemijn Ruberg, "The Letter as Medicine: Studying Health and Illness in Dutch Daily Correspondence, 1770–1850," *Social History of Medicine* 23 (2010): 492–508.
122. MdP. vol. 6081, dated 25 Jan. 1617: "l'occhio dell'intelletto stia affiso sempre in Dio come fa l'aquila al sole, e questo sguardo interno apporta pace e dolcezza al cuore"; MdP. vol. 6081, dated 16 Dec. 1621.
123. MdP. vol. 6081, dated 17 May 1620, 9 Sept. 1622, and 7 Dec. 1616, quote at latter: "La mia indispositione è stata un' poco di visita che m'ha fatta

il Signore per purgatione de miei difetti e peccati"; vol. 6101, fol. 585, dated 4 June 1617.

124. MdP. vol. 6081, dated 15 April 1616: "Io à dirglielo in secreto, procurerei di farlo distorre da i medici dicendogli che è troppo debole si come veramente gli è se gli ha bisogno di bagno si può far condur l'acque. Io Signora mia, non posso credere che questa volontà gli habbia a ire innanzi se vuole andare a pigliare aria può andare sulle montagnie e altri luoghi simili dove sono buon'arie da portargli giovamento."

125. MdP. vol. 6081, dated 12 Feb. 1618: "Penso che i medici sappino quel che bisogna, il Signore gli dia gratia che eleghino quei rimedi che sono più convenienti. [. . .] Io non ci ho esperientia e non so queste cose faccino quello che la lor prudentia e dottrina gli insegna."

126. MdP. vol. 6081, dated 12 Feb. 1618: "Quanto al pugner la vena in questi tempi se la furia del male non ne sforza non è bene cavarne massimo a un corpo cosi indebolito come è il suo. Credo che habbia più presto bisogno di mettersene che di cavarne."

127. Hannah Newton, "'She Sleeps Well and Eats an Egg': Convalescent Care in Early Modern England," in *Conserving Health in Early Modern Culture: Bodies and Environments in Italy and England,* ed. Sandra Cavallo and Tessa Storey (Manchester, UK: Manchester University Press, 2017), 104–132; Hannah Newton, *Misery to Mirth: Recovery from Illness in Early Modern England* (Oxford: Oxford University Press, 2018).

128. Francesco Rondinelli, *Relazione del contagio stato in Firenze l'anno 1630 e 1633* (Florence: Giovanbattista Landini, 1714), 22–23, 26. Important studies of the 1630 plague include Giulia Calvi, *Histories of a Plague Year: The Social and the Imaginary in Baroque Florence,* trans. Dario Biocca and Bryant T. Ragan (Berkeley: University of California Press, 1989); John Henderson, "'La schifezza, madre della corruzione': Peste e società nella Firenze della prima età moderna, 1630–31," *Medicina e storia* 2 (2001): 23–56; and Carlo M. Cipolla, *Cristofano and the Plague: A Study in the History of Public Health in the Age of Galileo* (Berkeley: University of California Press, 1973).

129. Nicholas A. Eckstein, "Florence on Foot: An Eye-Level Mapping of the Early Modern City in Time of Plague," *Renaissance Studies* 30 (2016): 273–297, at 293.

130. Colin Rose, "Plague and Violence in Early Modern Italy," *RQ* 71 (2018): 1000–1035.

131. Rondinelli, *Relazione,* 29.

132. Christiane Nockels Fabbri, "Treating Medieval Plague: The Wonderful Virtues of Theriac," *Early Science and Medicine* 12 (2007): 247–283.

133. Galilei, *Letters to Father,* 56, dated 2 Nov. 1630.

134. Jane Stevens Crawshaw, "Families, Medical Secrets and Public Health in Early Modern Venice," *Renaissance Studies* 28 (2014): 597–618.

135. Fabbri, "Treating Medieval Plague," 275–283.

136. Galilei, *Letters to Father,* 56–57, dated 8 Nov. 1630.
137. Peter Burke, "How to Be a Counter-Reformation Saint," in Peter Burke, *The Historical Anthropology of Early Modern Italy* (Cambridge, UK: Cambridge University Press, 1987), 48–62; Adriano Prosperi, *Tribunali della coscienza. Inquisitori, confessori, missionari* (Torino: Einaudi, 1996).
138. Anne Jacobsen Schutte, *Aspiring Saints: Pretense of Holiness, Inquisition, and Gender in the Republic of Venice, 1618–1750* (Baltimore, MD: Johns Hopkins University Press, 2001), 132.
139. Belardini, "'Piace molto a Giesù,'" 363, 365.
140. BNCF. Ms. Rossi Cassigoli. 134, fol. 100r: "per ordine della medesima Congregazione si abbruccarono in piazza tutte le scritte che ritrovarono appresso da Suor Orsola, le quali furono in numero grande come ancora si interdire tutti i brevi corone ed imagini benedetti per detta suora." See also Malena, *L'eresia dei perfetti,* 60–61; Belardini, "'Piace molto a Giesù,'" 362–363, 379–380.
141. BNCF. Ms. Rossi Cassigoli. 134, fol. 26r: "non dava contrasegno veruno di santità come da molti se aspettava."
142. Davis, *The Gift,* 78.
143. MdP. vol. 6081, dated 11 Aug. 1618 and 23 Sept. 1618. On this image-relic, see Alice E. Sanger, *Art, Gender, and Religious Devotion in Grand Ducal Tuscany* (Aldershot, UK: Ashgate, 2014), 88.
144. MdP. vol. 6081, dated 12 Feb. 1620 and 9 Sept. 1622.
145. Calvi, *Histories of a Plague Year,* 205; Harness, *Echoes of Women's Voices,* 282–317. Domenica's religiosity is analyzed by Lorenzo Polizzotto, "When Saints Fall Out: Women and the Savonarolan Movement in Early Sixteenth-Century Florence," *RQ* 46 (1993): 486–525.
146. Calvi, *Histories of a Plague Year,* 228–235.
147. Gianna Pomata, "Malpighi and the Holy Body: Medical Experts and Miraculous Evidence in Seventeenth-Century Italy," *Renaissance Studies* 21 (2007): 568–586, quote at 569.

CHAPTER THREE ~ THE BUSINESS OF HEALTH

1. Sandra Cavallo and David Gentilcore, "Introduction: Spaces, Objects and Identities in Early Modern Italian Medicine," *Renaissance Studies* 21 (2007): 473–479; Filippo De Vivo, "Pharmacies as Centres of Communication in Early Modern Venice," *Renaissance Studies* 21 (2007): 505–521.
2. Joanna Kostylo, "Pharmacy as a Centre for Protestant Reform in Renaissance Venice," *Renaissance Studies* 30 (2016): 236–253.
3. Evelyn Welch, "Space and Spectacle in the Renaissance Pharmacy," *Medicina e storia* 15 (2008): 127–158; Evelyn Welch, *Shopping in the Renaissance: Consumer Cultures in Italy, 1400–1600* (New Haven, CT: Yale University Press, 2005), 151–158; Kostylo, "Pharmacy."

4. David Gentilcore, *Medical Charlatanism in Early Modern Italy* (Oxford: Oxford University Press, 2006); William Eamon, *The Professor of Secrets: Mystery, Medicine, and Alchemy in Renaissance Italy* (Washington, DC: National Geographic Society, 2010).
5. Sharon T. Strocchia, "Introduction: Women and Healthcare in Early Modern Europe," *Renaissance Studies* 28 (2014): 496–514.
6. Elizabeth Cohen, "Miscarriages of Apothecary Justice: Un-separate Spaces of Work and Family in Early Modern Rome," *Renaissance Studies* 21 (2007): 480–504; Eleonora Carinci, "Una 'speziala' padovana: *Lettere di philosophia naturale* di Camilla Erculiani (1584)," *Italian Studies* 68 (2013): 202–229; Laurence Brockliss and Colin Jones, *The Medical World of Early Modern France* (Oxford: Oxford University Press, 1997), 263; F. H. Rawlings, "Two 17th Century Women Apothecaries," *Pharmaceutical Historian* 14 (1984): 7. De Vivo, "Pharmacies," 516, notes that five out of eight-five Venetian apothecaries registered in 1569 were women.
7. Arte dei Medici e Speziali. vol. 13, fols. 125v, 291v, dated 5 May 1576 and 29 Jan. 1592, respectively.
8. Sheila Barker, "The Contributions of Medici Women to Medicine in Grand Ducal Tuscany and Beyond," in *The Grand Ducal Medici and Their Archive (1537–1743),* ed. Alessio Assonitis and Brian Sandberg (Turnhout, Belgium: Brepols, 2016), 101–116; Arte dei Medici e Speziali. vol. 14, fols. 142v ("lava cappelli in piazza"), 240v, 254v ("per poter vendere polvere da rogna"). Remarking on the intrafamilial transfer of certain guild privileges, Giulia Calvi notes that a woman named Santa, the daughter of the Florentine barber Giovanni Peretoli, ran a barbershop on Piazza Pitti in 1633, and that her mother, Margherita, opened another shop on Ponte Vecchio; Giulia Calvi, *Histories of a Plague Year: The Social and the Imaginary in Baroque Florence,* trans. Dario Biocca and Bryant T. Ragan (Berkeley: University of California Press, 1989), 229–230.
9. Tessa Storey, "Face Waters, Oils, Love Magic and Poison: Making and Selling Secrets in Early Modern Rome," in *Secrets and Knowledge in Medicine and Science, 1500–1800,* ed. Elaine Leong and Alisha Rankin (Aldershot, UK: Ashgate, 2011), 143–163.
10. Marinella Franchi, "La spezieria: gestione e funzionamento," in *Una farmacia preindustriale in Valdelsa. La spezieria e lo spedale di Santa Fina nella città di San Gimignano, secc. XIV–XVIII,* ed. Gabriele Borghini (San Gimignano: Città di San Gimignano, 1981), 126.
11. Jutta Gisela Sperling, *Convents and the Body Politic in Late Renaissance Venice* (Chicago: University of Chicago Press, 1999), 18. In 1563, Florence counted forty-seven convents housing 3,823 nuns, who represented 5.5 to 6.5 percent of the urban population; Sharon T. Strocchia, *Nuns and Nunneries in Renaissance Florence* (Baltimore, MD: Johns Hopkins University Press, 2009), 200n.

12. Quoted in Gianna Pomata, "Practicing between Heaven and Earth: Women Healers in Seventeenth-Century Bologna," *Dynamis* 19 (1999): 119–143, at 133–134.
13. Katharine Park, *Secrets of Women: Gender, Generation, and the Origins of Human Dissection* (New York: Zone Books, 2006), 39–76.
14. *Erbe e speziali. I laboratori della salute,* ed. Margherita Breccia Fratadocchi and Simonetta Buttò (Sansepolcro: Aboca, 2007), 283.
15. Strocchia, *Nuns and Nunneries,* 28–38, 111–151, which informs the remainder of the paragraph.
16. Lorenzo Polizzotto, *The Elect Nation: The Savonarolan Movement in Florence* (Oxford: Oxford University Press, 1994), 397, 409.
17. OSMN. vol. 192, fol. 173r, dated 28 Feb. 1558. The number of Florentine apothecaries dropped from 113 heads of household in 1427 to sixty in 1480; Antonella Astorri, "Appunti sull'esercizio dello speziale a Firenze nel Quattrocento," *Archivio storico italiano* 147 (1989): 31–62. That number declined further in the first half of the Cinquecento. A 1553 guild census put the number of apothecaries resident in the Medici dominion at 200, with forty-eight living in the city (OSMN. vol. 192, fol. 439r). However, a ducal tax survey taken in the 1530s reported sixty-eight Florentine apothecaries, although perhaps not all of them practiced their trade; Esther Diana, "Medici, speziali e barbieri nella Firenze della prima metà del '500," *Rivista di Storia della Medicina,* n.s. 4 (1994): 13–27.
18. By 1617, the number of Venetian apothecary shops was on the rise: more than a hundred pharmacies serviced roughly 142,000 people, giving a ratio of one shop per 1,420 residents. Richard Palmer, "Pharmacy in the Republic of Venice in the Sixteenth Century," in *The Medical Renaissance of the Sixteenth Century,* ed. Andrew Wear, Roger French, and I. M. Lonie (Cambridge, UK: Cambridge University Press, 1985), 100–117. The ratio of Venetian pharmacists reported by De Vivo, "Pharmacies," 505, should be adjusted.
19. Ninety pharmacies catered to a Roman populace of 53,000 people; Leonardo Colapinto, "L'arte degli speziali a Roma nel Rinascimento," *AMAISF* 12 (1995): 59–63.
20. Sandra Giovannini and Gabriella Mancini, *L'Officina profumo-farmaceutica di Santa Maria Novella in Firenze. Sette secoli di storia e di arte* (Rome: Chitarrini, 1994), 16–24.
21. *Farmacie storiche in Toscana* (Florence: Polistampa, 1998), 19, 32; Convento di Camaldoli. Appendice. Vol. 500.
22. Giovannini and Mancini, *L'Officina,* 21–24.
23. Monachus Aromatarius, "Il contributo degli ordini monastici alla botanica in Italia," in *Di sana pianta. Erbaria e taccuini di sanità,* ed. Rolando Bussi (Modena: Panini, 1988), 29–32.
24. Strocchia, *Nuns and Nunneries,* 89.

25. Maria Celeste Galilei's pharmaceutical skills are noted by Dava Sobel, *Galileo's Daughter: A Historical Memoir of Science, Faith and Love* (New York: Walker & Co., 1999), 9, 85, 112, 325.
26. Guido Vannini, "La spezieria: formazione e dotazione," in *Una farmacia preindustriale,* 37–38.
27. Burr Litchfield, *Florence Ducal Capital, 1530–1630,* Paragraph 222, and Table 5.1.
28. Saundra Weddle, "Identity and Alliance: Urban Presence, Spatial Privilege, and Florentine Renaissance Convents," in *Renaissance Florence: A Social History,* ed. Roger Crum and John Paoletti (Cambridge, UK: Cambridge University Press, 2006), 394–412.
29. Nicholas Terpstra, "Locating the Sex Trade in the Early Modern City: Space, Sense, and Regulation in Sixteenth-Century Florence," in *Mapping Space, Sense, and Movement in Florence: Historical GIS and the Early Modern City,* ed. Nicholas Terpstra and Colin Rose (New York: Routledge, 2016), 107–124.
30. These activities brought in 500 florins annually by the 1450s. Kate Lowe, "Women's Work at the Benedictine Convent of Le Murate in Florence: Suora Battista Carducci's Roman Missal of 1509," in *Women and the Book: Assessing the Visual Evidence,* ed. J. Taylor and L. Smith (London: British Library, 1997), 133–146, at 142; Strocchia, *Nuns and Nunneries,* 114, 121–122.
31. BNCF. II. II. 509, fol. 27r: "sendo che con l'infermeria che si concluse in tre anni, feciono attorno la cucina per poter cuocere all' inferme con le stanze basse della spetieria accomodate per stillare, e altre cose necessarie co' lor mobili et masseritie da usarsi in que' servitii sopra la qual cucina fece fare una camera per le sue figliuole e altre della sua famiglia per quando ci venissino ad habitare che però ancora la chiamiamo la Cella de' Benci." Sister Giustina Niccolini, *The Chronicle of Le Murate,* ed. and trans. Saundra Weddle (Toronto: University of Toronto Press, 2011), 88–89, discusses some of the site problems associated with the infirmary.
32. BNCF. II. II. 509, fols. 88r, 98v–99v: "fece alcuni miglioramenti nella n[os]tra spezieria, che furono certi vasi di terra per fare le confettione, et altre cose appartenente et il proposito a quel mestiero." See also *Chronicle of Le Murate,* 179, 195.
33. *Suor Plautilla Nelli (1523–1588): The First Woman Painter of Florence,* ed. Jonathan Nelson (Florence: Cadmo, 2000).
34. CRSGF. 106. vol. 30, fol. 49r.
35. David Gentilcore, "'For the Protection of Those Who Have Both Shop and Home in This City': Relations between Italian Charlatans and Apothecaries," *Pharmacy in History* 45 (2003): 108–121; CRSGF. 106. vol. 13, fols. 51r, 81v.
36. CRSGF. 106. vol. 51, fol. 47v, and vol. 52.

37. CRSGF. 106. vol. 13, fols. 51r, 81v; vol. 30, fol. 95r; Patrick Macey, *Bonfire Songs: Savonarola's Musical Legacy* (Oxford: Oxford University Press, 1998), 124–129.
38. CRSGF. 106. vol. 30, fols. 90r, 95r; Strocchia, *Nuns and Nunneries,* 34.
39. CRSGF. 106. vol. 35, fol. 207v.
40. Carlo Borromeo, *Instructiones fabricae et suppellectilis ecclesiasticae* (1577): *A Translation with Commentary and Analysis,* ed. Evelyn C. Voelker (Ann Arbor: University Microfilms International, 1993), bk. I, ch. 33. http://www.evelynvoelker.com.
41. *La spezieria di San Benedetto a Montefiascone. Dalle collezioni di Palazzo Venezia in Roma,* ed. Maria Selene Sconci and Romualdo Luzi (Ferrara: Belriguardo, 1994), 50–51.
42. Julia Delancey, "Dragonsblood and Ultramarine: The Apothecary and Artists' Pigments in Florence," in *The Art Market in Italy (15th–17th Centuries),* ed. Louisa Matthew and Sara Matthews-Grieco (Modena: F.C. Panini, 2003), 141–150; Louisa Matthew, "'Vendecolori a Venezia': The Reconstruction of a Profession," *Burlington Magazine* 144 (2002): 680–686.
43. Catherine Turrill, "*Compagnie* and *Discepole:* The Presence of Other Women Artists at Santa Caterina da Siena," in *Suor Plautilla Nelli,* 83–102.
44. Sheila Barker, "Painting and Humanism in Early Modern Florentine Convents," in *Artiste nel chiostro. Produzione artistica nei monasteri femminili in età moderna,* ed. Sheila Barker with Luciano Cinelli (Florence: Nerbini, 2015), 105–139.
45. CRSGF. 98. vol. 25, fols. 10r, 60v.
46. Barker, "Painting and Humanism," 120–121, 129–131.
47. Barker, "Painting and Humanism," 121–122.
48. Astorri, "Appunti," 54–55.
49. CRSGF. 105. vol. 27, fol. 30v; vol. 61, fols. 128v–129r, 149v–150r. Andrea Del Garbo hailed from a dynasty of physician-pharmacists; he also supplied medicinals to the convent pharmacy at San Francesco. Another supplier at the Annalena was Bernarbo Bicci, descendant of the painters Bicci di Lorenzo and Lorenzo di Bicci. For the Del Garbo family, see Katharine Park, *Doctors and Medicine in Early Renaissance Florence* (Princeton, NJ: Princeton University Press, 1985).
50. Giuseppe Richa, *Notizie istoriche delle chiese fiorentine* (Florence: P.G. Viviani, 1754–1762), 10: 156–157.
51. CRSGF. 105. vols. 29 and 30. In the second half of the century, textile work accounted for roughly 35 percent of earned income, boarding fees for 50–55 percent.
52. CRSGF. 105. vol. 29, fol. 46r; vol. 35, fols. 10r, 15v, 25r, 55v. The ducat was roughly equivalent to the florin after 1530.
53. Giovannini and Mancini, *L'Officina,* 15–16.
54. MSS. vol. 176, insert 1; S. Agata. vol. 16, fols. 54v, 56r, 103v–104r.

55. Auditore. vol. 4894, fol. 68r/v: "respetto alla conservatione della loro spetieria, la quale dicano essere il sostentamento loro [. . .] et che dette monache non hanno altro assegnamento più vivo di detto spetieria, et che non viveno d'altro che di quella et di elemosine."
56. Elissa Weaver, *Convent Theater in Early Modern Italy: Spiritual Fun and Learning for Women* (Cambridge, UK: Cambridge University Press, 2002), 72–73, 134–135.
57. *S. Caterina de' Ricci: Cronache, diplomatica, lettere varie,* ed. Guglielmo Di Agresti (Florence: L.S. Olschki, 1969), lxxxvi–lxxxvii; Turrill, "Compagnie e Discepole," 100–101.
58. *S. Caterina,* lxxviii, 5, 80, quotes at 13; Silvestro Bardazzi and Eugenio Castellani, *Il monastero di S. Vincenzo in Prato* (Prato: Edizioni del Palazzo, 1982), 35–38; *Farmacie storiche,* 81–82. Anna Vittoria Laghi discusses the later development of the pharmacy in "Di tre 'spezierie' pratesi," in *Il Settecento a Prato,* ed. Renzo Fantappiè (Prato: CariPrato, 1999), 345–352.
59. CRSGF. 81. vol. 8, fols. 251v, 264v.
60. Convent records indicate that nuns kept separate ledgers for distinct craft activities. For instance, the Santa Felicita nuns made repeated reference to "the account book of our pharmacy" in 1597, although this ledger no longer exists; CRSGF. 83. vol. 77, fols. 310v–311r. For the Giglio inventories, Shaw and Welch, *Making and Marketing Medicine,* 65–67.
61. John Henderson, *The Renaissance Hospital: Healing the Body and Healing the Soul* (New Haven, CT: Yale University Press, 2006), 311; Giovannini and Mancini, *L'officina,* 18.
62. By decree dated 17 May 1577, the Florentine College of Physicians restricted annual production of theriac to the period between 1 June and 31 August; Giovannini and Mancini, *L'officina,* 60–61, 89–90.
63. Corey Tazzara, "Capricious Demands: Artisanal Goods, Business Strategies, and Consumer Behavior in Seventeenth-Century Florence," in *Early Modern Things: Objects and Their Histories, 1500–1800,* ed. Paula Findlen (New York: Routledge, 2013), 204–224.
64. Shaw and Welch, *Making and Marketing Medicine,* 159.
65. Giuseppe Franchi, "Apparecchi e metodi per 'lambiccare' secondo Mattioli," in *I Giardini dei Semplici e gli Orti Botanici della Toscana,* ed. Sara Ferri and Francesca Vannozzi (Perugia: Quattroemme, 1993), 201–204.
66. Antonio Clericuzio, "Chemical Medicine and Paracelsianism in Italy, 1550–1650," in *The Practice of Reform in Health, Medicine and Science, 1500–2000. Essays for Charles Webster,* ed. Margaret Pelling and Scott Mandelbrote (Aldershot, UK: Ashgate, 2005), 59–79.
67. BNCF. Ms. Targioni Tozzetti. 56. 3 vols, I: fols. 33r, 188v–191v.
68. Douglas Biow, *The Culture of Cleanliness in Renaissance Italy* (Ithaca, NY: Cornell University Press, 2006); Sandra Cavallo and Tessa Storey, *Healthy*

Living in Late Renaissance Italy (Oxford: Oxford University Press, 2013), esp. 240–269.

69. LCF. vol. 1264, accounts of the soap-maker Luca di Antonio Castrucci from 1593 to 1603.
70. Calvi, *Histories of a Plague Year*, 283n; Girolamo Maccioni, *Risposta al parere del Sig. Gasparo Marcucci intorno alla qualità del sapon molle* (Florence: Sermartelli, 1630).
71. John Styles, "Product Innovation in Early Modern London," *Past and Present* 168 (2000): 124–169; *L'Officina*, 18.
72. Aud. Rif. vol. 21, fols. 653r–654r. In the 1520s, San Pier Maggiore marked urban property holdings with its corporate emblem; Strocchia, *Nuns and Nunneries*, 58–59. For institutional interiors and objects, see Vannini, "La spezieria," 42n.
73. The petition from the Pisan soap-maker Giovanni Celli to Duke Cosimo illustrates many of these points. Celli had marketed his product for eighteen years under the name "Pisan soap" *(sapone pisano)*. He complained that recently certain soap from the Genoese Riviera was being sold widely throughout the dominion under the same name. Because this soap was "bad" *(cattivo)*, it damaged the reputation of the genuine article. After sampling the soap, however, guild officials did not find it to be "bad"; moreover, the foreign soap bore a different trademark. Responding to this case, Cosimo declared that all soap entering Florence should be "good" *(buoni)* and could be sold by anyone, but instructed customs officers to appoint apothecaries to sample all such products before being cleared for sale. OSMN. vol. 194. no. 13, dated 17 Dec 1566; no. 16, dated 23 Dec. 1566.
74. CRSGF. 108. vol. 16, fol. 14v (pills); vol. 18, fols. 9v–10r (rose unguent); vol. 15, fol. 205r (press).
75. See for example CRSGF. 108. vol. 43, fol. 27r; vol. 45, fol. 22v.
76. Strocchia, *Nuns and Nunneries*, 126.
77. Cavallo and Storey, *Healthy Living*, 247–249.
78. *S. Caterina*, cv. Sick nuns were often prescribed a diet of capons, sugared almonds, wine, and refined bread.
79. Fanny Kieffer, "La Confiserie des Offices: Art, Sciences et Magnificence à la Cour des Medicis," *Predella* 33 (2013): 89–106.
80. CRSGF. 81. vol. 9, fols. 71v, 73r, 74r / v.
81. Sandra Cavallo, "Health, Beauty, and Hygiene," in *At Home in Renaissance Italy*, ed. Marta Ajmar-Wollheim and Flora Dennis (London: V & A, 2006), 174–187; Luke Demaitre, "Skin and the City: Cosmetic Medicine as an Urban Concern," in *Between Text and Patient: The Medical Enterprise in Medieval and Early Modern Europe*, ed. Florence Eliza Glaze and Brian Nance (Florence: SISMEL, 2011), 97–120; Mariacarla Gadebusch Bondio, "La *Carne di Fuori*: Discorsi medici sulla natura e l'estetica della pelle nel '500," *Micrologus* 13 (2005): 537–570.

82. Monachus Aromatarius, "Il contributo," 30–31.
83. Enrica Viviani della Robbia, *Nei monasteri fiorentini* (Florence: Sansoni, 1946), 21; Richa, *Notizie* 7: 51.
84. Gianna Pomata, "Medicina delle monache. Pratiche terapeutiche nei monasteri femminili di Bologna in età moderna," in *I monasteri femminili come centri di cultura fra Rinascimento e Barocco,* ed. Gianna Pomata and Gabriella Zarri (Rome: Edizioni di storia e letteratura, 2005), 331–363. Candida Carrino, *Le monache ribelli raccontate da Suor Fulvia Caracciolo* (Naples: Intra Moenia, 2013), 38–39, notes that S. Gregorio Armeno was renowned for making "tablets of damask rose and the best quince jelly"; the Sapienza concocted bean-shaped lozenges from sugar and amber to settle coughs; S. Maria Donnaromita specialized in candied rose petals.
85. *Tariffa et prezzi del medicinale et mercanzie attenente alli speziali della città di Firenze* (Florence: Giorgio Marescotti, 1593) [LCF. vol. 164], unnumbered front flyleaf; p. 72. This list incorporated small price protections. Convents legally could sell their restorative chicken broth for half as much as apothecary shops (two soldi versus four soldi per ounce).
86. CRSGF. 81. vol. 5, fol. 44v: "per limosina del convento et ricompenso delle fatiche della spetieria per le sue infirmita."
87. Shaw and Welch, *Making and Marketing Medicine,* 62–67.
88. CRSGF. 105. vol. 64, fols. 172v–173r; vol. 65, fols. 184v–185r; vol. 66, fols. 136v–137r.
89. MdP. vol. 370, fol. 551; Aud. Rif. vol. 11, fasc. 136, dated 11 April 1573.
90. For instance, the nuns of San Benedetto near Viterbo grew most simples in their garden, but purchased powdered coral, rubies, and sapphires from druggists and Jewish merchants in Rome, Viterbo, and Civitavecchia; *La spezieria di San Benedetto,* 51–54.
91. For the Annalena, CRSGF. 105. vol. 68, fol. 194r; for San Vincenzo, see *Vita di S. Caterina de' Ricci,* ed. Gugliemo Di Agresti (Florence: L.S. Olschki, 1965), 78.
92. CRSGF. 81. vol. 100, #326, dated 25 Sept. 1499.
93. K. J. P. Lowe, *Nuns' Chronicles and Convent Culture in Renaissance and Counter-Reformation Italy* (Cambridge, UK: Cambridge University Press, 2003), 139–140.
94. *Una farmacia preindustriale,* 126.
95. CRSGF. 106. vol. 33, fols. 6r, 19r, 47v, 50r, 83r, 84r.
96. CRSGF. 106. vol. 35, fols. 69v, 101v, 130v–132r.
97. Evelyn Welch, "Scented Buttons and Perfumed Gloves: Smelling Things in Renaissance Italy," in *Ornamentalism: The Art of Renaissance Accessories,* ed. Bella Mirabella (Ann Arbor: University of Michigan Press, 2011), 13–39; Cavallo and Storey, *Healthy Living,* 98–105.
98. Dietrich Heikamp, "Agostino del Riccio. Del giardino di un re," in *Il giardino storico italiano. Problemi e indagine sulle fonti letterarie e storiche,* ed.

Giovanna Ragionieri (Florence: L.S. Olschki, 1981), 59–64; Utzima Benzi, "Le frère dominicain Agostino del Riccio et son *Arte della memoria locale* (1595)," *Bulletin de l'Association des Historiens de l'Art Italien* 13 (2007): 25–43.

99. BNCF. Ms. Targioni Tozzetti. 56. I, fol. 126v: "così usano fare le nostre monache di S. Domenico, che stanno appresso S. Marco in Firenze; altre si fanno il medesimo le accorte e prudenti donne fiorentine che stanno fuori alla Porta della Croce così hanno tante vivuole, e gherofani, e fioralisi, et altri fiori, che ne cavano assai denari ogni anno. Io voglio pur dire che le nostre monache cavano più di 50 scudi l'anno, e più e meno, secondo che vogliono, imperocchè i fiori la derrata danno a Bologna, et a Ferrara, et all'altre città."

100. CRSGF. 108. vol. 15. Don Luis began creating an impressive pleasure garden on the site in 1551; after losing possession, the land reverted to the convent in 1580; Anatole Tchikine, "Horticultural Differences: The Florentine Garden of Don Luis de Toledo and the Nuns of San Domenico del Maglio," *Studies in the History of Gardens and Designed Landscapes: An International Quarterly* 30 (2010): 224–240.

101. Aud. Rif. vol. 21, fols. 609–611, dated 23 Nov. 1596: those opposing this privilege were "alcuni monasterii che attendono a questi fiori et molti contadini." The patent is recorded in Pratica Segreta. vol. 190, fol. 27r/v.

102. Strocchia, *Nuns and Nunneries.*

103. BNCF. Ms. Targioni Tozzetti. 56, 1: fol. 52r.

104. This information is culled from CRSGF. 108. vols. 15 and 16.

105. Giovan Vettorio Soderini, *Il trattato degli arbori,* ed. Alberto Bacchi della Lega (Bologna: Romagnoli Dall'Acqua, 1904), vol. 3: 330: "sono frutti che temono il freddo e massimamente il vento da tramontana."

106. Mar Rey Bueno, "The Health of Philip II, A Matter of State: Medicines and Medical Institutions in the Spanish Court (1556–1598)," in *Être médecin à la cour (Italie, France, Espagne, XIIIe–XVIIIe siècle),* ed. Elisa Andretta and Marilyn Nicoud (Florence: SISMEL, 2013), 149–159, at 155n.

107. MdP. vol. 613, fol. 21, dated 7 March 1551: "li melaranci hanno patito qualche poco che talvolta dubitamo di maggior perdita, et spero che presto rimetteranno con questi tempi le foglie fresche."

108. MdP. vol. 220, fol. 76, dated 27 Jan. 1564: "Si ricevettono le 119 piante di frutti et aspettiamo che voi et Andrea ci mandiate tutto il restante di quelle che vi si sono domandate che qua è buono piantare. Però sollecitate di mandarcele tutte, et di più vi commettiamo che ci provediate 80 piante d'aranci in pentole et 25 piante di limoni pure in pentole, che in cotesti monasteri di S. Felicita et nel Paradiso et altri ne troverrete, però subito cercategli et comperateli ingegnandovi d'havere de più grossi che si truovi, et subito con diligentia mandateceli."

109. May Woods and Arete Swartz Warren, *Glass Houses: A History of Greenhouses, Orangeries and Conservatories* (New York: Rizzoli, 1988), 10–15.

110. BNCF. Ms. Targioni Tozzetti. 56, 1: fol. 163r/v; Valentina Fornaciai, *'Toilette,' Perfumes and Make-Up at the Medici Court: Pharmaceutical Recipe Books, Florentine Collections and the Medici Milieu Uncovered* (Livorno: Sillabe, 2007), 41–59.
111. BNCF. Ms. Targioni Tozzetti. 56, 1: fol. 129v: "le nostre amorevole sorelle [. . .] che se ne prestieno in prigione nelle stanze serrate con poco aria in questo modo starebbono più sane"; Cavallo and Storey, *Healthy Living.*
112. Shaw and Welch, *Making and Marketing Medicine,* 81, 100–102.
113. Sharon Strocchia and Julia Rombough, "Women behind Walls: Tracking Nuns and Socio-Spatial Networks in Sixteenth-Century Florence," in *Mapping Space, Sense, and Movement,* 87–106.
114. CRSGF. 105. vol. 67, fol. 202r: "medicine, sciloppi, et altri medicamenti avuti dalla nostra spetieria"; Strocchia, *Nuns and Nunneries,* 142–143.
115. CRSGF. 105. vol. 29, fol. 82v; vol. 64, 54v–55r; vol. 67, fols. 201v–202r, quote at 184v: "per medicine, sciloppi, et altri medicamenti avuti dalla nostra spetieria per insino l'anno 1587"; vol. 68, fol. 12r.
116. CRSGF. 106. vol. 36, fols. 106v, 168v.
117. Strocchia, *Nuns and Nunneries,* 142–143.
118. CRSGF. 111. vol. 67, unfol.; Monastero di S. Luca. vol. 89, fols. 149v, 160r.
119. Auditore. vol. 4892, fol. 144r/v.
120. CRSGF. 106. vol. 71, unnumb. fasc. dated 26 April 1604; Turrill, "*Compagnie* and *Discepole,*" 88–89.
121. CRSGF. 111. vol. 67, unfoliated; second receipt dated 1571.
122. Paolo Zacchia, *De' mali hipochondriaci libri tre* (Rome: Vitale Mascardi, 1644), 181–190.
123. CRSGF. 105. vol. 67, fol. 19r.
124. Monica Green, "Bodies, Gender, Health, Disease: Recent Work on Medieval Women's Medicine," *Studies in Medieval and Renaissance History,* ser. 3, 5 (2005): 1–46.
125. Shaw and Welch, *Making and Marketing Medicine,* 57–58, 84–85.
126. BNCF. Ms. Landau Finlay. 72, fol. 55v: "che visse sempre con loro praticò confabulo e confidentemente trattò."
127. CRSGF. 112. vols. 85 and 97 for lists of women who supported the Pietà financially. On the Pietà, see Nicholas Terpstra, "Mothers, Sisters, and Daughters: Girls and Conservatory Guardianship in Late Renaissance Florence," *Renaissance Studies* 17 (2003): 201–229, and *Lost Girls: Sex and Death in Renaissance Florence* (Baltimore, MD: Johns Hopkins University Press, 2010).
128. Incurabili. vol. 102, fols. 45v–46r.
129. Monastero di S. Luca. vols. 83 and 89; Monastero dell'Arcangelo Raffaello. vol. 14. The latter account book, covering the period 1652 to 1719, belonged to the convent of San Luca but has been archived incorrectly.

130. Nicholas Terpstra, "Competing Visions of the State and Social Welfare: The Medici Dukes, the Bigallo Magistrates, and Local Hospitals in Sixteenth-Century Tuscany," *RQ* 54 (2001): 1319–1355.
131. *Legislazione Toscana raccolta e illustrata,* ed. Lorenzo Cantini (Florence: Albizziniana, 1800), 1: 260–264.
132. Gilberto Aranci, *Formazione religiosa e santità laicale a Firenze tra Cinque e Seicento* (Florence: Pagnini, 1997), 29–44.
133. *Statuti dell'Arte dei Medici e Speziali,* ed. Raffaele Ciasca (Florence: Vallecchi, 1922), 547–611. The guild decreed that all medicinal wares must be "buone, nette realmente e senza fraude alcuna." OSMN. vol. 192, fols. 556–557, dated 30 Dec. 1547. Other pharmaceutical legislation is published in *Legislazione Toscana,* 2: 292–294; 3: 95–102, 128–131, 369–375, 378–387, 392–399.
134. Aud. Rif. vol. 8, fols. 502r–503r, dated 12 Oct. 1560, following first round of guild inspections.
135. Colapinto, "Arte speziali a Roma," 61.
136. Auditore. vol. 4894, fol. 68r/v. Sant'Orsola petitioned "che qualunche persona potessi andare al Monasterio loro senza altra licentia del ordinario, respetto alla conservatione della loro spetieria, la quale dicano essere il sostentamento loro." After investigation, the officials found that "molte persone si astengano dall'andare et mandare per quelle, remosse dall'incommodo et briga che hanno d'andare ogn'hora a provedersi della licentia." Considering that the nuns "non hanno altro assegnamento più vivo di detto spetieria," the deputies concluded that "crederemo fussi bene di permettere [. . .] che tutte quelle persone, le quali o con la ricetta d'alcun medico della città di Firenze, o con il consenso della Badessa o spetiala di detto Monasterio anderanno a quello, possino liberamente andarvi senza altra licentia." The verso notes that this provision was not intended to support any convent "industria di spetieria."
137. OSMN. vol. 193, fol. 186r/v, dated 30 April 1562. Inspectors praised the dispensary at Santa Maria Nuova, "nella quale trovano abbondantia si di drogherie di composti, et le celebrano eccelentissime et con ordine et modo del ministrare raro; et similmente lodano se di Santa Orsola et di Santa Caterina di ben condotte et ministrate con diligentia, ne vi hanno trovato cosa da danarla." However, they lambasted Fra Baccio "per haver egli mostro poco medicinale et buona parte mal ministrato et condotto."
138. OSMN. vol. 193, fol. 229r/v, reply dated 30 Dec. 1562: "expongono a quella essere loro di grande interesse per quel poco di spetieria fanno per lo più per la povera casa loro l'aver a ragunare a ogni loro compositione tutto il collegio de revisori; et per tal causa supplicono adunque a quella che per gratia li vogli concedere basti a tal acto Messer Francesco Gamberelli, nostro medico fisico, el quale è del collegio, e Stefano Rosselli, tutti a dua di detto offitio; et è interessato detto Stefano al nostro monastero

per averci monache tre sorelle." Prior inspection had shown that the pharmacy maintained "buon medicinale et tenuto con assai diligentia."

139. OSMN. vol. 193, fol. 242r/v, registered 6 April 1563. "fu [. . .] trovata detta spetieria fornite di buone et bene composte medicine; et perchè intendono che ogni volta che vorrano comporre medicine sono tenute chiamare i detti deputati visitatori del medicinale. Et havendo dette monache la loro spetieria nel mezzo del convento dispiace loro assai che detti visitatori secolari habbino a passare quasi per tutto il convento." Consequently, the nuns requested "che il visitatore della loro spetieria sia sempre il loro medico fisico."
140. OSMN. vol. 193, fols. 229v, 242v, dated July 1572.
141. OSMN. vol. 197, no. 169, dated 10 Nov. 1586; no. 173, dated 21 Jan. 1587; BNCF. II. II. 509, fols. 141v–142r.
142. Borromeo, *Instructiones fabricae et supellectilis ecclesiasticae* (1577), bk I, ch. 33; BNCF. Codice Panciatichi. 119. II, pp. 87, 95–97.
143. Quoted in Giuliana Boccadamo, "Una riforma impossibile? I papi e i primi tentativi di riforma dei monasteri femminili a Napoli nel '500," *Campania sacra. Rivista di storia sociale e religiosa del Mezzogiorno* 21 (1990): 96–122, at 121: "per fare cose di zucchero, saponetti et altre cose di cocina a diverse persone, oltre la moltitudine di frati et di preti, dalli quali sotto varie spetie di pratiche e di divotione erano frequentate et visitate spesso, in modo che quando detti monasterij si vedevano, non parevano monasterij chiusi di donne, ma più presto publici mercati per la multitudine di gente che con esse haveva da negoziare."
144. Carrino, *Monache ribelli,* 37–38. The ordinances prohibited local nuns from compensating confessors with "cose di zuccaro o vero sciroppate per venderle o donare."
145. CRSGF. 105. vol. 35, fols. 196r–212v; vol. 71.
146. Colapinto, "Arte speziali a Roma," 61.
147. Quoted in Pomata, "Practicing between Earth and Heaven," 131.
148. CRSGF. 106. vol. 71, no. 23; Monachus Aromatarius, "Il contributo," 32.

CHAPTER FOUR ~ AGENTS OF HEALTH

1. David Gentilcore, "Apothecaries, 'Charlatans,' and the Medical Marketplace in Italy, 1400–1750," *Pharmacy in History* 45 (2003): 91–94.
2. Ursula Klein and E. C. Spary, "Introduction: Why Materials?," in *Materials and Expertise in Early Modern Europe: Between Market and Laboratory,* ed. Ursula Klein and E. C. Spary (Chicago: University of Chicago Press, 2010), 1–23, at 2.
3. Alisha Rankin, *Panaceia's Daughters: Noblewomen as Healers in Early Modern Germany* (Chicago: University of Chicago Press, 2013), 129, 138.

4. Pamela H. Smith, "Making Things: Techniques and Books in Early Modern Europe," in *Early Modern Things: Objects and Their Histories, 1500–1800,* ed. Paula Findlen (New York: Routledge, 2013), 173–203.
5. Bruce T. Moran, *Distilling Knowledge: Alchemy, Chemistry, and the Scientific Revolution* (Cambridge, MA: Harvard University Press, 2005), 114.
6. The following discussion is culled from CRSGF. 105. vols. 33, 35.
7. AAF. Professioni di monache. filza II, unfol., dated December 1585.
8. The painter Cosimo Rosselli began training in Neri di Bicci's workshop in 1453 at age fourteen; other apprentices there were eleven and eighteen years old. Neri di Bicci, *Le ricordanze (10 marzo 1453–24 aprile 1475),* ed. Bruno Santi (Pisa: Edizioni Marlin, 1976), 3, 51–52, 101, 262.
9. Dionisio Pulinari da Firenze, *Cronache dei Frati Minori della Provincia di Toscana,* ed. Saturnino Mencherini (Arezzo: Cooperativa tipografica, 1913), 263–264: "amata discepola e compagna"; "non buttava fetore alcuna."
10. Auguste Baudot, *Études historiques sur la pharmacie en Bourgogne avant 1803* (Paris: A. Maloine, 1905), 337–338.
11. Janine Christina Maegraith, "Nun Apothecaries and the Impact of the Secularization in Southwest Germany," *Continuity and Change* 25 (2010): 313–344.
12. BNCF. II. II. 509, fol. 121v; BNCF. Ms. Landau Finlay. 72, fol. 91r; *Le lettere spirituali e familiari di S. Caterina de' Ricci, fiorentina,* ed. Cesare Guasti (Prato: R. Guasti, 1861), LXV.
13. CRSGF. 105. vols. 64, 65, 66.
14. San Jacopo. vol. 23, fol. 150v: "Fece l'ofitio del pane e 4 anni quello della spetieria nel quale avea particolare intendere e se ben non sapeva leggere, conosceva e intendeva tutte le cose di detto ofitio che doppo fu uscita, tutte le spetiale monache sempre la chiamavano e non mai a rieno fatto un lattovario o cosa d'importanza senza che lei vi fussi stata presente."
15. BNCF. Ms. Landau Finlay. 72, fols. 90v–92r; CRSGF. 106. vol. 12, fol. 2r/v.
16. Guglielmo Di Agresti, *S. Caterina de' Ricci: Cronache, diplomatica, lettere varie* (Florence: L.S. Olschki, 1969), 28, 29, 32, 88; Nicholas Terpstra, "Mothers, Sisters, and Daughters: Girls and Conservatory Guardianship in Late Renaissance Florence," *Renaissance Studies* 17 (2003): 201–229.
17. A. Nannizzi, "L'arte degli speziali in Siena," *Bullettino senese di storia patria* 46 (1939): 93–131, 214–260; Leonardo Colapinto, "L'arte degli speziali a Roma nel Rinascimento," *AMAISF* 12 (1995): 59–63.
18. BNCF. Ms. Landau Finaly. 72, fol. 91r/v: "le suore istesse fra di loro determinorono non volergli conferire altri ofizzi del monastero, che quello di speziala quantunque capacissima per il suo buono iudicio e grande spirito."
19. BNCF. Ms. Landau Finaly. 72, fol. 91r/v. On Gondi charitable connections, see Terpstra, "Mothers," 209.

20. Following Raffaella Gondi's death in 1599, Sister Maria Cleofe succeeded her as head pharmacist. Gostanza Gondi died in 1601. CRSGF. 106. vol. 2, fol. 116v; vol. 36, fols. 8v–9r.
21. BNCF. Ms. Landau Finaly. 72, fol. 91r/v; BNCF. II. II. 509, fols. 150r, 161v–162r.
22. BNCF. II. II. 509, fols. 149v–150r, 151v, 152r.
23. Pulinari, *Cronache,* 258–264, quotes at 262–263.
24. CRSGF. 100. vol. 90, bks. 12, 14, 15; Giuseppe Richa, *Notizie istoriche delle chiese fiorentine* (Florence: Albizziniana, 1754–1762), 7: 51; Enrica Viviani della Robbia, *Nei monasteri fiorentini* (Florence: Sansoni, 1946), 21.
25. *Il Ricettario medicinale necessario a tutti i Medici e Speziali* (Florence: Giunti, 1567), 2: "Il buono speziale debbe essere d'ingegno, e di corpo destro, di buoni costumi, non avaro, diligente, fedele, esercitato da giovane nella cognizione delle medicine semplici e delle composte, haver cercato tutti i luoghi atti a produrre l'herbe, e l'altre medicine, che nascono nel nostro paese: sapere tanto della lingua Latina, che egli possa leggere Dioscoride, Galeno, Plinio, Serapione, Mesuè, Avicenna e gl'altri, che parlano della materia dello Speziale, ò vero non ne sapendo, debbe essere instruito da uno intelligente maestro & esercitarsi in leggere i moderni, i quali hanno tradotto, ò scritto in tal materia in lingua volgare."
26. Cristina Bellorini, *The World of Plants in Renaissance Tuscany: Medicine and Botany* (Burlington, VT: Ashgate, 2016), 87–122; Paolo Luzzi and Fernando Fabbri, "I tre orti botanici di Firenze," in *I giardini dei semplici e gli orti botanici della Toscana,* ed. Sara Ferri and Francesca Vannozzi (Perugia: Quattroemme, 1993), 49–68; Paula Findlen, "The Formation of a Scientific Community: Natural History in Sixteenth-Century Italy," in *Natural Particulars: Nature and the Disciplines in Renaissance Europe,* ed. Anthony Grafton and Nancy Siraisi (Cambridge, MA: Harvard University Press, 1999), 369–400.
27. Baudot, *Études historiques,* 334, plate IX.
28. Diodata Malvasia, *Writings on the Sisters of San Luca and Their Miraculous Madonna,* ed. and trans. Danielle Callegari and Shannon McHugh (Toronto: University of Toronto Press, 2015), 91.
29. Elissa Weaver, entry on Fiammetta Frescobaldi, http://www.lib.uchicago.edu/efts/IWW/BIOS/A0160.html, accessed April 2018.
30. Elaine Leong, "'Herbals she peruseth': Reading Medicine in Early Modern England," *Renaissance Studies* 28 (2014): 556–578.
31. Victoria Primhak, "Women in Religious Communities: The Benedictine Convents in Venice, 1400–1550," PhD dissertation, University of London, 1991.
32. Carmela Compare, "Inventari di biblioteche monastiche femminili alla fine del XVI secolo," *Genesis: rivista della società italiana delle storiche* 2 (2003): 220–232, at 226; Danilo Zardin, "Libri e biblioteche negli ambienti mo-

nastici dell'Italia del primo Seicento," in *Donne, filosofia e cultura nel Seicento,* ed. Pina Totaro (Rome: Consiglio nazionale delle ricerche, 1999), 347–383, at 376–377. Gianna Pomata, "Medicina delle monache. Pratiche terapeutiche nei monasteri femminili di Bologna in età moderna," in *I monasteri femminili come centri di cultura fra Rinascimento e Barocco,* ed. Gianna Pomata and Gabriella Zarri (Rome: Edizioni di storia e letteratura, 2005), 331–363, discusses convent medical texts and health manuals at 333–334.

33. Zardin, "Libri e biblioteche," 376–377; Guardaroba Medicea. vol. 463, fol. 1v.
34. Zardin, "Libri e biblioteche," 376–377.
35. Monica Green, "Books as a Source of Medical Education for Women in the Middle Ages," *Dynamis* 20 (2000): 331–369.
36. Smith, "Making Things," 178–179.
37. *Nuovo receptario composto dal famossisimo chollegio degli ex.mi dottori della arte et medicina della inclita cipta di Firenze* (reprinted Florence: L.S. Olschki, 1968), bk. 1, seconda dottrina; Teresa Huguet-Termes, "Standardising Drug Therapy in Renaissance Europe? The Florence (1499) and Nuremberg Pharmacopoeia (1546)," *Medicina e storia* 8 (2008): 77–101.
38. Mattioli's pamphlet was titled *Del modo di distillare le acque da tutte le piante* (Venice: Valgrisi, 1568). Prospero Borgarucci, *La fabrica de gli spetiali, partita in dodici distintioni* (Venice: Vincenzo Valgrisio, 1566).
39. *Ricettario medicinale* (1567), 6, 8, 97.
40. Sandra Cavallo and Tessa Storey, *Healthy Living in Late Renaissance Italy* (Oxford: Oxford University Press, 2013), 25–26.
41. Leong, "'Herbals she peruseth.'"
42. Pulinari, *Cronache,* 263: "in un armadio in spezieria."
43. Steven Shapin, *A Social History of Truth: Civility and Science in Seventeenth-Century England* (Chicago: University of Chicago Press, 1994); Deborah E. Harkness, *The Jewel House: Elizabethan London and the Scientific Revolution* (New Haven, CT: Yale University Press, 2007).
44. OSP. vol. 771, fols. 88r, 112r.
45. CRSGF. 105. vol. 66, fols. 167v–168r, 225v–226r; vol. 67, fols. 26v–27r.
46. Sandra Cavallo, *Artisans of the Body in Early Modern Italy: Identities, Families and Masculinities* (Manchester, UK: Manchester University Press, 2007).
47. Auditore. vol. 4892, fols. 33r/v, 95r–96v, 141r.
48. CRSGF. 105. vol. 31, fol. 75r; vol. 33, fol. 77r; vol. 65, fols. 84v–85r, 341v–342r.
49. Sister Orsola and Fra Agostino (b. 1541, secular name Pietro) were the offspring of Francesco Del Riccio. Since she was already head pharmacist at the Annalena by 1558, when her brother took monastic vows, most likely she was the older sibling.
50. Suzanne Butters, *The Triumph of Vulcan: Sculptors' Tools, Porphyry, and the Prince in Ducal Florence* (Florence: L.S. Olschki, 1996), I: 220–223; Marco

Beretta, "Material and Temporal Powers at the Casino di San Marco (1574–1621)," in *Laboratories of Art: Alchemy and Art Technology from Antiquity to the 18th Century,* ed. Sven Dupré (Berlin: Springer, 2014), 129–156, at 144–145.

51. Auditore. vol. 4892, fol. 141r (S. Orsola); CRSGF. 81. vol. 8, fols. 151v, 276r (Murate).
52. *Ricettario medicinale* (1567), unpaginated front matter. Bettini's appointment as convent operaio is recorded in Auditore. vol. 4896, fol. 513r/v.
53. *S. Caterina de' Ricci,* 34–35, 85, 88, 104. CRSGF. 106. vol. 12, fols. 3v, 81r, 133v; Rosselli del Turco. vol. 16, fols. 54v–55r; vol. 26, fol. 39r. Maria Angela di Piero Bettini (secular name Caterina) took the habit on 19 May 1521; Silvestro Bardazzi and Eugenio Castellani, *Il monastero di S. Vincenzo in Prato* (Prato: Edizioni del Palazzo, 1982), 30.
54. CRSGF. 108. vol. 44, fols. 5v, 15r.
55. AAF. Religiosi e religiose. Professioni di monache. filza 1, unfol., under 25 Oct. 1577.
56. Guardaroba Medicea. vol. 1166 bis.
57. Carlo Borromeo, *Instructiones fabricae et suppellectilis ecclesiasticae* (1577): *A Translation with Commentary and Analysis,* ed. Evelyn C. Voelker (Ann Arbor: University Microfilms International, 1993), bk. I, ch. 33. http://www.evelynvoelker.com. Borromeo also recommended that the pharmacy have its own courtyard and well.
58. Evelyn Welch, *Shopping in the Renaissance: Consumer Cultures in Italy, 1400–1600* (New Haven, CT: Yale University Press, 2005), 151–158.
59. LCF. vol. 1570, fol. 42 bis r.
60. Leonardo Colapinto, "Monachesimo e spezierie conventuale in Italia tra XII al XVII secolo," *AMAISF* 12 (1995): 107–113.
61. *La spezieria di San Benedetto a Montefiascone. Dalle collezioni di Palazzo Venezia in Roma,* ed. Maria Selene Sconci and Romualdo Luzi (Ferrara: Belriguardo, 1994).
62. In the 1570s and 1580s, the San Marco friars commissioned new pharmacy jars depicting Dominican saints such as Peter Martyr and the recently canonized Antoninus. Several vessels are illustrated in *Catalogo delle Maioliche. Museo Nazionale di Firenze. Palazzo del Bargello,* ed. Giovanni Conti (Florence: Centro Di, 1971), nos. 78, 83, 85.
63. K. J. P. Lowe, *Nuns' Chronicles and Convent Culture in Renaissance and Counter-Reformation Italy* (Cambridge, UK: Cambridge University Press, 2003), 337–338; BNCF. II. II. 509, fol. 121v: "dove si ordinava le compositioni de' lattovari et medicine."
64. Alisha Rankin, "Exotic Materials and Treasured Knowledge: The Valuable Legacy of Noblewomen's Remedies in Early Modern Germany," *Renaissance Studies* 28 (2014): 533–555; Leong, "'Herbals she peruseth.'"

65. AAB. Monache in città, busta 261, unfol.: "molte volte si va pericolo di cadere, come ci è accaduto più volte."
66. BNCF. II. II. 509, fols. 139v–140r: "fu necessario fare una stanza in alto et accomodarle sicure."
67. BNCF. II. II. 509, fols. 141r–142r.
68. BNCF. II. II. 509, fol. 141v: "calculare e' lattovari et altre cose medicinali"; CRSGF. 81. vol. 8, fols. 21v, 125v.
69. CRSGF. 81. vol. 8, fol. 264v: "dalla vendita di lattovari, di stillature d'acque et simile per un anno"; BNCF. II. II. 509, fols. 141v–142r: "con molta sua industria et diligentia."
70. Giuseppe Franchi, "Apparecchi e metodi per 'lambiccare' secondo Mattioli," in *I giardini dei semplici,* 201–204. See R. Burr Litchfield, *Florence Ducal Capital,* para. 267 on the development of the local glass industry from 1561 to 1642.
71. BLF. Ms. San Marco. vol. 903, fol. 134r: "ci sono poi di molti vetri cioè boccie orinali ampolline coppelli storte et recipienti di più sorte che di questi non se ne può dar conto: perché ogni dì se ne rompe et ogni dì se ne compra."
72. BNCF. II. II. 509, fol. 141r/v: "disse che l'herbe stillate nel vetro conferivono assai più a' corpi di quelle stillate nel piombo."
73. Franchi, "Apparecchi," 203.
74. This image is published and briefly described by Catherine Monbeig Goguel, "Cristofano Allori, 1577–1621," *Prospettiva* 39 (1984): 76–80. The portrait is not listed in any studies of Gherardini's oeuvre, but is signed and dated on the book pictured in the lower left quadrant. My thanks to Sheila Barker for calling this image to my attention.
75. Archivio Provinciale del Convento di S. Marco, Florence. vol. 53, fol. 2v.
76. Among these portraits was a likeness of the apothecary Benedecta Smidt (d. 5 Feb. 1623). The inscription in the upper-left-hand corner noted that she had served the hospital pharmacy for forty years prior to her death at age sixty-five. Evelien Vanden Berghe, "The Pharmacy of the St. John's Hospital in Bruges," http://farmasihistorie.com/w/index.php?title=The_pharmacy_of_the_St._John%E2%80%99s_hospital_in_Bruges, accessed April 2019.
77. Erika Monahan, "Locating Rhubarb: Early Modernity's Relevant Obscurity," in *Early Modern Things,* 227–251.
78. Pamela O. Long, *Artisan/Practitioners and the Rise of the New Sciences, 1400–1600* (Corvallis: Oregon State University Press, 2011).
79. Elaine Leong and Alisha Rankin, "Introduction: Secrets and Knowledge," in *Secrets and Knowledge in Medicine and Science, 1500–1800,* ed. Elaine Leong and Alisha Rankin (Aldershot, UK: Ashgate, 2011), 1–20; William Eamon, *The Professor of Secrets: Mystery, Medicine and Alchemy in Renaissance Italy* (Washington, DC: National Geographic Society, 2010).

80. William Eamon, *Science and the Secrets of Nature: Books of Secrets in Medieval and Early Modern Culture* (Princeton, NJ: Princeton University Press, 1994).
81. Pamela O. Long, "Invention, Authorship, 'Intellectual Property,' and the Origin of Patents: Notes toward a Conceptual History," *Technology and Culture* 32 (1991): 846–884; Luca Molà, "Inventors, Patents and the Market for Innovations in Renaissance Italy," *History of Technology* 32 (2014): 7–34; Francesca Trivellato, "Guilds, Technology and Economic Change in Early Modern Venice," in *Guilds, Innovation, and the European Economy, 1400–1800,* ed. Steven R. Epstein and Maarten Prak (Cambridge, UK: Cambridge University Press, 2008), 199–231.
82. Tessa Storey, "Face Waters, Oils, Love Magic and Poison: Making and Selling Secrets in Early Modern Rome," in *Secrets and Knowledge,* 143–163.
83. Eamon, *Science and the Secrets of Nature,* 164–165. Jo Wheeler maintains that Giovanni Bariletto, Curtio Troiano di Navò, and Mario Caboga supervised its publication. http://renaissancesecrets.blogspot.com/2013/04/.
84. Meredith K. Ray, *Daughters of Alchemy: Women and Scientific Culture in Early Modern Italy* (Cambridge, MA: Harvard University Press), 73–93.
85. Jo Wheeler, *Renaissance Secrets: Recipes and Formulas* (London: V & A, 2009), 13; Montserrat Cabré, "Women or Healers? Household Practices and the Categories of Health Care in Late Medieval Iberia," *BHM* 82 (2008): 18–51.
86. Jane Stevens Crawshaw, "Families, Medical Secrets and Public Health in Early Modern Venice," *Renaissance Studies* 28 (2014): 597–618. Similarly, Gianna Pomata, "Practicing between Earth and Heaven: Women Healers in Seventeenth-Century Bologna," *Dynamis* 19 (1999): 119–143, cites examples of women holding state patents on secrets.
87. Semidea Poggi (secular name Ginevra) was the daughter of Countess Lodovica Pepoli and Cristoforo di Cristoforo Poggi. Virginia Cox, *The Prodigious Muse: Women's Writing in Counter-Reformation Italy* (Baltimore, MD: Johns Hopkins University Press, 2011), 266, provides a short biography. Semidea's uncle, the future cardinal Giovanni Poggi, began building the family palace in 1549. Poggi's mother Lodovica Pepoli was among the noble widows praised in verse by Lodovico Frati, *Poesia in lode di alcune dame vedove bolognesi (1615)* (Bologna: Nicola Zanichelli, 1889), 19.
88. Caroline P. Murphy, "In Praise of the Ladies of Bologna: The Image and Identity of the Sixteenth-Century Bolognese Female Patriciate," *Renaissance Studies* 31 (1999): 440–454.
89. Cox, *Prodigious Muse,* 266, states that Poggi became a canoness in 1579 along with her sister, who took the religious name Cleria (or Clelia). However, Gian Lodovico Masetti Zanini, "Una canonichessa erborista: Semidea Poggi (sec. XVI–XVII)," *Strenna storica bolognese* 45 (1995): 395–401, places Poggi's entrance date around 1583.

90. *Abiti e lavori delle monache di Bologna in una serie di disegni del secolo XVIII,* ed. Mario Fanti (Bologna: Tamari, 1972), 48: "Di cottogne fan gelo delicate da dame e cavaglieri assai stimato"; Pomata, "Medicina delle monache," 333.
91. Bartolomeo Scappi, *Opera* (Venice: Michele Tramezzino, 1570), 391, 393–396.
92. The following paragraph draws on Craig Monson, *Nuns Behaving Badly: Tales of Music, Magic, Art and Arson in the Convents of Italy* (Chicago: University of Chicago Press, 2010), 25–62.
93. Virginia Cox, *Women's Writing in Italy, 1400–1650* (Baltimore, MD: Johns Hopkins University Press, 2008), 212, 214, 365n. Cox maintains that "some of her verse hints at periods of depression or spiritual crisis"; *Lyric Poetry by Women of the Renaissance* (Baltimore, MD: Johns Hopkins University Press, 2013), 398.
94. Masetti Zannini, "Una canonichessa erborista," 397.
95. This phrase is borrowed from Suzanne Butters, "The Use and Abuse of Gifts in the World of Ferdinando de' Medici (1549-1609)," *I Tatti Studies in the Italian Renaissance* 11 (2007): 243-354, at 245.
96. Rankin, *Panaceia's Daughters,* 39–41.
97. Mazetti Zanini, "Una canonichessa erborista," quote at 397: "Il rimedio è sicuro ed è questo: faccia lei raccogliere il mese di ottobre la semenza di viole zoppe, quelle che spontano fuora la primavera, le semenze sono in certe vesighete verde, le quali vesiche ánno drento le dette semenze sono simile a queste che li mando per mostra le quali senne furno raccolte, avertisca farne raccogliere in grandissima quantità e la sera le faccia porre in infusione in vino o in brodo un cucchiaio colmo, la mattina poi se le magna e non dubita nulla che con l'aiuto del Signore Iddio se spezzaranno le pietre e le orinarà senza dolore di sorte alcuna e restarà libero risanato affatto' non creda a medici che le faranno danno in contra di aiuto."
98. Two Paracelsian remedies for stone developed by Don Antonio de' Medici used these ingredients; Filippo Luti, "Don Antonio de' Medici 'professore de' secreti,'" *Medicea* 1 (2008): 34–47.
99. Florike Egmond, "Apothecaries as Experts and Brokers in the Sixteenth-Century Network of the Naturalist Carolus Clusius," *History of Universities* 23 (2008): 59–91.
100. Gian Lodovico Mazetti Zanini, *Motivi storici della educazione femmnile. Scienza, lavoro, giuochi* (Naples: D'Auria, 1981), 372, dated 31 Oct. 1621.
101. Leong, "'Herbals she peruseth.'"
102. Moran, *Distilling Knowledge,* 111.
103. Stefano Rosselli, *Mes Secrets: A Florence au temps des Médicis, 1593: patisserie, parfumerie, medicine,* ed. Rodrigo de Zayas (Paris: J.-M. Place, 1996).
104. Butters, *Triumph of Vulcan,* I: 219.
105. Auditore. vol. 4892, fols. 151r–153r; Strocchia, *Nuns and Nunneries,* 20–21.

106. Suzanne G. Cusick, *Francesca Caccini at the Medici Court: Music and the Circulation of Power* (Chicago: University of Chicago Press, 2009), 54–59. The four Medici princesses were Caterina di Giulio, discussed below, and Leonora, Maria, and Giovanna, the illegitimate daughters of Don Piero de' Medici and his Spanish mistress, Antonia de Carvajal.
107. Caterina's birth date is unknown. Her father Giulio was the illegitimate son of Duke Alessandro; her mother likely was Giulio's wife Lucrezia Gaetani of Pisa.
108. CRSGF. 81. vol. 9, fols. 39r, 42v, 45v, 51v.
109. CRSGF. 81. vol. 27, fol. 62v.
110. Pierfilippo Covoni, *Don Antonio de' Medici al Casino di San Marco* (Florence: Tipografia cooperativa, 1892); Filippo Luti, *Don Antonio de' Medici e i suoi tempi* (Florence: L.S. Olschki, 2006); Beretta, "Material and Temporal Powers."
111. Antonio's correspondence is conserved as MdP. vols. 5128 and 5130. Molly Bourne discusses the medical and botanical interests of these Medici women in "Medici Women at the Gonzaga Court, 1584–1627," in *Italian Art, Society, and Politics. A Festschrift in Honor of Rab Hatfield,* ed. Barbara Deimling, Jonathan K. Nelson, and Gary Radke (Syracuse, NY: Syracuse University Press, 2007), 223–243.
112. *Apparato della Fonderia dell'Illustrissimo et Eccellentissimo Sig. D. Antonio Medici:* BNCF. Magl. XVI. 63. Bellorini discusses the Paracelsian content of these manuscripts in *The World of Plants,* 165–166.
113. Luti, "Don Antonio de' Medici," 37.
114. *La fonderia dell'Ill.mo et Ecc.mo Sig. Don Antonio Medici Principe di Capistrano, etc. Nella quale si contiene tutta l'arte spagirica di Teofrasto Paracelso, & sue medicine, etc.* Florence: Stampata nel Palazzo del Casino di S.E. Illustrissima, 1604.
115. *Segreti sperimentati dall'Illustrissimo et Eccellentissimo Sig.r Principe D. Antonio de Medici nella sua Fonderia del Casino:* BNCF. Magl. XV. 140.
116. Luti, "Don Antonio de' Medici," 37.
117. BNCF. Magl. XV. 140, fols. 5v–6r, 13r/v; Luti, "Don Antonio de' Medici," n43.
118. Beretta, "Material and Temporal Powers," 147.
119. Wellcome Library. Ms. 485. "Segreti di Don Antonio de' Medici principe di Toscana esperimentati di sua propria mano tutti verissimi dati, e donati a me Caterina Medici nel Convento delle Murate di Firenze, l'anno 1598." Storey, "Face Waters," 150, mentions this volume but was unable to identify its recipient.
120. Sheila Barker, "Christine of Lorraine and Medicine at the Medici Court," in *Medici Women: The Making of a Dynasty in Grand Ducal Tuscany,* ed. Giovanna Benadusi and Judith C. Brown (Toronto: University of Toronto Press, 2015), 155–181.

121. Wellcome Library. Ms. 485, fols. 209r/v, 216r–217r: "l'ambra sia tartufata, non nera, e che habbia odor soave, perché la nera ha troppo dell'hodore del zibetto e svapora presto."
122. Wellcome Library. Ms. 485, fols. 166r–168r, 169r–176r, 213v–216r.
123. Moran, *Distilling Knowledge,* 46.
124. Wellcome Library. Ms. 485, third leaf: "Trattato copioso diviso in due parti, nella prima si insegna il modo da fare tutti i lisci senza offesa delle carni et ogni acqua, olio, et altre sorte di belletti per far bianche, morvide, e belle le carni, et i lisci per i capelli, barba e peli."
125. Wellcome Library. Ms. 485, fols. 1r/v, 6r, 29r–31r.
126. Wellcome Library. Ms. 218, unnumbered first folio: "fa la carne polita et chiara com'un[o] specc[h]io."
127. Gentilcore, *Medical Charlatanism,* 206. Skin products also "represented the most important single remedy category amongst medicinal recipes" in the books of secrets studied by Storey, "Face Waters," 154.
128. Will Fischer, "The Renaissance Beard: Masculinity in Early Modern England," *RQ* 54 (2001): 155–187; Douglas Biow, *On the Importance of Being an Individual in Renaissance Italy: Men, Their Professions, and Their Beards* (Philadelphia: University of Pennsylvania Press, 2015), 181–224. See Wellcome Library. Ms. 5139, no. 1, travel pass dated 22 June 1630.
129. Carte Bardi. ser. 2. vol. 40, fol. 4v: "a fare la barba e capelli neri o castagnuoli"; Wheeler, *Renaissance Secrets,* 35; BMF. Cod. C. CXLV, fol. 44r; Wellcome Library. Ms. 485, fols. 137r–150v, 158v–162r.
130. Wellcome Library. Ms. 485, third leaf: "Nella seconda [parte] si insegnano i modo di fare l'acque et olii, e balsami, odoriferi, profumi, paste odorifere, manteche, pomate, grassetti odoriferi, polveri profummate di cipri, e di monpolieri di varii colori, guancioletti alla lucchese, saponetti odoriferi, e conce di guancie d'ogni sorte e colori e tutte le saponi muschiati teneri e sodi all'uso di Bologna."
131. Heiki Mikkeli, *Hygiene in the Early Modern Medical Tradition* (Helsinki: Finnish Academy of Science and Letters, 1999), 58.
132. Wheeler, *Renaissance Secrets,* 33; Giovanventura Rosetti, *Notandissimi secreti de l'arte profumatoria* (Venice: Francesco Rampazetto, 1555).
133. Ivan Day, *Pomanders, Washballs and Other Scented Articles* (London: The Herb Society, n.d.), 20–21.
134. Strocchia, *Nuns and Nunneries,* 117–126.
135. Wheeler, *Renaissance Secrets,* 33.
136. Wellcome Library. Ms. 218, unfol.
137. Christine's interest in chemical remedies is discussed by Barker, "Christine of Lorraine."
138. Barker, "Christine of Lorraine."
139. Caterina Eletta died 4 Nov. 1634; CRSGF. 81. vol. 31, fols. 89r, 110r. Cusick, *Francesca Caccini,* 51–54, describes the turmoil wracking the Murate.

140. See for example MdP. vol. 6113, fol. 542, dated 6 Feb. 1628.
141. Miscellanea Medicea. vol. 30, ins. 25, fols. 1r, 4r, dated 8 and 9 July 1614.
142. Antonio Zobi, *Manuale storico degli ordinamenti economici vigenti in Toscana* (Florence: Accademia de'Georgofili, 1858), 408.
143. MdP. vol. 1351, fol. 156, dated 31 March 1613: "del quale dui podagrosi ne sono subito stati guariti, cioè ne cessò in sei hore il dolore a fatto."
144. MdP. vol. 1350, fol. 77, dated 19 April 1613: "Et quando di nuovo gli verranno a far'dolore havendo dismesso il Cerotto, bisognerà reiterare l'istesso medicamento et modo, che non molto tempo seguira poi che la se ne vedra a fatto liberata, si come è arrivato in me et in persone di molte altri che ancora loro hanno usato il medesimo rimedio."
145. MdP. vol. 1350, fol. 89, dated 20 April 1613; fol. 189, dated 11 May 1613.
146. MdP. vol. 1704, dated 1 Aug. 1618: "Mi ha prima risposto che ne scriverebbe al Signore Capitano et acciò lei uscisse di sospetto. Io le ho poi alla fine mostrato la ricetta et pregatala che la riscontrassimo con la sua. M'ha risposto che lei non ha ricetta del fare il cerotto, ma si bene che il monasterio ha fatto stampare gli usi del cerotto, et dell'estratto o bollitura per potere distribuire a chi danno il cerotto accio possino fare l'estratto da l'oro per non dare spesa al monasterio del grecho et brigha che vi occorre; et che il cerotto lo fecie fare il capitano allo spetiale."
147. See for instance OSMN. vol. 194, no. 65, dated 27 March 1568.
148. *Collezionismo mediceo e storia artistica,* vol. 1: tomo 1, *Da Cosimo I a Cosimo II, 1540–1621,* ed. Paola Barocchi and Giovanna Gaeta Bertelà (Florence: Studio per edizioni scelte, 2002), 475.
149. AAB. Monache in città, busta 261, undated, convent of Corpus Domini.
150. MdP. vol. 1704, dated 1 Aug. 1618: "credo che lei senz'altro habbia la ricetta del cerotto, et nella ricetta delli usi, hanno intitolato il monasterio e non il capitano, ma non so se lo vendono o se lo donano. [. . .] se io ho a dire il vero all'orrecchio di Vostra Signoria, non stimo nulla questo suo segreto inpirico, et alla giornata suanirà come suani presto il medesimo cerotto due anni sono che ne predicava miracoli per levare la lagrimatione et infiamatione et il dolore da denti che mai fecie effetto buono a nessuno, et hora ha trovato nuovo modo di servirsene a diversi mali in diversi modi per il cio a dio che facci anche maggiore effetti che lui dicie."
151. The special issue of *BHM* 91 (2017), edited by Elaine Leong and Alisha Rankin, is devoted to early modern drug trials.
152. Alisha Rankin, "On Anecdote and Antidotes: Poison Trials in Sixteenth-Century Europe," *BHM* 91 (2017): 274–302.
153. MdP. vol. 1704, dated 21 Nov. 1619.

CHAPTER FIVE ~ RESTORING HEALTH

1. Katharine Park and John Henderson, "'The First Hospital among Christians': The Ospedale di Santa Maria Nuova in Early Sixteenth-Century Florence," *Medical History* 35 (1991): 164–188.
2. Peregrine Horden, "'A Discipline of Relevance': The Historiography of the Later Medieval Hospital," *Social History of Medicine* 1 (1988): 359–374; Paolo Savoia, "The *Book of the Sick* of Santa Maria della Morte in Bologna and the Medical Organization of a Hospital in the Sixteenth Century," *Nuncius* 31 (2016): 163-235.
3. R. Burr Litchfield, *Florence Ducal Capital, 1530–1630* (ACLS e-book, 2008), para. 131.
4. Brian Pullan, *Tolerance, Regulation and Rescue: Dishonoured Women and Abandoned Children in Italy, 1300–1800* (Manchester, UK: Manchester University Press, 2016); Nicholas Terpstra, *Lost Girls: Sex and Death in Renaissance Florence* (Baltimore, MD: Johns Hopkins University Press, 2010).
5. The vast literature on Italian hospitals tends to be local or regional in nature. Francesco Bianchi, "Italian Renaissance Hospitals: An Overview of the Recent Historiography," *Mitteilungen des Instituts für Österreichische Geschichtsforschung* 115 (2007): 394–403, provides a good general survey. Several studies have explored Renaissance hospital architecture and visual programs in relation to their spiritual objectives, since these institutions aimed at healing both body and soul. See *La bellezza come terapia. Arte e assistenza nell'ospedale di Santa Maria Nuova a Firenze*, ed. Enrico Ghidetti and Esther Diana (Florence: Polistampa, 2005), and John Henderson, *The Renaissance Hospital: Healing the Body and Healing the Soul* (New Haven, CT: Yale University Press, 2006).
6. On the educational role of Islamic hospitals, see Ahmed Ragab, *The Medieval Islamic Hospital: Medicine, Religion, and Charity* (Cambridge, UK: Cambridge University Press, 2015). Henderson, *Renaissance Hospital,* 232, notes that senior physicians at the Florentine civic hospital of Santa Maria Nuova advised junior colleagues as part of an apprenticeship arrangement, but this was not part of the formal curriculum. Paolo Savoia shows how a surgeon's artisanal know-how could be integrated into learned medical culture, despite status differences between hospital practitioners, in "Skills, Knowledge, and Status: The Career of an Early Modern Italian Surgeon," *BHM* 93 (2019): 27-54, quote at 36.
7. Susan E. Dinan, *Women and Poor Relief in Seventeenth-Century France: The Early History of the Daughters of Charity* (Aldershot, UK: Ashgate, 2006); Nicholas Terpstra, "Making a Living, Making a Life: Work in the Orphanages of Florence and Bologna," *Sixteenth Century Journal* 31 (2000): 1063–1079. Initially the Daughters of Charity visited house-bound patients and only embarked upon hospital nursing in 1640.

8. Mary E. Fissell, "Introduction: Women, Health, and Healing in Early Modern Europe," *BHM* 82 (2008): 1–17.
9. Jon Arrizabalaga, John Henderson, and Roger French, *The Great Pox: The French Disease in Renaissance Europe* (New Haven, CT: Yale University Press, 1997), 1–3; Claude Quétel, *History of Syphilis,* trans. Judith Braddock and Brian Pike (Baltimore, MD: Johns Hopkins University Press, 1990), 11–16.
10. Jon Arrizabalaga, "Medical Responses to the 'French Disease' in Europe at the Turn of the Sixteenth Century," in *Sins of the Flesh: Responding to Sexual Disease in Early Modern Europe,* ed. Kevin Siena (Toronto: University of Toronto Press, 2005), 33–55.
11. This phrase comes from Steven Shapin, *The Social History of Truth: Civility and Science in Seventeenth-Century England* (Chicago: University of Chicago Press, 1994).
12. Anna Foa, "The New and the Old: The Spread of Syphilis (1494–1530)," in *Sex and Gender in Historical Perspective,* ed. Edward Muir and Guido Ruggiero (Baltimore, MD: Johns Hopkins University Press, 1990), 26–45.
13. Quétel, *History of Syphilis,* 11–16. In this chapter I use the terms French disease, pox, and syphilis interchangeably.
14. Quétel, *History of Syphilis,* 66–67.
15. Pullan, *Tolerance,* 48–56; Laura J. McGough, *Gender, Sexuality, and Syphilis in Early Modern Venice: The Disease that Came to Stay* (Basingstoke: Palgrave, 2011), 45–70; and Winfried Schleiner, "Moral Attitudes towards Syphilis and Its Prevention in the Renaissance," *BHM* 68 (1994): 389–410.
16. Arrizabalaga, Henderson, and French, *Great Pox,* 145–170. German pox hospitals have been studied by Robert Jütte, "Syphilis and Confinement: Hospitals in Early Modern Germany," in *Institutions of Confinement: Hospitals, Asylums, and Prisons in Western Europe and North America, 1500–1950,* ed. Norbert Finzsch and Robert Jütte (Cambridge, UK: Cambridge University Press, 1996), 97–115. Kevin Siena discusses the piecemeal approach to institutional care for London pox patients in "The Clean and the Foul: Paupers and the Pox in London Hospitals, c. 1550–c. 1700," in *Sins of the Flesh,* 261–284.
17. Cristian Berco, "The Great Pox, Symptoms, and Social Bodies in Early Modern Spain," *Social History of Medicine* 28 (2014): 225–244.
18. Quétel, *History of Syphilis,* 9–32; Siena, "Clean and the Foul"; MSS. vol. 173, no. 5, unfol.
19. Francesco Guicciardini, *The History of Italy,* trans. Sidney Alexander (London: Macmillan, 1969), 109.
20. MSS. vol. 173, no. 5, unfol.: "falsi unguenti."
21. Luca Landucci, *Diario fiorentino dal 1450 al 1516* (Florence: G.C. Sansoni, 1883), 132, reports the appearance in May 1496 of "una certa infermità, che le chiamavano bolle franciose, ch'erano come un vagiuolo grosso."

22. Arrizabalaga, Henderson, and French, *Great Pox,* 160–162. Incurabili. vol. 102, fols. 45v–46r; Sharon T. Strocchia, "The Nun Apothecaries of Renaissance Florence: Marketing Medicines in the Convent," *Renaissance Studies* 25 (2011): 627–647.
23. MSS. vol. 173, no. 5, unfol.: "diligentemente nutriti, curati e medicate finche guarissero."
24. Andrea Nordio, "Presenze femminili nella nascita dell'ospedale degli Incurabili di Venezia," *Regnum dei* 120 (1994): 11–39; Guido Donatone, *La farmacia degli Incurabili e la maiolica napoletana del Settecento* (Naples: Edizioni del Delfino, 1972), 11.
25. Incurabili. vol. 102, fol. 19r. Between 1522 and 1529, the confraternity collected 1089 lire in dues from women, compared to 839 lire from men.
26. Incurabili. vol. 59, fol. 13v, dated March 1550.
27. Luigi Passerini, *Storia degli stabilimenti di beneficenza e d'istruzione elementare gratuita della città di Firenze* (Florence: Le Monnier, 1853), 210–211.
28. Jütte, "Syphilis and Confinement," 101.
29. Vivian Nutton, "The Reception of Fracastoro's Theory of Contagion: The Seed that Fell Among Thorns?," *Osiris,* ser. 2, vol. 6 (1990): 196–234.
30. Pullan, *Tolerance,* 55.
31. Pullan, *Tolerance,* 55.
32. Eunice D. Howe, "The Architecture of Institutionalism: Women's Space in Renaissance Hospitals," in *Architecture and the Politics of Gender in Early Modern Europe,* ed. Helen Hills (Aldershot, UK Ashgate, 2003), 63–82. Renaissance hospitals segregated male and female patients in one of several ways: by placing their beds along opposite walls; by using chapels to partition wards; or by constructing separate floors or wings. In contrast to Florence, the Incurabili hospitals in Rome and Venice had two parallel wards for women and men; Arrizabalaga, Henderson, and French, *Great Pox,* 185.
33. The much larger Neapolitan pox hospital, housing about 1500 patients and staff, cared for persons afflicted by pox, tuberculosis, cancer, and madness well into the seventeenth century; Giulio Cesare Capaccio, *Descrizione di Napoli ne' principii del secolo XVII* (Naples: Società di storia patria, 1882), 34.
34. Maria Luz Lopez Terrada, "El tratamiento de la sifilis en un hospital renascentista: la sala del *mal de siment* del hospital general de Valencia," *Asclepio* 41 (1989): 19–50.
35. Incurabili. vol. 1, pp. 36–37; Jütte, "Syphilis and Confinement."
36. Incurabili. vol. 1, pp. 29, 55–56.
37. Claudia Stein, *Negotiating the French Pox in Early Modern Germany* (Burlington, VT: Ashgate, 2009), 136–145; Siena, "Clean and the Foul," 269, 274. Venereal patients in London also were treated gratis until 1667, when

they became the first hospital patients in England required to pay admission fees.

38. Incurabili. vol. 2, fol. 47v, dated 29 Aug. 1596; Daniela Lombardi, "Poveri a Firenze. Programmi e Realizzazioni della Politica Assistenziale dei Medici tra Cinque e Seicento," in *Timore e Carità. I Poveri nell'Italia Moderna,* ed. Giorgio Politi, Maria Rosa, and Franco della Peruta (Cremona: Libreria del convegno, 1982), 165–184.
39. Stein, *Negotiating the French Pox,* 138–145, analyzes the rich petitions submitted by applicants for admission to the Augsburg pox hospital.
40. Montserrat Cabré, "Women or Healers? Household Practices and the Categories of Health Care in Late Medieval Iberia," *BHM* 82 (2008): 18–51, at 25n. Bernice J. Trexler, "Hospital Patients in Florence: San Paolo, 1567–68," *BHM* 48 (1974): 41–59, notes that probably twenty of San Paolo's thirty-five beds belonged to the men's ward. However, Jütte, "Syphilis and Confinement," 110, reports that the Strasbourg syphilis hospital treated more women than men in the mid-sixteenth century.
41. Incurabili. vols. 61, 62.
42. Arrizabalaga, Henderson, and French, *Great Pox,* 193.
43. Incurabili. vol. 61, fols. 4r, 5r, 6r–10r, 14v–17v, 34v–35r, 44r. Terpstra, *Lost Girls.*
44. Samuel K. Cohn Jr., *Cultures of Plague: Medical Thinking at the End of the Renaissance* (Oxford: Oxford University Press, 2010), esp. 238–263.
45. For the early history of San Paolo, see Richard A. Goldthwaite and W. R. Rearick, "Michelozzo and the Ospedale di San Paolo in Florence," *Mitteilungen des Kunsthistorischen Instituts in Florenz* 21 (1977): 221–306; Passerini, *Storia degli stabilimenti,* 163–188.
46. OSP. vol. 905, fol. 18v: "dove fussino con carità ricevuti e poveri convalescenti fino a tanto che havessero recuperate le pristine forze et sanità."
47. Nicholas Terpstra, *Abandoned Children of the Italian Renaissance: Orphan Care in Florence and Bologna* (Baltimore, MD: Johns Hopkins University Press, 2005), argues that early modern Italians regularly relied on "constructed families" that extended notions of kinship out from the biological household unit. Jane Stevens Crawshaw examines the influence of familial models on Venetian hospital practice in "Families, Medical Secrets and Public Health in Early Modern Venice," *Renaissance Studies* 28 (2014): 597–618. For the German context, see Annemarie Kinzelbach, "Women and Healthcare in Early Modern German Towns," *Renaissance Studies* 28 (2014): 619–638; and Merry Wiesner, *Working Women in Renaissance Germany* (New Brunswick, NJ: Rutgers University Press, 1986).
48. Pullan, *Tolerance,* 86–105.
49. Incurabili. vol. 1, pp. 11–12 (1521 hospital statutes): "oppressi da alcuna infermita o da morbo gallico o altra malattia incurabile excepto che di lebbra o peste."

50. Richelle Munkhoff, "Searchers of the Dead: Authority, Marginality, and the Interpretation of Plague in England, 1574–1665," *Gender & History* 11 (1999): 1–29.
51. Quétel, *History of Syphilis,* 21–22. Roger French and Jon Arrizabalaga note the perceived parallels with leprosy in "Coping with the French Disease: University Practitioners' Strategies and Tactics in the Transition from the Fifteenth to the Seventeenth Century," in *Medicine from the Black Death to the French Disease,* ed. Roger French, Jon Arrizabalaga, Andrew Cunningham, and Luis Garcia-Ballester (Aldershot, UK: Ashgate, 1998), 248–287.
52. Ceppo. vol. 1 bis, fol. 12v: "li guardino bene il capo et gola et il resto del corpo"; "et trovando si infetta di male incurabili o pericolosi di infettare le altre," then the girl was denied admission.
53. Roy Porter, "The Rise of Physical Examination," in *Medicine and the Five Senses,* ed. W. F. Bynum and Roy Porter (Cambridge, UK: Cambridge University Press, 1993), 179–197.
54. Incurabili. vol. 1, p. 31; vol. 49, fol. 36v; vol. 59, fols. 80r, 93r, 97r. The hospital employed Carlo Cortesi as surgeon from 1549 to 1551. During that period, he was ill for three months and was replaced by another practitioner. In 1552 the hospital "hired" the ducal physician Baccio Baldini, who treated patients gratis as a charitable gesture. Yet another physician had taken over by 1563.
55. Henderson, *Renaissance Hospital,* 230–238. At Augsburg pox hospitals, physicians exercised enormous authority, whereas the Florentine Incurabili vested administrative power in the hospital warden, who was a priest rather than a medical professional. Stein, *Negotiating the French Pox,* 107–145; Incurabili. vol. 1, pp. 26–28.
56. ASR. Ospedale di S. Giacomo. vol. 366, no. 41, fol. 56r. Mario Vanti, *S. Giacomo degl'Incurabili di Roma nel Cinquecento* (Rome: Pustet, 1938), remains a valuable study.
57. Incurabili. vol. 1, pp. 30–32; vol. 60, fols. 178v–179r, 180v, dated 1557, recording monthly salaries of three and a half lire, or roughly six scudi annually. Henderson, *Renaissance Hospital,* 210.
58. For changing perceptions of household service, see Diane Wolfthal, "Household Help: Early Modern Portraits of Female Servants," *Early Modern Women* 8 (2013): 5–52.
59. Incurabili. vol. 3, fol. 78r/v; Henderson, *Renaissance Hospital,* 83.
60. Their names and salaries are recorded in ASR. Ospedale di S. Giacomo. vol. 366 and 367. During April and May, when the guaiac cure was administered, the number of female and male assistants temporarily doubled; Vanti, *S. Giacomo,* 48.
61. Carole Rawcliffe, "Hospital Nurses and Their Work," in *Daily Life in the Late Middle Ages,* ed. Richard Britnell (Gloucestershire: Sutton, 1998), 43–64.

62. Anita Malamani, "L'Ospedale degli Incurabili di Pavia dalle origini al suo assorbimento nel P. L. Pertusati," *Archivio storico lombardo* 100 (1974): 145–170, at 158; Henderson, *Renaissance Hospital,* 199, 218.
63. Henderson, *Renaissance Hospital,* 217; Deborah E. Harkness, "A View from the Streets: Women and Medical Work in Elizabethan London," *BHM* 82 (2008): 52–85, notes that some Elizabethan nursing sisters were reprimanded for brawling and public intoxication.
64. Incurabili. vol. 1, pp. 33–34. This age profile most closely resembled the nursing sisters called *soeurs grises,* whose 1483 statutes required that entrants be seventeen to thirty years old. The group flourished in Flanders and northern France but was never active in Italy. Henri Lemaitre, "Statuts des religieuses du tiers ordres franciscain dites soeurs grises hospitalières (1483)," *Archivum franciscanum historicum* 4 (1911): 713–731, at 721.
65. Incurabili. vol. 1, pp. 32–33, 35; Harkness, "View from the Streets."
66. Michael Rocke, *Forbidden Friendships: Homosexuality and Male Culture in Renaissance Florence* (New York: Oxford University Press, 1996), 53.
67. Pullan, *Tolerance,* 94.
68. Incurabili. vol. 2, fol. 2r/v, dated 6 July 1570: "atteso alla gran povertà della Francesca figliuola fu (blank) da Prato e come lei si trova senza padre e madre e altri stretti parenti . . . e d'esser già d'età d'anni xvi e di buona indole e costumi." Terpstra, *Lost Girls,* notes that most girls were admitted to conservatories around age eleven or twelve.
69. Incurabili. vol. 2, fol. 2r/v: "e atteso quanto pericolo porti una così fatta fanciulla . . . accettate per suora in detto spedale e congregatione per far quanto di sopra e conservarsi la pudicitia e far vita celibe."
70. Incurabili. vol. 2, fols. 56r/v, 100r: "buona fanciulla, sana, et obediente, e dura fatica volontieri"; "essendo fanciulla grande di vita e robusta." Pullan, *Tolerance,* 180, notes cultural attitudes toward country girls.
71. Incurabili. vol. 2, fol. 32r; vol. 61, fol. 3r/v.
72. OSP. vol. 616, pp. 1467–68, dated 22 Oct. 1592: "essendo persona di poca vita e gentile di complessione, dubitiamo che non habbia a potere reggere le molte fatiche che si fanno in detta casa."
73. Incurabili. vol. 2, fols. 48r, 51r: "rimasta senza padre et abbandonata dalla madre quale si è rimaritata"; "aver visto la sopradetta Solomea sana et atta al servitio di detto nostro arciospedale." Zerbinelli's appointment to the Collegio in 1614 is recorded in OSMN. vol. 199, fol. 32r.
74. Incurabili. vol. 49, fol. 256v, dated 12 Dec. 1574: "per essere inhabile alle fatiche di questo spedale et non essere a proposito per quando se havuto informatione dalla maggior parte di tutte le altre servigiale examinate con diligentia."
75. Incurabili. vol. 2, fol. 5v: "l'oppinione loro a una per una."
76. Incurabili. vol. 1, pp. 33–34; vol. 2 notes the rotation of prioresses.

77. Incurabili. vol. 1, pp. 26–28, 33–36.
78. Incurabili. vol. 2, fols. 59r/v, 65r/v.
79. Incurabili. vol. 2, fols. 119v, 125r, 127v, 128v, 129r, 137r, 140v, 142r/v, 143v; vol. 3, fols. 3r, 5r, 8r, 9r, 10v, 11v, 13r/v, 14r.
80. Incurabili. vol. 1, pp. 39–56. Richard A. Goldthwaite, *The Economy of Renaissance Florence* (Baltimore, MD: Johns Hopkins University Press, 2009), 577, puts annual earnings of an unskilled construction worker in the 1580s at thirty-three scudi, or 231 lire. No comparable figures are available for the 1640s.
81. Incurabili. vol. 49, fols. 14r, 82v, 101r; vol. 58, fols. 60v, 102r; vol. 62, fols. 75v, 79r, 89r, 99r; vol. 68, fols. 17r, 25v. Other pharmacy receipts documenting similar medical commerce are contained in Monastero dell'Arcangelo Raffaello, vol. 14. This volume has been miscataloged; it belonged to the convent of San Luca. Spanning the years from 1652 to 1719, the register reveals the web of commercial medical exchanges between Santa Caterina da Siena, Santa Lucia, San Luca, the Incurabili, the hospital of Bonifazio, and lay apothecaries. Zerbinelli also contributed 100 scudi from personal funds in 1655 to refurbish the hospital chapel and part of the women's ward; Incurabili. vol. 3, fol. 11v.
82. Margaret Pelling, "Nurses and Nursekeepers: Problems of Identification in the Early Modern Period," in *The Common Lot: Sickness, Medical Occupations and the Urban Poor in Early Modern England* (London: Longman, 1998), 179–202.
83. BRF. Ms. 1133, fols. 48r–52r; BNCF. Magl. XV. 92, fols. 44r/v, 171v, 190v.
84. Henderson, *Renaissance Hospital,* 187–188.
85. See the 1615 hospital statutes in OSP. vol. 637, Cap. 2.
86. On the cultural ambivalence and disgust triggered by household work, see Wolfthal, "Household Help."
87. Carlo M. Cipolla, *Miasmas and Disease: Public Health and the Environment in the Pre-Industrial Age* (New Haven, CT: Yale University Press, 1992).
88. Sandra Cavallo, "Health, Air and Material Culture in the Early Modern Italian Domestic Environment," *Social History of Medicine* 29 (2016): 695–716.
89. Sandra Cavallo and Silvia Evangelisti, "Introduction," in *Domestic Institutional Interiors in Early Modern Europe,* ed. Sandra Cavallo and Silvia Evangelisti (Burlington, VT: Ashgate, 2009), 1–23.
90. OSP. vol. 616, pp. 589–90: "ripieno continuamente di puzzolente fetore."
91. Lopez Terrada, "Tratamiento de la sifilis," 26.
92. Incurabili. vol. 2, fol. 68v, dated 27 April 1604.
93. Biagio Palma, *Regole della congregatione dell'Umiltà di S. Carlo* (Rome: Giacomo Mascardi, 1629), 12: "visitar le povere donne inferme particolarmente negli Hospidali e di aiutarle quanto sarà possibile."

94. Biagio Palma, *Dialogo tra il padre rettore della congregatione dell'Humiltà di S. Carlo e la sorella della medesima congregatione* (Rome: Giacomo Mascardi, 1629), 23. This dialogue was published together with the preceding work but has a separate title and pagination. For more about this congregation, see Pamela M. Jones, *Altarpieces and Their Viewers in the Churches of Rome from Caravaggio to Guido Reni* (Aldershot, UK: Ashgate, 2008), 165–168. On the hospital as a sensorium, see Jonathan Reinarz, "Learning to Use Their Senses: Visitors to Voluntary Hospitals in Eighteenth-Century England," *Journal for Eighteenth-Century Studies* 35 (2012): 505-520.
95. Palma, *Dialogo,* 4–9: "un poco schifoso"; "l'opera mi pare molto fatigosa"; "per lo più suole rendere nausea, massime alle donne che sono naturalmente delicate;" "vermiglie e odorifere rose."
96. Jones, *Altarpieces and Their Viewers,* 169–170.
97. Incurabili. vol. 1, pp. 32–33: "E l'uffitio di tutte sia con diligentia e carità governare le donne ammalate in quel luogo, e cuocere il legno, e ogn'altra cosa che occorressi così per lo spedale degli uomini cioè per quello delle donne, e generalmente per tutti i ministri et famiglia di casa, e per tutti quelli che alloggeranno in esso."
98. Incurabili. vol. 1, p. 35; vol. 59, fol. 93v; vol.102, fol. 189r.
99. Rawcliffe, "Hospital Nurses," 62.
100. Auguste Baudot, *Études historiques sur la pharmacie en Bourgogne avant 1803* (Paris: A. Maloine, 1905), 334-338.
101. BNCF. Magl. XV. 92; OSP. vol. 771, fol. 25r; vol. 912, fols. 9v–10r; vol. 932, fols. 63v–64r. See also Rawcliffe, "Hospital Nurses," 58–59, 61.
102. AOI. vol. 4496, unfol. The role of the physician-pharmacists was to "ordinare e tener conto della spetieria e medicine e medicare secondo l'arte loro"; the nurses were tasked with "dar le medicine e governar li infermi con carita e diligentia." See also Philip Gavitt, *Gender, Honor, and Charity in Late Renaissance Florence* (Cambridge, UK: Cambridge University Press, 2011), 167–168.
103. Lucia Sandri, "Fanciulle e balie. La presenza femminile nell'Ospedale," in *Madri, figlie, balie. Il coretto della chiesa e la comunità femminile degli Innocenti,* ed. Stefano Filipponi and Eleonora Mazzocchi (Florence: Nardini, 2010), 12–23.
104. AOI. vol. 6855, capitolo 14: "che alla spetieria si tenga il libro de medici, ove dì per dì si scriva le medicine et syroppi et tutto quello che si consuma in casa et per lo spedale, che servira per vedere ove si consumano le cose, et ove vanno: oltre al libro ove si nota il venduto et comperato."
105. Gavitt, *Gender, Honor, and Charity,* 168–171.
106. Incurabili. vol. 49, contains numerous references to making remedies for ringworm as well as various distillates; vol. 102, fols. 259v–260r ("per fatura di una manaia da tagliare erbe per la medicheria"), 262v–263r, 266r.

107. Nicholas Terpstra, "Competing Visions of the State and Social Welfare: The Medici Dukes, the Bigallo Magistrates, and Local Hospitals in Sixteenth-Century Tuscany," *RQ* 54 (2001): 1319–1355.
108. OSP. vol. 616, p. 140.
109. Francisco Delicado, *El modo de adoperare el legno de India occidentale: Salutifero remedio a ogni piaga e mal incurabili* (Venice: n.p., 1525); Max H. Frisch, *Nicolas Pol, doctor, 1494, with a critical text of his guaiac tract* (New York: Herbert Reichner, 1947).
110. Incurabili. vol. 1, p. 31; MSS. vol. 173, no. 5, unfol.
111. Stein, *Negotiating the French Pox,* 147–154; Lorenzo Gaetano Fabbri, *Dell'uso del mercurio sempre temerario in medicina. Della fondazione e del medicamento dell'Arcispedale degl'Incurabili nella città di Firenze* (Cologne: Federigo Tirbien, 1749), 44.
112. Passserini, *Stabilimenti,* 209–210.
113. MSS. vol. 173, no. 5, unfol.; Incurabili. vol. 49, fols. 3v, 67r; 150r.
114. CRSGF. 111. vol. 67, unfol., under date 25 Sept. 1567: "tagliato da pezzo grosso."
115. LCF. vol. 164, pp. 15, 35.
116. Stein, *Negotiating the French Pox,* 148.
117. Goldthwaite, *Economy of Renaissance Florence,* 577.
118. For comparative data, see the detailed financial records kept by a German merchant: John L. Flood and David J. Shaw, "The Price of the Pox in 1527: Johannes Sinapius and the Guaiac Cure," *Bibliothèque d'Humanisme et Renaissance* 54 (1992): 691–707.
119. Incurabili. vol. 68, fol. 152r.
120. Incurabili. vol. 49, fol. 95v, dated Sept. 1564.
121. Arrizabalaga, Henderson, and French, *Great Pox,* 187–188; Alfonso Corradi, *L'acqua del legno e le cure depurative nel Cinquecento* (Milan: Rechiedei, 1884); Stein, *Negotiating the French Pox,* 149–152.
122. Incurabili. vol. 48, fols. 1v, 3v; vol. 102, fols. 269v–270r. By contrast, the smaller Fugger-run pox hospital in Augsburg consumed only 166 pounds of guaiac annually from 1583 to 1631; Stein, *Negotiating the French Pox,* 147.
123. Incurabili. vol. 1, pp. 34–35.
124. Strocchia, "Nun Apothecaries."
125. See for example the 1551 statutes of Ceppo. vol. 1 bis, fols. 14r–15r.
126. Elizabeth D. Harvey, "Introduction: The 'Sense of All Senses,'" in *Sensible Flesh: On Touch in Early Modern Culture,* ed. Elizabeth D. Harvey (Philadelphia: University of Pennsylvania Press, 2003), 1–21, at 2.
127. Michael R. McVaugh, "Smell and the Medieval Surgeon," *Micrologus* 10 (2002): 113–132.
128. Harvey, "Introduction," 1–2.
129. Keith Thomas, "Magical Healing: The King's Touch," in *The Book of Touch,* ed. Constance Classen (Oxford: Berg, 2005), 354–362.

130. Harvey, "Introduction," 16.
131. Jerome Bylebyl, "The Manifest and the Hidden in the Renaissance Clinic," in *Medicine and the Five Senses,* 40–60.
132. *Historia: Empiricism and Erudition in Early Modern Europe,* ed. Gianna Pomata and Nancy G. Siraisi (Cambridge, MA: Harvard University Press, 2005).
133. Sharon T. Strocchia, "Introduction: Women and Healthcare in Early Modern Europe," *Renaissance Studies* 28 (2014): 496–514.
134. Fissell, "Introduction," 1.
135. For the extensive tacit knowledge underpinning care routines, see Cordula Nolte, "Domestic Care in the Sixteenth Century: Expectations, Experiences, and Practices from a Gendered Perspective," in *Gender, Health, and Healing, 1250-1550,* ed. Sara Ritchey and Sharon Strocchia (Amsterdam: Amsterdam University Press, forthcoming).
136. Sandra Cavallo and Tessa Storey, *Healthy Living in Late Renaissance Italy* (Oxford: Oxford University Press, 2013).
137. *La prattica dell'infermiero di Fra Francesco dal Bosco di Valdebiadene, detto il Castagnaro, Minorita Cappuccino* (Bologna: Gioseffo Longhi, 1677), 42: "ben governar gl'infermi circa le sei cose non naturali."
138. Filippo Festini and Angelica Nigro, *Prima di Florence Nightingale. La letteratura infermieristica italiana, 1676–1846* (Padua: Libreriauniversitaria, 2012), 11–27.
139. *La prattica,* unnumbered prefatory page, 15 ("il calo febrile"), 35, 37, 68, 245.
140. *La prattica,* 40, 42–43: "ogni minimo errore commesso dal'infermiero."
141. Rawcliffe, "Hospital Nurses," 54.
142. Incurabili. vol. 14, fol. 5v.
143. Vanti, *S. Giacomo,* 42.
144. Girolamo Fracastoro, *Syphilis or the French Disease: A Poem in Latin Hexameters,* trans. Heneage Wynne-Finch (London: William Heinemann, 1935), 209.
145. Jane Stevens Crawshaw, *Plague Hospitals: Public Health for the City in Early Modern Venice* (Aldershot, UK: Ashgate, 2012), 143, 172, 177.
146. Stein, *Negotiating the French Pox,* 158.
147. "Regole di monache per le citelle di Santa Croce di Verona, capitolo 38, Dell'ufficio dell'infermeria," in Gian Lodovico Masetti Zannini, *Motivi storici della educazione femminile. II: Scienza, lavoro, giuochi* (Naples: D'Auria, 1982), 375; Cavallo, "Health, Air, and Material Culture," 710–712.
148. Georges Vigarello, *Concepts of Cleanliness: Changing Attitudes in France since the Middle Ages,* trans. Jean Birrell (Cambridge, UK: Cambridge University Press, 1988), 7–37; Cavallo and Storey, *Healthy Living,* 250–257.
149. Cavallo and Storey, *Healthy Living,* 250; Vanti, *S. Giacomo,* 42.
150. Vivian Nutton, "The Seeds of Disease: An Explanation of Contagion and Infection from the Greeks to the Renaissance," *Medical History* 27 (1983): 1–34.

151. Cavallo, "Health, Air, and Material Culture," 700.
152. Stein, *Negotiating the French Pox,* 19–20, 43.
153. Henderson, *Renaissance Hospital,* 209–210.
154. Douglas Biow, *The Culture of Cleanliness in Renaissance Italy* (Ithaca, NY: Cornell University Press, 2006), 99.
155. Aud. Rif. vol. 25, fols. 200r–208r, 285r–290r.
156. Fanciulle di Santa Caterina. vol. 7, quaderno A, capitolo 16 and 16 bis: "Habitando tante fanciulle in cosi stretto luogo e con tanta povertà, si ricerca per benefitio della sanità una gran pulitezza per tutta la casa"; Masetti Zannini, *Motivi storici,* 375.
157. Cavallo and Storey, *Healthy Living,* 179–181, 189.
158. Vanti, *S. Giacomo,* 52–53. No written ordinances have surfaced for the Florentine Incurabili, but its internal regulations probably resembled the Roman model.
159. The following discussion is based on Palma, *Dialogo,* 34–107, subsection headed "Dialogo tra la sorella della congregatione e l'inferma."
160. Hannah Newton, *Misery to Mirth: Recovery from Illness in Early Modern England* (Oxford: Oxford University Press, 2018), 139.
161. Palma, *Dialogo,* 40 ("Iddio guarisce le ulcere delle anime nostre"), 45 ("me ne stò in questo letto come un'animale, senza un sentimento di Dio"), 58 ("sono hor mai due mesi ch'io stò in questo letto tutto pista e rotta, ne ci giovano rimedij, tanti medicamenti, io sono martirizzata"), 61 ("il medico non ha conosciuto, ne credo che ancor conosca la mia infirmità"), 79 ("mi trovo pur inchiodata in questo letto").

BIBLIOGRAPHY

A. PRIMARY SOURCES

A1. Archival Manuscripts

Archivio Arcivescovile, Bologna

Monache in città. Busta 261, convent of Corpus Domini, undated

Archivio Arcivescovile, Florence

Religiosi e religiose. Professioni di monache.

Filza I. 1577–1582

Filza II. 1582–1585

Archivio Provinciale del Convento di S. Marco, Florence

53. Notizie di quelle cose che sono contengono nella spezieria di S. Marco di Firenze dei padri predicatori. Quaderno sec. XVIII, e inventario degli arredi della farmacia, sec. XIX

Archivio della Quiete, Florence

A VII, n. 46. Eleonora Ramirez di Montalvo, *Istruzione alle maestre,* circa 1656–1657

Archivio di Stato, Florence

Arte de' Medici e Speziali

13. Matricole, May 1566–May 1592

14. Matricole, May 1592–May 1614

Auditore dei Benefici Ecclesiastici poi Segreteria del Regio Diritto

4892. Negozi della deputazione sopra i monasteri, 1548–1552

4894. Negozi della deputazione sopra i monasteri, 1561–1617

4896. Negozi della deputazione sopra i monasteri, 1562–1581

Auditore delle Riformagioni

8. Filza di informazioni di Messer Francesco Vinta, 1560–1564

11. Filza di relazioni di Messer Paolo Vinta, 1571–1575

21. Filza ottava di relazione di Messer Jacopo Dani, 1595–1596

25. Filza terza de ministri delle Riformagioni del 1604 et 1605

Carte Bardi

Ser. 2. Vol. 40. Ricettario della famiglia Bardi, sec. XVII

Ceppo
Vol. 1 bis. Statutes, 1551–1598
Convento di Camaldoli. Appendice.
500. Entrata & uscita dell'infermeria, 1466–1493
Corporazioni Religiose Soppresse dal Governo Francese
81. SS. Annunziata delle Murate
5. Libro giornale, 1569–1575
8. Libro giornale, 1587–1592
9. Libro giornale, 1592–1598
27. Entrate & uscite, 1598–1606
31. Entrate & uscite, 1628–1636
100. Filza di lettere, 1500–1700
83. S. Felicita
77. Debitori & creditori, 1582–1600
98. S. Maria a Monticelli
25. Entrate & Uscite, 1460–1478
100. S. Orsola
90. Dieci libri di entrate & uscite, 1563–1602
105. S. Vincenzo d'Annalena
27. Entrate & uscite, 1530–1536
29. Entrate & uscite, 1557–1566
30. Entrate & uscite, 1566–1574
31. Entrate & uscite, 1574–1582
33. Entrate & uscite, 1582–1593
35. Entrate & uscite, 1593–1603
61. Campione, 1530–1540
64. Campione, 1557–1570
65. Campione, 1566–1574
66. Campione, 1574–1582
67. Campione, 1582–1593
68. Campione, 1593–1603
106. S. Caterina da Siena
2. Giornale, 1593–1614
12. Entrate & uscite, 1496–1509
13. Entrate & uscite, 1513–1525
30. Debitori & creditori, testamenti, ricordi, 1510–1525
33. Debitori & creditori, e ricordi, 1531–1535
35. Debitori & creditori, e ricordi, 1555–1598
36. Campione, Debitori & creditori segnato E, 1599–1641
51. Entrate & uscite di procuratrice, 1517–1523
52. Entrate & uscite di procuratrice, 1524–1525
71. Filza prima di negozi diversi

108. S. Domenico nel Maglio
15. Entrata, 1566–1575
16. Uscita segnato B, 1578–1589
18. Uscita segnato D, 1601–1607
43. Quaderno di cassa della sindaca, segnato A, 1577–1578
44. Entrata di cassa segnato B, 1578–1589
45. Giornale di cassa, 1582–1584
111. S. Lucia
67. Spese di vitto, spezierie, sagrestia, 1500–1600
112. La Pietà
85. Memoriale, 1555–1563
97. Registro delle donne di compagnia, 1554–1556
Fanciulle di S. Caterina
7. Quaderno A. Capitoli
Guardaroba Medicea
463. Libro della Fonderia, 1633–1641
1166 bis. Inventario della spezzieria, 1708
Libri di Commercio e di Famiglia
164. Tariffario dei medicinali di Firenze, 1593
1264. Giornale e ricevute segnato A, Luca di Antonio Castrucci, saponaio, 1595–1603
1570. Convento di Chiarito. Giornale, 1592–1634
3374. Libro di ricette farmacologiche di Natale Martini, speziale, 1620
Manoscritti
173, no. 5. Memoria e documenti, confraternity and hospital of SS. Trinita
176. Memorie ecclesiastiche della diocesi fiorentina
321. Medici court payrolls, extracted from other ledgers
Mediceo avanti il Principato
Filze 56, 85, 106. Lettere
Mediceo del Principato
4, 5, 220, 269, 345, 346, 358, 361, 370, 401, 413a, 613, 1169, 1170, 1170a, 1171, 1172, 1173, 1175, 1350, 1351, 1704, 4959, 5128, 5130, 5926, 6081, 6101, 6107, 6110, 6113
Miscellanea Medicea
30, insert 25. Ricevuta di Orazio Tornabuoni, 1614
Monastero dell'Arcangelo Raffaello
14. Libro delle ricevute dello speziale per le monache di S. Luca, 1652–1719
Monastero di S. Luca
83. Debitori & creditori, campione, 1581–1590
89. Debitori & creditori, campione segnato A secondo, 1643–1655

Ospedale della SS. Trinita detto degli Incurabili
1. Capitoli, 1521–1645
2. Libri dei partiti e decreti, 1570–1649
3. Libri dei partiti e decreti, 1649–1687
14. Entrate & uscite, giornale e ricordi, Spedale dello Ceppo, 1489–1547
48. Giornale E, Spedale degli Incurabili, 1588–1607
49. Giornale C, Spedale degli Incurabili, 1562–1593
58. Entrate & uscite, 1542–1545
59. Entrate & uscite, 1548–1554
60. Entrate & uscite, 1554–1561
61. Entrate & uscite, e quaderno di cassa A, 1571–1574
62. Entrate & uscite, e quaderno di cassa B, 1575–1577
68. Entrate & uscite, e quaderno di cassa H, 1635–1645
102. Debitori & creditori, 1520–1542
Ospedale di S. Maria Nuova
192. Memoriali e negozi, 1554–1559
193. Memoriali e negozi, 1559–1566
194. Memoriali e negozi, 1565–1568
197. Memoriali, 1582–1591
199. Memoriali, 1613–1627
Ospedale di S. Paolo
616. Raccolta di alcuni residui di scritture
637. Capitoli e regole dello spedale, 1614–1615
771. Entrate & uscite segnato K, 1568–1571
905. Inventari, relazione, copia di lettere, 1590–1593
912. Riforma per lo spedale e monastero di S. Paolo, 1569
932. Debitori & creditori segnato K, 1568–1574
Pratica Segreta
190. Libro V di privilegi, 1594–1615
Rosselli del Turco
16. Debitori & creditori segnato A, Stefano di Maestro Romolo Rosselli, 1556–1570
26. Giornale segnato C, Stefano di Maestro Romolo Rosselli, 1583–1597
San Jacopo di Ripoli
23. Cronache, segnato A, 1508–1778
Sant'Agata di Bibbiena
16. Ricordanze, 1545–1776
Archivio di Stato, Rome
Ospedale di S. Giacomo
366. Personale laico, sec. XVI–XIX
367. Personale laico, sec. XVI–XVIII
Archivio storico dell'Ospedale degli Innocenti, Florence
4496. Nota delli ufitii et arte delle donne, 1582

6855. Negozi diversi di Mons. Borghini riguardanti tutti i monasteri dello stato, specialmente circa la clausura
12720, 12721, 12722. Forniture granducali. Debitori & creditori, 1546–1568
12954, 12955, 12956. Forniture granducali. Debitori & creditori, 1537–1546
Biblioteca comunale dell' Archiginnasio, Bologna
Ms. B. 4231. Religiosi claustrali d'ambo i sessi che trovavansi in Bologna al finire del secolo XVIII
Biblioteca Laurenziana, Florence
Ms. Antinori. 51. Stefano Rosselli, Segreti diversi, 1569–1588
Ms. San Marco. 903. Libro di ricordi, 1495–1532
Biblioteca Marucelliana, Florence
Codice C. CXLV. Stefano Rosselli, Zibaldone di diversi segreti, 1572
Biblioteca Nazionale Centrale, Florence
Ms. II. II. 509. Suora Giustina Niccolini, Cronache del monastero delle Murate di Firenze, 1597
Codice Panciatichi. 119. II. Memorie di chiese e luoghi pii della città e diocesi fiorentina, 1737
Ms. Landau Finaly. 72. Memorie e notizie istoriche del monastero di S. Caterina da Siena della città di Firenze, 1744
Ms. Landau Finaly. 268. Ricette mediche e segreti vari, sec. XV
Ms. Magliabechiana. XV. 14. Segreti diversi of Caterina Sforza, Duchess of Forlì. Copy by Lucantonio Cuppano, segnato B
Ms. Magliabechiana. XV. 92. Ricettario of S. Maria Nuova, 1515
Ms. Magliabechiana. XV. 140. Segreti sperimentati dall'Illustrissimo et Eccellentissimo Sig. Principe D. Antonio de Medici nella sua Fonderia del Casino
Ms. Magliabechiana. XVI. 63. Apparato della Fonderia dell'Illustrissimo et Eccellentissimo Sig. D. Antonio Medici. 4 vol.
Ms. Palatino. 1021. Miscellanea di medicina e farmacologia
Ms. Rossi Cassigoli. 134. Felice Dondori, Selva di varie cose antiche e moderne della città di Pistoia, 1639
Ms. Targioni Tozzetti. 56. Agostino Del Riccio, Agricoltura sperimentale, 1595. 3 vol.
Biblioteca Riccardiana, Florence
Ms. 1133. Due lettere di Brigida Baldovinetti, sec. XV
Ms. 2712. Lettere di uomini eruditi
Wellcome Library, London
Ms. 218. Recette pour la baute, 1596
Ms. 485. Segreti di Don Antonio de' Medici principe di Toscana, 1598
Ms. 5139. Italian Health Passes, 1630–1633

A2. Printed Primary Sources

Alberti, Leon Battista. *The Family in Renaissance Florence.* Trans. Renée Neu Watkins. Columbia: University of South Carolina Press, 1969.

Anne of France. *Les enseignements d'Anne de France, duchesse de Bourbonnois et d'Auvergne, à sa fille Susanne de Bourbon.* Marseille: Lafitte, 1978.

Borgarucci, Prospero. *La fabrica de gli spetiali, partita in dodici distintioni.* Venice: Vincenzo Valgrisio, 1567.

Borromeo, Carlo. *Instructiones fabricae et supellectilis ecclesiasticae* (1577): *A Translation with Commentary and Analysis.* Ed. Evelyn C. Voelker. Ann Arbor: University Microfilms International, 1993. http://www.evelynvoelker.com.

Capaccio, Giulio Cesare. *Descrizione di Napoli ne' principii del secolo XVII.* Naples: Società di storia patria, 1882.

Champier, Symphorien. *La nef des dames vertueuses.* Ed. Judy K. Kem. Paris: Honoré Champion, 2007.

Christine of Lorraine. *Lettere alla figlia Caterina de' Medici Gonzaga duchessa di Mantova (1617–1629).* Ed. Beatrice Biagioli and Elisabetta Stumpo. Florence: Firenze University Press, 2015.

Dal Bosco, Francesco. *La prattica dell'infermiero di Fra Francesco dal Bosco di Valdebiadene, detto il Castagnaro, Minorita Cappuccino.* Bologna: Gioseffo Longhi, 1677.

Delicado, Francisco. *El modo de adoperare el legno de India occidentale: Salutifero remedio a ogni piaga e mal incurabili.* Venice, 1525.

Fracastoro, Girolamo. *Syphilis or the French Disease: A Poem in Latin Hexameters.* Trans. Heneage Wynne-Finch. London: William Heinemann, 1935.

Frati, Lodovico. *Poesia in lode di alcune dame vedove bolognesi (1615).* Bologna: Nicola Zanichelli, 1889.

Frisch, Max H. *Nicolas Pol, doctor, 1494, with a critical text of his guaiac tract.* Ed. and trans. Dorothy M. Schullian. New York: Herbert Reichner, 1947.

Galilei, Maria Celeste. *Letters to Father: Suor Maria Celeste to Galileo (1623–1633).* Trans. Dava Sobel. New York: Penguin, 2001.

Guicciardini, Francesco. *The History of Italy.* Trans. Sidney Alexander. London: Macmillan, 1969.

Landucci, Luca. *Diario fiorentino dal 1450 al 1516.* Florence: G.C. Sansoni, 1883.

Legislazione Toscana raccolta e illustrata (1532–1775). Ed. Lorenzo Cantini. 32 vols. Florence: Albizziniana, 1800–1808.

Maccioni, Girolamo. *Risposta al parere del Sig. Gasparo Marcucci intorno alla qualità del sapon molle.* Florence: Sermartelli, 1630.

Malvasia, Diodata. *Writings on the Sisters of San Luca and Their Miraculous Madonna.* Ed. and trans. Danielle Callegari and Shannon McHugh.Toronto: University of Toronto Press, 2015.

Mattioli, Pietro Andrea. *Del modo di distillare le acque da tutte le piante.* Venice: Valgrisi, 1568.

Medici, Antonio de'. *La fonderia dell'Ill.mo et Ecc.mo Sig. Don Antonio Medici Principe di Capistrano, etc. Nella quale si contiene tutta l'arte spagirica di Teofrasto Paracelso, & sue medicine.* Florence: Stampata nel Palazzo del Casino di S.E. Illustrissima, 1604.

Medici, Giovanni de'. "Lettere di Giovanni de' Medici detto delle Bande Nere." *Archivio Storico Italiano* n.s. Vol. 9, part 2° (1859): 109–147.

Neri di Bicci. *Le ricordanze (10 marzo 1453–24 aprile 1475).* Ed. Bruno Santi. Pisa: Edizioni Marlin, 1976.

Niccolini, Sister Giustina. *The Chronicle of Le Murate.* Ed. and trans. Saundra Weddle. Toronto: University of Toronto Press, 2011.

Nuovo receptario composto dal famossisimo chollegio degli ex.mi dottori della arte et medicina della inclita cipta di Firenze. Florence, 1499; reprinted Florence: L.S. Olschki, 1968.

Palma, Biagio. *Dialogo tra il padre rettore della congregatione dell'Humiltà di S. Carlo e la sorella della medesima congregatione.* Rome: Giacomo Mascardi, 1629.

Palma, Biagio. *Regole della congregatione dell'Umiltà di S. Carlo.* Rome: Giacomo Mascardi, 1629.

[Pietro Ispano]. *Il Tesoro dei poveri. Ricettario medico del XIII secolo.* Ed. Luca Pesante. Arezzo: Aboca, 2007.

Pulinari, Dionisio (da Firenze). *Cronache dei Frati Minori della Provincia di Toscana.* Ed. Saturnino Mencherini. Arezzo: Cooperativa tipografica, 1913.

"Regole di monache per le citelle di Santa Croce di Verona." In Gian Lodovico Masetti Zannini, *Motivi storici della educazione femminile. Scienza, lavoro, giuochi.* Naples: D'Auria, 1982.

[Ricci, Caterina de']. *Le lettere spirituali e familiari di S. Caterina de' Ricci fiorentina.* Ed. Cesare Guasti. Prato: R. Guasti, 1861.

[Ricci, Caterina de']. *Santa Caterina de' Ricci: Cronache, diplomatica, lettere varie.* Ed. Guglielmo Di Agresti. Florence: L.S. Olschki, 1969.

El ricettario dell'arte, et universita de medici, et spetiali della città di Firenze. Riveduto dal collegio de medici per ordine dello illustrissimo et eccellentissimo signore duca di Firenze. Florence: Lorenzo Torrentino, 1550.

Il Ricettario medicinale necessario a tutti i Medici e Speziali. Florence: Giunti, 1567.

Rondinelli, Francesco. *Relazione del contagio stato in Firenze l'anno 1630 e 1633.* Florence: Giovanbattista Landini, 1714.

Rosetti, Giovanventura. *Notandissimi secreti de l'arte profumatoria.* Venice: Francesco Rampazetto, 1555.

Rosselli, Stefano. *Mes Secrets: A Florence au temps des Medici, 1593: patisserie, parfumerie, medecine.* Ed. Rodrigo de Zayas. Paris: J.-M. Place, 1996.

Savonarola, Michele. *Il trattato ginecologico-pediatrico in volgare (Ad mulieres Ferrarienses De regimine pregnantium et noviter natorum usque ad septennium).* Ed. Luigi Belloni. Milan: Stucchi, 1952.

Scappi, Bartolomeo. *Opera.* Venice: Michele Tramezzino, 1570.

[Sforza, Caterina]. *Experimenti de la Ex.ma S.r Caterina da Furlj Matre de lo Illux.mo S.r Giouanni de Medici.* Ed. Pier Desiderio Pasolini. Imola: Ignazio Galeati e Figlio, 1894.

Soderini, Giovan Vettorio. *Le opere. Vol. 3: Il trattato degli arbori.* Ed. Alberto Bacchi della Lega. Bologna: Romagnoli Dall'Acqua, 1904.

Statuti dell'Arte dei Medici e Speziali. Ed. Raffaele Ciasca. Florence: Vallecchi, 1922.

Tariffa et prezzi dei medicinali et mercanzie attenente alli speziali della città di Firenze. Florence: Giorgio Marescotti, 1593.

Vallambert, Simon de. *Cinq Livres, de la maniere de nourrir et gouverner les enfans dès leur naissance.* Ed. Colette H. Winn. Geneva: Droz, 2005.

Vita di Santa Caterina de' Ricci. Ed. Guglielmo Di Agresti. Florence: L.S. Olschki, 1965.

Vives, Juan Luis. *The Education of a Christian Woman: A Sixteenth-Century Manual.* Ed. and trans. Charles Fantazzi. Chicago: University of Chicago Press, 2000 [1524].

Zacchia, Paolo. *De' mali hipochondriaci libri tre.* Rome: Vitale Mascardi, 1644.

B. SECONDARY SOURCES

Abiti e lavori delle monache di Bologna in una serie di disegni del secolo XVIII. Ed. Mario Fanti. Bologna: Tamari, 1972.

Acerboni, Carolina. "L'infanzia dei principi di Casa Medici: Saggio storico sulla vita privata fiorentina nel Cinquecento," *Rassegna nazionale,* ser. 2, anno 38, vol. 5 (1916): 108–125.

Acerboni, Carolina. "L'infanzia dei principi di Casa Medici. Saggio storico sulla vita privata fiorentina nel Cinquento. Balocchi, giuochi, feste tradizionali di famiglia, onoranze funebri," *Rassegna nazionale,* ser. 2, anno 39, vol. 9 (1917): 34–43.

Acerboni, Carolina. "L'infanzia dei principi di Casa Medici. Saggio storico sulla vita privata fiorentina nel Cinquecento. Vestiti ed appartamenti dei Principi Medicei," *Rassegna nazionale,* ser. 2, anno 39, vol. 8 (1917): 202–211.

Acerboni, Carolina. "L'infanzia dei principi di Casa Medici. Saggio storico sulla vita privata fiorentina nel Cinquecento. Vita di famiglia e vita pubblica," *Rassegna nazionale,* ser. 2, anno 39, vol. 7 (1917): 301–316.

Aranci, Gilberto. *Formazione religiosa e santità laicale a Firenze tra Cinque e Seicento. Ippolito Galantini fondatore della Congregazione di San Francesco della dottrina cristiana di Firenze (1565–1620).* Florence: Pagnini, 1997.

Arrizabalaga, Jon. "Medical Responses to the 'French Disease' in Europe at the Turn of the Sixteenth Century." In *Sins of the Flesh: Responding to Sexual Disease in Early Modern Europe.* Ed. Kevin Siena. Toronto: University of Toronto Press, 2005, 33–55.

Arrizabalaga, Jon, John Henderson, and Roger French. *The Great Pox: The French Disease in Renaissance Europe.* New Haven, CT: Yale University Press, 1997.

Astorri, Antonella. "Appunti sull'esercizio dello speziale a Firenze nel Quattrocento," *Archivio storico italiano* 147 (1989): 31–62.

Bardazzi, Silvestro and Eugenio Castellani. *Il monastero di S. Vincenzo in Prato.* Prato: Edizioni del Palazzo, 1982.

Barker, Sheila. "Christine of Lorraine and Medicine at the Medici Court." In *Medici Women: The Making of a Dynasty in Grand Ducal Tuscany.* Ed. Giovanna Benadusi and Judith C. Brown. Toronto: University of Toronto Press, 2015, 155–181.

Barker, Sheila. "The Contributions of Medici Women to Medicine in Grand Ducal Tuscany and Beyond." In *The Grand Ducal Medici and Their Archive (1537–1743).* Ed. Alessio Assonitis and Brian Sandberg. Turnhout, Belgium: Brepols, 2016, 101–116.

Barker, Sheila. "Malaria and the Search for Its Cure in Granducal Tuscany," *Medicea* 5 (2010): 54–59.

Barker, Sheila. "Painting and Humanism in Early Modern Florentine Convents." In *Artiste nel chiostro. Produzione artistica nei monasteri femminili in età moderna.* Ed. Sheila Barker with Luciano Cinelli. Florence: Nerbini, 2015, 105–139.

Barker, Sheila, and Sharon Strocchia, "Household Medicine for a Renaissance Court: Caterina Sforza's *Ricettario* Reconsidered." In *Gender, Health, and Healing, 1250–1550.* Ed. Sara Ritchey and Sharon Strocchia. Amsterdam: Amsterdam University Press, forthcoming.

Battistini, Mario. "Il medico Andrea Pasquali," *Rivista di storia delle scienze mediche e naturali,* anno 17, vol. 8 (1926): 231–233.

Baudot, Auguste. *Études historiques sur la pharmacie en Bourgogne avant 1803.* Paris: Librairie A. Maloine, 1905.

Belardini, Manuela. "'Piace molto a Giesù la nostra confidanza:' Suor Orsola Fontebuoni a Maria Maddalena d'Austria." In *Per lettera. La scrittura epistolare femminile tra archivio e tipografia secoli XV–XVII.* Ed. Gabriella Zarri. Rome: Viella, 1999, 359–383.

La bellezza come terapia. Arte e assistenza nell'ospedale di Santa Maria Nuova a Firenze. Ed. Enrico Ghidetti and Esther Diana. Florence: Polistampa, 2005.

Bellorini, Cristina. *The World of Plants in Renaissance Tuscany: Medicine and Botany.* Burlington, VT: Ashgate, 2016.

Benzi, Utzima. "Le frère dominicain Agostino del Riccio et son Arte della memoria locale (1595)." *Bulletin de l'Association des Historiens de l'Art Italien* 13 (2007): 25–43.

Berco, Cristian. "The Great Pox, Symptoms, and Social Bodies in Early Modern Spain," *Social History of Medicine* 28 (2014): 225–244.

Beretta, Marco. "Material and Temporal Powers at the Casino di San Marco (1574–1621)." In *Laboratories of Art: Alchemy and Art Technology from Antiquity to the 18th Century.* Ed. Sven Dupré. Berlin: Springer, 2014, 129–156.

Bernstein, William J. *A Splendid Exchange: How Trade Shaped the World.* New York: Grove / Atlantic, 2008.

Bianchi, Francesco. "Italian Renaissance Hospitals: An Overview of the Recent Historiography," *Mitteilungen des Instituts für Ősterreichische Geschichtsforschung* 115 (2007): 394–403.

Biow, Douglas. *The Culture of Cleanliness in Renaissance Italy.* Ithaca, NY: Cornell University Press, 2006.

Biow, Douglas. *On the Importance of Being an Individual in Renaissance Italy: Men, Their Professions, and Their Beards.* Philadelphia: University of Pennsylvania Press, 2015.

Blumenthal, Debra. "Domestic Medicine: Slaves, Servants and Female Medical Expertise in Late Medieval Valencia," *Renaissance Studies* 28 (2014): 515–532.

Boccadamo, Giuliana. "Una riforma impossibile? I papi e i primi tentativi di riforma dei monasteri femminili a Napoli nel '500," *Campania sacra. Rivista di storia sociale e religiosa del Mezzogiorno* 21 (1990): 96–122.

Bourne, Molly. "From Court to Cloister and Back Again: Margherita Gonzaga, Caterina de' Medici and Lucrina Fetti at the Convent of Sant'Orsola in Mantua." In *Domestic Institutional Interiors in Early Modern Europe.* Ed. Sandra Cavallo and Silvia Evangelisti. Burlington, VT: Ashgate, 2009, 153–179.

Bourne, Molly. "Medici Women at the Gonzaga Court, 1584–1627." In *Italian Art, Society, and Politics: A Festschrift for Rab Hatfield.* Ed. Barbara Deimling, Jonathan K. Nelson, and Gary M. Radke. Syracuse, NY: Syracuse University Press, 2007, 223–243.

Brockliss, Laurence, and Colin Jones. *The Medical World of Early Modern France.* Oxford: Oxford University Press, 1997.

Broomhall, Susan. *Women's Medical Work in Early Modern France.* Manchester, UK: Manchester University Press, 2004.

Brown, Judith C. "Introduction." In *Medici Women: The Making of a Dynasty in Grand Ducal Tuscany.* Ed. Giovanna Benadusi and Judith C. Brown. Toronto: University of Toronto Press, 2015, 17–57.

Burke, Peter. "How to Be a Counter-Reformation Saint." In Peter Burke, *The Historical Anthropology of Early Modern Italy.* Cambridge, UK: Cambridge University Press, 1987, 48–62.

Butters, Suzanne. *The Triumph of Vulcan: Sculptors' Tools, Porphyry, and the Prince in Ducal Florence.* Florence: L.S. Olschki, 1996.

Butters, Suzanne. "The Use and Abuse of Gifts in the World of Ferdinando de' Medici (1549-1609)," *I Tatti Studies in the Italian Renaissance* 11 (2007): 243–354.

Bylebyl, Jerome. "The Manifest and the Hidden in the Renaissance Clinic." In *Medicine and the Five Senses.* Ed. W. F. Bynum and Roy Porter. Cambridge, UK: Cambridge University Press, 1993, 40–60.

Cabré, Montserrat. "Beautiful Bodies." In *A Cultural History of the Human Body in the Medieval Age.* Ed. Linda Kalof. Oxford: Berg, 2010, 121–139.

Cabré, Montserrat. "Keeping Beauty Secrets in Early Modern Iberia." In *Secrets and Knowledge in Medicine and Science, 1500–1800.* Ed. Elaine Leong and Alisha Rankin. Aldershot, UK: Ashgate, 2011, 167–190.

Cabré, Montserrat. "Women or Healers? Household Practices and the Categories of Health Care in Late Medieval Iberia," *BHM* 82 (2008): 18–51.

Calderara, Augusto. *Abraxas: Glossario dei termini di sostanze, formule e oggetti usati in pratiche magiche o terapeutiche, citati nei documenti di ABRATASSA*. Lucca: M. Pacini Fazzi, 1989.

Calonaci, Stefano. "Gli angeli del testamento. Donne fedecommissarie e fedecommittenti nella Toscana moderna." In *Nobildonne, monache e cavaliere dell'ordine di Santo Stefano. Modelli e strategie femminili nella vita pubblica della Toscana granducale*. Ed. Marcella Aglietti. Pisa: ETS, 2009, 79–96.

Calvi, Giulia. *Histories of a Plague Year: The Social and the Imaginary in Baroque Florence*. Trans. Dario Biocca and Bryant T. Ragan. Berkeley: University of California Press, 1989.

Cantini, Lorenzo. *Vita di Cosimo de' Medici primo granduca di Toscana*. Florence: Albizziniana, 1805.

Capecchi, Stefania. "Il ricettario della molto reverenda Madre Donna Maddalena Favilla, spetiale nel monastero di San Mercuriale della città di Pistoia nell'anno 1750," *AMAISF* 18 (2001): 55–60.

Cardini, Franco. "Il 'breve' (secoli XIV-XV): tipologia e funzione," *La ricerca folklorica* 5 (1982): 63–73.

Carinci, Eleonora. "Una 'speziala' padovana: *Lettere di philosophia naturale* di Camilla Erculiani (1584)," *Italian Studies* 68 (2013): 202–229.

Carneiro de Sousa, Ivo. *A Rainha D. Leonor (1458–1525). Poder, Misericórdia, Religiosidade e Espiritualidade no Portugal do Renascimento*. Lisbon: Fundação Calouste Gulbenkian, 2002.

Carneiro de Sousa, Ivo. "A Rainha D. Leonor e as Murate de Florença (Notas de Investigação), *Revista da Faculdade de Letras do Porto,* serie historia, ser. 2, 4 (1987): 119–133.

Carrino, Candida. *Le monache ribelli raccontate da Suor Fulvia Caracciolo*. Naples: Intra Moenia, 2013.

Casini, Alfonso. *Passitea Crogi, donna senese*. Siena: Edizioni Cantagalli, 1991.

Castiglione, Caroline. "Peasants at the Palace: Wet Nurses and Aristocratic Mothers in Early Modern Rome." In *Medieval and Renaissance Lactations: Images, Rhetorics, Practices*. Ed. Jutta Gisela Sperling. Burlington, VT: Ashgate, 2013, 79–99.

Catalogo delle Maioliche. Museo Nazionale di Firenze. Palazzo del Bargello. Ed. Giovanni Conti. Florence: Centro Di, 1971.

Cavallo, Sandra. *Artisans of the Body in Early Modern Italy: Identities, Families and Masculinities*. Manchester, UK: Manchester University Press, 2007.

Cavallo, Sandra. "Health, Air and Material Culture in the Early Modern Italian Domestic Environment," *Social History of Medicine* 29 (2016): 695–716.

Cavallo, Sandra. "Health, Beauty, and Hygiene." In *At Home in Renaissance Italy*. Ed. Marta Ajmar-Wollheim and Flora Dennis. London: V & A, 2006, 174–187.

Cavallo, Sandra. "Pregnant Stones as Wonders of Nature." In *Reproduction: From Antiquity to the Present Day.* Ed. Nick Hopwood, Rebecca Flemming, and Lauren Kassell. Cambridge, UK: Cambridge University Press, 2018, E17.

Cavallo, Sandra. "Secrets to Healthy Living: The Revival of the Preventive Paradigm in Late Renaissance Italy." In *Secrets and Knowledge in Medicine and Science, 1500–1800.* Ed. Elaine Leong and Alisha Rankin. Aldershot, UK: Ashgate, 2011, 191–212.

Cavallo, Sandra, and Silvia Evangelisti. "Introduction." In *Domestic Institutional Interiors in Early Modern Europe.* Ed. Sandra Cavallo and Silvia Evangelisti. Burlington, VT: Ashgate, 2009, 1–23.

Cavallo, Sandra, and David Gentilcore. "Introduction: Spaces, Objects and Identities in Early Modern Italian Medicine," *Renaissance Studies* 21 (2007): 473–79.

Cavallo, Sandra, and Tessa Storey. *Healthy Living in Late Renaissance Italy.* Oxford: Oxford University Press, 2013.

Cipolla, Carlo M. *Cristofano and the Plague: A Study in the History of Public Health in the Age of Galileo.* Berkeley: University of California Press, 1973.

Cipolla, Carlo M. *Miasmas and Disease: Public Health and the Environment in the Pre-Industrial Age.* New Haven, CT: Yale University Press, 1992.

Clericuzio, Antonio. "Chemical Medicine and Paracelsianism in Italy, 1550–1650." In *The Practice of Reform in Health, Medicine and Science, 1500–2000. Essays for Charles Webster.* Ed. Margaret Pelling and Scott Mandelbrote. Aldershot, UK: Ashgate, 2005, 59–79.

Cohen, Elizabeth. "Miscarriages of Apothecary Justice: Un-separate Spaces of Work and Family in Early Modern Rome," *Renaissance Studies* 21 (2007): 480–504.

Cohn, Samuel K. Jr. *Cultures of Plague: Medical Thinking at the End of the Renaissance.* Oxford: Oxford University Press, 2010.

Colapinto, Leonardo. "L'arte degli speziali a Roma nel Rinascimento," *AMAISF* 12 (1995): 59–63.

Colapinto, Leonardo. "Monachesimo e spezierie conventuale in Italia tra XII al XVII secolo," *AMAISF* 12 (1995): 107–113.

Collezionismo mediceo e storia artistica. Vol. 1: Tomo 1, *Da Cosimo I a Cosimo II, 1540–1621.* Ed. Paola Barocchi and Giovanna Gaeta Bertelà. Florence: Studio per edizioni scelte, 2002.

Compare, Carmela. "Inventari di biblioteche monastiche femminili alla fine del XVI secolo," *Genesis: rivista della società italiana delle storiche* 2 (2003): 220–232.

Conklin, Beth A. *Consuming Grief: Compassionate Cannibalism in an Amazonian Society.* Austin: University of Texas Press, 2001.

Corradi, Alfonso. *L'acqua del legno e le cure depurative nel Cinquecento.* Milan: Rechiedei, 1884.

Corradi, Alfonso. *Le prime farmacopee italiane, ed in particolare: Dei Ricettari Fiorentini. Memoria.* Milan: Rechiedei, 1887.

Correia, Fernando da Silva. "O Julgamento da Rainha D. Leonor, seguido de tres relatorios medicos (I: A historia clinica d'El Rei D. Joao II; II: A historia clinica da Infanta Santa Joana; III: Rainha D. Leonor, pelo dr. Julio Dantas)," sep. da *Revista Ocidente,* Lisbon, 1943.

Covoni, Pierfilippo. *Don Antonio de' Medici al Casino di San Marco.* Florence: Tipografia cooperativa, 1892.

Cox, Virginia. *Lyric Poetry by Women of the Renaissance.* Baltimore, MD: Johns Hopkins University Press, 2013.

Cox, Virginia. *The Prodigious Muse: Women's Writing in Counter-Reformation Italy.* Baltimore, MD: Johns Hopkins University Press, 2011.

Cox, Virginia. *Women's Writing in Italy, 1400–1650.* Baltimore, MD: Johns Hopkins University Press, 2008.

Crisciani, Chiara. "Cura e educazione a corte: note su medici e giovani principi a Milano (sec. XV)." In *I bambini di una volta.* Ed. Monica Ferrari. Milan: Franco Angeli, 2006, 41–50.

Criscuolo, Vincenzo. "Documenti vaticani su Passitea Crogi clarissa cappuccina senese (1564–1615)," *Collectanea franciscana* 62 (1992): 651–683.

Cusick, Suzanne G. *Francesca Caccini at the Medici Court: Music and the Circulation of Power.* Chicago: University of Chicago Press, 2009.

Davis, Natalie Z. *The Gift in Sixteenth-Century France.* Madison: University of Wisconsin Press, 2000.

Day, Ivan. *Pomanders, Washballs and Other Scented Articles.* London: The Herb Society, n.d.

Delancey, Julia. "Dragonsblood and Ultramarine: The Apothecary and Artists' Pigments in Florence." In *The Art Market in Italy (15th–17th Centuries).* Ed. Marcello Fantoni, Louisa C. Matthew, and Sara F. Matthews-Grieco. Modena: F.C. Panini, 2003, 141–150.

Demaitre, Luke. "Skin and the City: Cosmetic Medicine as an Urban Concern." In *Between Text and Patient: The Medical Enterprise in Medieval and Early Modern Europe.* Ed. Florence Eliza Glaze and Brian Nance. Florence: SISMEL, 2011, 97–120.

De Renzi, Silvia. "Tales from Cardinals' Deathbeds: Medical Hierarchy, Courtly Etiquette and Authority in the Counter Reformation." In *Être médicin à la cour (Italie, France, Espagne, XIIIe–XVIIIe siècle).* Ed. Elisa Andretta and Marilyn Nicoud. Florence: SISMEL, 2013, 235–258.

De Vivo, Filippo. "Pharmacies as Centres of Communication in Early Modern Venice," *Renaissance Studies* 21 (2007): 505–521.

Diana, Esther. "Medici, speziali e barbieri nella Firenze della prima metà del '500," *Rivista di Storia della Medicina,* n.s. 4 (1994): 13–27.

Dinan, Susan E. *Women and Poor Relief in Seventeenth-Century France: The Early History of the Daughters of Charity.* Aldershot, UK: Ashgate, 2006.

Donatone, Guido. *La farmacia degli Incurabili e la maiolica napoletana del Settecento.* Naples: Edizioni del Delfino, 1972.

Dubost, Jean-Francois. "Liberalità calcolate: politiche del dono tra corte di Francia e corti italiane al tempo di Maria de' Medici." In *Medici Women as Cultural Mediators (1533–1743). Le donne di casa Medici e il loro ruolo di mediatrici culturali fra le corti d'Europa.* Ed. Christina Strunck. Milan: Silvana, 2011, 207–225.

Eamon, William. *The Professor of Secrets: Mystery, Medicine, and Alchemy in Renaissance Italy.* Washington, DC: National Geographic Society, 2010.

Eamon, William. *Science and the Secrets of Nature: Books of Secrets in Medieval and Early Modern Culture.* Princeton, NJ: Princeton University Press, 1994.

Eckstein, Nicholas A. "Florence on Foot: An Eye-Level Mapping of the Early Modern City in Time of Plague," *Renaissance Studies* 30 (2016): 273–297.

Edelstein, Bruce. "Eleonora di Toledo e la gestione dei beni familiari: una strategia economica?" In *Donne di potere nel Rinascimento.* Ed. Letizia Arcangeli and Susanna Peyronel. Rome: Viella, 2008, 733–764.

Edelstein, Bruce. "*La fecundissima Signora Duchessa:* The Courtly Persona of Eleonora di Toledo and the Iconography of Abundance." In *The Cultural World of Eleonora di Toledo, Duchess of Florence and Siena.* Ed. Konrad Eisenbichler. Aldershot, UK: Ashgate, 2004, 71–97.

Egmond, Florike. "Apothecaries as Experts and Brokers in the Sixteenth-Century Network of the Naturalist Carolus Clusius," *History of Universities* 23 (2008): 59–91.

Erbe e speziali. I laboratori della salute. Ed. Margherita Breccia Fratadocchi and Simonetta Buttò. Sansepolcro: Aboca, 2007.

Evangelisti, Silvia. "To Find God in Work? Female Social Stratification in Early Modern Italian Convents," *European History Quarterly* 38 (2008): 398–416.

Fabbri, Christiane Nockels. "Treating Medieval Plague: The Wonderful Virtues of Theriac," *Early Science and Medicine* 12 (2007): 247–283.

Fabbri, Lorenzo Gaetano. *Dell'uso del mercurio sempre temerario in medicina. Della fondazione e del medicamento dell'Arcispedale degl'Incurabili nella città di Firenze.* Cologne: Federigo Tirbien, 1749.

Fantoni, Marcello. "Feticci di prestigio: il dono alla corte medicea." In *Rituale cerimoniale etichetta.* Ed. Sergio Bertelli and Giuliano Crifò. Milan: Bompiani, 1985, 141–161.

Farmacie storiche in Toscana. Florence: Polistampa, 1998.

Ferrari, Monica. "*Ordini da servare nella vita* ed *Emploi du temps.* Il ruolo pedagogico del medico in due corti europee tra '400 e '600," *Micrologus* 16 (2008): 295–313.

Festini, Filippo and Angelica Nigro. *Prima di Florence Nightingale. La letteratura infermieristica italiana, 1676–1846.* Padua: Libreriauniversitaria, 2012.

ffolliott, Sheila. "Catherine de' Medici as Artemisia: Figuring the Powerful Widow." In *Rewriting the Renaissance: The Discourses of Sexual Difference in Early*

Modern Europe. Ed. Margaret W. Ferguson, Maureen Quilligan, and Nancy Vickers. Chicago: University of Chicago Press, 1986, 227–241.

Findlen, Paula. "The Formation of a Scientific Community: Natural History in Sixteenth-Century Italy." In *Natural Particulars: Nature and the Disciplines in Renaissance Europe*. Ed. Anthony Grafton and Nancy Siraisi. Cambridge, MA: Harvard University Press, 1999.

Finucci, Valeria. *The Prince's Body: Vincenzo Gonzaga and Renaissance Medicine*. Cambridge, MA: Harvard University Press, 2015.

Fischer, Will. "The Renaissance Beard: Masculinity in Early Modern England," *RQ* 54 (2001): 155–187.

Fissell, Mary E. "Introduction: Women, Health, and Healing in Early Modern Europe," *BHM* 82 (2008): 1–17.

Fissell, Mary E. "The Marketplace of Print." In *Medicine and the Market in England and Its Colonies, c. 1450–c. 1850*. Ed. Mark S. R. Jenner and Patrick Wallis. New York: Palgrave Macmillan, 2007, 108–132.

Fiumi, Fabrizia and Giovanna Tempesta, "Gli 'experimenti' di Caterina Sforza." In *Caterina Sforza: Una Donna del Cinquecento*. Imola: La mandragora, 2000, 139–146.

Flood, John L., and David J. Shaw, "The Price of the Pox in 1527: Johannes Sinapius and the Guaiac Cure," *Bibliothèque d'Humanisme et Renaissance* 54 (1992): 691–707.

Foa, Anna. "The New and the Old: The Spread of Syphilis (1494–1530)." In *Sex and Gender in Historical Perspective*. Ed. Edward Muir and Guido Ruggiero. Baltimore, MD: Johns Hopkins University Press, 1990, 26–45.

Fornaciai, Valentina. *"Toilette," Perfumes and Make-Up at the Medici Court: Pharmaceutical Recipe Books, Florentine Collections and the Medici Milieu Uncovered*. Livorno: Sillabe, 2007.

Forster, Elborg. "From the Patient's Point of View: Illness and Health in the Letters of Liselotte von der Pfalz (1652–1722)," *BHM* 60 (1986): 297–320.

Franceschini, Chiara. "*Los scholares son cosa de su excelentia, como lo es toda la Compañia:* Eleonora di Toledo and the Jesuits." In *The Cultural World of Eleonora di Toledo, Duchess of Florence and Siena*. Ed. Konrad Eisenbichler. Aldershot, UK: Ashgate, 2004, 181–206.

Franchi, Giuseppe. "Apparecchi e metodi per 'lambiccare' secondo Mattioli." In *I giardini dei semplici e gli orti botanici della Toscana*. Ed. Sara Ferri and Francesca Vannozzi. Perugia: Quattroemme, 1993, 201–204.

Franchi, Marinella. "La spezieria: gestione e funzionamento." In *Una farmacia preindustriale in Valdelsa. La spezieria e lo spedale di Santa Fina nella città di San Gimignano, secc. XIV–XVIII*. Ed. Gabriele Borghini. San Gimignano: Città di San Gimignano, 1981, 126.

French, Roger, and Jon Arrizabalaga. "Coping with the French Disease: University Practitioners' Strategies and Tactics in the Transition from the Fifteenth to the Seventeenth Century." In *Medicine from the Black Death to the French Disease.* Ed. Roger French, Jon Arrizabalaga, Andrew Cunningham, and Luis Garcia-Ballester. Aldershot, UK: Ashgate, 1998, 248–287.

Gadebusch Bondio, Mariacarla. "La *Carne di Fuori:* Discorsi medici sulla natura e l'estetica della pelle nel '500," *Micrologus* 13 (2005): 537–570.

Galasso Calderara, Estella. *La Granduchessa Maria Maddalena D'Austria: Un amazzone tedesca nella Firenze medicea del '600.* Genoa: SAGEP, 1985.

Gavitt, Philip. *Gender, Honor, and Charity in Late Renaissance Florence.* Cambridge, UK: Cambridge University Press, 2011.

Geary, Patrick. "Sacred Commodities: The Circulation of Medieval Relics." In *The Social Life of Things: Commodities in Cultural Perspective.* Ed. Arjun Appadurai. Cambridge, UK: Cambridge University Press, 1986, 169–194.

Gentilcore, David. "Apothecaries, 'Charlatans,' and the Medical Marketplace in Italy, 1400–1750," *Pharmacy in History* 45 (2003): 91–94.

Gentilcore, David. "'For the Protection of Those Who Have Both Shop and Home in this City': Relations between Italian Charlatans and Apothecaries," *Pharmacy in History* 45 (2003): 108–121.

Gentilcore, David. *Medical Charlatanism in Early Modern Italy.* Oxford: Oxford University Press, 2006.

Giovannini, Sandra, and Gabriella Mancini. *L'Officina profumo-farmaceutica di Santa Maria Novella in Firenze. Sette secoli di storia e di arte.* Rome: Chitarrini, 1994, 16–24.

Giuffra, Valentina, and Gino Fornaciari. "Breastfeeding and Weaning in Renaissance Italy: The Medici Children," *Breastfeeding Medicine* 8 (2013): 1–6.

Goguel, Catherine Monbeig. "Cristofano Allori, 1577–1621," *Prospettiva* 39 (1984): 76–80.

Goldberg, Edward L. "Artistic Relations between the Medici and the Spanish Courts, 1587–1621: Part 1," *Burlington Magazine* 138 (1996): 105–114.

Goldthwaite, Richard A. *The Economy of Renaissance Florence.* Baltimore, MD: Johns Hopkins University Press, 2009.

Goldthwaite, Richard A., and W. R. Rearick. "Michelozzo and the Ospedale di San Paolo in Florence," *Mitteilungen des Kunsthistorischen Instituts in Florenz* 21 (1977): 221–306.

Graziani, Natale, and Gabriella Venturelli. *Caterina Sforza.* Milan: Mondadori, 2001.

Green, Monica. "Bodies, Gender, Health, Disease: Recent Work on Medieval Women's Medicine," *Studies in Medieval and Renaissance History,* ser. 3, 5 (2005): 1–46.

Green, Monica. "Books as a Source of Medical Education for Women in the Middle Ages," *Dynamis* 20 (2000): 331–369.

Green, Monica. "Gendering the History of Women's Healthcare," *Gender & History* 20 (2008): 487–518.

Green, Monica. "Integrative Medicine: Incorporating Medicine and Health into the Canon of Medieval European History," *History Compass* 7 (2009): 1218–1245.

Green, Monica. "Women's Medical Practice and Health Care in Medieval Europe," *Signs* 14 (1989): 434–473.

Grieco, Allen J. "Medieval and Renaissance Wines: Taste, Dietary Theory, and How to Choose the 'Right' Wine (14th–16th centuries)," *Medievalia* 30 (2009): 15–42.

Guidi Bruscoli, Francesco. *Bartolomeo Marchionni, 'homem de grossa fazenda,' (ca. 1450–1530). Un mercante fiorentino a Lisbona e l'impero portoghese.* Florence: L.S. Olschki, 2014.

Hairston, Julia L. "The Economics of Milk and Blood in Alberti's *Libri della famiglia:* Maternal versus Wet-Nursing." In *Medieval and Renaissance Lactations: Images, Rhetorics, Practices.* Ed. Jutta Gisela Sperling. Burlington, VT: Ashgate, 2013, 187–212.

Hariguchi, Jennifer. "*Istruzione alle maestre* (Instruction for Teachers): A Model Text for Women's Lay Conservatories in Seventeenth and Eighteenth-Century Tuscany," *Early Modern Women* 10 (2016): 3–21.

Harkness, Deborah E. *The Jewel House: Elizabethan London and the Scientific Revolution.* New Haven, CT: Yale University Press, 2007.

Harkness, Deborah E. "A View from the Streets: Women and Medical Work in Elizabethan London," *BHM* 82 (2008): 52–85.

Harness, Kelly. *Echoes of Women's Voices: Music, Art, and Female Patronage in Early Modern Florence.* Chicago: University of Chicago Press, 2006.

Harvey, Elizabeth D. "Introduction: The 'Sense of All Senses.'" In *Sensible Flesh: On Touch in Early Modern Culture.* Ed. Elizabeth D. Harvey. Philadelphia: University of Pennsylvania Press, 2003, 1–21.

Heikamp, Dietrich. "Agostino del Riccio. Del giardino di un re." In *Il giardino storico italiano. Problemi e indagine sulle fonti letterarie e storiche.* Ed. Giovanna Ragionieri. Florence: L.S. Olschki, 1981, 59–64.

Henderson, John. *The Renaissance Hospital: Healing the Body and Healing the Soul.* New Haven, CT: Yale University Press, 2006.

Henderson, John. "'La schifezza, madre di corruzione.' Peste e società nella Firenze della prima età moderna, 1630–31." *Medicina e storia* 2 (2001): 23–56.

Herbert, Amanda E. *Female Alliances: Gender, Identity and Friendship in Early Modern Britain.* New Haven, CT: Yale University Press, 2014.

Historia: Empiricism and Erudition in Early Modern Europe. Ed. Gianna Pomata and Nancy G. Siraisi. Cambridge, MA: MIT Press, 2005.

Horden, Peregrine. "'A Discipline of Relevance': The Historiography of the Later Medieval Hospital," *Social History of Medicine* 1 (1988): 359–374.

Howe, Eunice D. "The Architecture of Institutionalism: Women's Space in Renaissance Hospitals." In *Architecture and the Politics of Gender in Early Modern Europe.* Ed. Helen Hills. Aldershot, UK: Ashgate, 2003, 63–82.

Huguet Termes, Teresa. "Standardising Drug Therapy in Renaissance Europe? The Florence (1499) and Nuremberg Pharmacopoeia (1546)," *Medicina e storia* 8 (2008): 77–101.

Jacquart, Danielle. "Naissance d'une pédiatrie en milieu de cour," *Micrologus* 16 (2008): 271–294.

Jenner, Mark S. R. "Follow Your Nose? Smell, Smelling, and Their Histories," *American Historical Review* 116 (2011): 335–351.

Jenner, Mark S. R., and Patrick Wallis, "The Medical Marketplace." In *Medicine and the Market in England and Its Colonies, c. 1450–c. 1850.* Ed. Mark S. R. Jenner and Patrick Wallis. New York: Palgrave Macmillan, 2007, 1–23.

Jones, Pamela M. *Altarpieces and Their Viewers in the Churches of Rome from Caravaggio to Guido Reni.* Aldershot, UK: Ashgate, 2008.

Jütte, Robert. "Syphilis and Confinement: Hospitals in Early Modern Germany." In *Institutions of Confinement: Hospitals, Asylums, and Prisons in Western Europe and North America, 1500–1950.* Ed. Norbert Finzsch and Robert Jütte. Cambridge, UK: Cambridge University Press, 1996, 97–115.

Kieffer, Fanny. "La Confiserie des Offices: Art, Sciences et Magnificence à la Cour des Medicis," *Predella* 33 (2013): 89–106.

Kinzelbach, Annemarie. "Women and Healthcare in Early Modern German Towns," *Renaissance Studies* 28 (2014): 619–638.

Klapisch-Zuber, Christiane. *Women, Family, and Ritual in Renaissance Italy.* Trans. Lydia G. Cochrane. Chicago: University of Chicago Press, 1985.

Klein, Ursula, and E. C. Spary, "Introduction: Why Materials?" In *Materials and Expertise in Early Modern Europe: Between Market and Laboratory.* Ed. Ursula Klein and E. C. Spary. Chicago: University of Chicago Press, 2010, 1–23.

Koslofsky, Craig. "Knowing Skin in Early Modern Europe, c. 1450–1750," *History Compass* 12 (2014): 794–806.

Kostylo, Joanna. "Pharmacy as a Centre for Protestant Reform in Renaissance Venice," *Renaissance Studies* 30 (2016): 236–253.

Laghi, Anna Vittoria. "Di tre 'spezierie' pratesi." In *Il Settecento a Prato.* Ed. Renzo Fantappiè and Sandro Bellesi. Prato: CariPrato, 1999, 345–352.

Lamberini, Daniela. "Il monastero di San Mercuriale a Pistoia. Lineamenti di storia dei secoli X–XIX dalla documentazione d'archivio." In *L'architettura del San Mercuriale a Pistoia. Un frammento di città.* Ed. Francesco Gurrieri. Florence: Alinea, 1989, 15–25.

Langdon, Gabrielle. *Medici Women: Portraits of Power, Love and Betrayal from the Court of Duke Cosimo I.* Toronto: University of Toronto Press, 2006.

Laroche, Rebecca. *Medical Authority and Englishwomen's Herbal Texts, 1550–1650.* Aldershot, UK: Ashgate, 2009.

LeJacq, Seth Stein. "The Bounds of Domestic Healing: Medical Recipes, Storytelling, and Surgery in Early Modern England," *Social History of Medicine* 26 (2013): 451–468.

Lemaitre, Henri. "Statuts des religieuses du tiers ordres franciscain dites soeurs grises hospitalières (1483)," *Archivum franciscanum historicum* 4 (1911): 713–731.

Leong, Elaine. "'Herbals she peruseth': Reading Medicine in Early Modern England," *Renaissance Studies* 28 (2014): 556–578.

Leong, Elaine. "Making Medicines in the Early Modern Household," *BHM* 82 (2008): 145–168.

Leong, Elaine, and Sara Pennell. "Recipe Collections and the Currency of Medical Knowledge in the Early Modern 'Medical Marketplace.'" In *Medicine and the Market in England and Its Colonies, c. 1450–c. 1850.* Ed. Mark S. R. Jenner and Patrick Wallis. New York: Palgrave Macmillan, 2007, 133–152.

Leong, Elaine, and Alisha Rankin. "Introduction: Secrets and Knowledge." In *Secrets and Knowledge in Medicine and Science, 1500–1800.* Ed. Elaine Leong and Alisha Rankin. Aldershot, UK: Ashgate, 2011, 1–20.

Lequain, Élodie. "Le bon usage du corps dans l'education des princesses à la fin du Moyen Âge." In *Cultures de cour, cultures du corps (XIVe–XVIIIe siècle).* Ed. Catherine Lanoë, Mathieu da Vinha, and Bruno Laurioux. Paris: Publications de l'université Paris-Sorbonne, 2011, 115–125.

Lev, Efraim. "Mediators between Theoretical and Practical Medieval Knowledge: Medical Notebooks from the Cairo Genizah and Their Significance," *Medical History* 57 (2013): 487–515.

Litchfield, R. Burr. "Demographic Characteristics of Florentine Patrician Families, Sixteenth to Nineteenth Centuries," *Journal of Economic History* 29 (1969): 191–205.

Litchfield, R. Burr. *Florence Ducal Capital, 1530–1630.* ACLS e-book, 2008.

Lombardi, Daniela. "Poveri a Firenze. Programmi e Realizzazioni della Politica Assistenziale dei Medici tra Cinque e Seicento." In *Timore e carità. I poveri nell'Italia moderna.* Ed. Giorgio Politi, Maria Rosa, and Franco della Peruta. Cremona: Libreria del convegno, 1982, 165–184.

Long, Pamela O. *Artisan/Practitioners and the Rise of the New Sciences, 1400–1600.* Corvallis: Oregon State University Press, 2011.

Long, Pamela O. "Invention, Authorship, 'Intellectual Property,' and the Origin of Patents: Notes toward a Conceptual History," *Technology and Culture* 32 (1991): 846–884.

Lopez Terrada, Maria Luz. "El tratamiento de la sifilis en un hospital renascentista: la sala del *mal de siment* del hospital general de Valencia," *Asclepio* 41 (1989): 19–50.

Lowe, K. J. P. *Nuns' Chronicles and Convent Culture in Renaissance and Counter-Reformation Italy.* Cambridge, UK: Cambridge University Press, 2003.

Lowe, Kate. "Rainha D. Leonor of Portugal's Patronage in Renaissance Florence and Cultural Exchange." In *Cultural Links between Portugal and Italy in the*

Renaissance. Ed. Kate Lowe. Cambridge, UK: Cambridge University Press, 2000, 225–248.

Lowe, Kate. "Understanding Cultural Exchange between Portugal and Italy in the Renaissance: Social and Institutional Relations." In *Cultural Links between Portugal and Italy in the Renaissance.* Ed. Kate Lowe. Cambridge, UK: Cambridge University Press, 2000, 1–18.

Lowe, Kate. "Women's Work at the Benedictine Convent of Le Murate in Florence: Suora Battista Carducci's Roman Missal of 1509." In *Women and the Book: Assessing the Visual Evidence.* Ed. Lesley Smith and Jane H. M. Taylor. London: British Library, 1997, 133–146.

Lubkin, Gregory. *A Renaissance Court: Milan under Galeazzo Maria Sforza.* Berkeley: University of California Press, 1994.

Luti, Filippo. *Don Antonio de' Medici e i suoi tempi.* Florence: L.S. Olschki, 2006.

Luti, Filippo. "Don Antonio de' Medici 'professore de' secreti,'" *Medicea* 1 (2008): 34–47.

Luzzi, Paolo, and Fernando Fabbri, "I tre orti botanici di Firenze." In *I giardini dei semplici e gli orti botanici della Toscana.* Ed. Sara Ferri and Francesca Vannozzi. Perugia: Quattroemme, 1993, 49–68.

Macey, Patrick. *Bonfire Songs: Savonarola's Musical Legacy.* Oxford: Oxford University Press, 1998.

Maegraith, Janine Christina. "Nun Apothecaries and the Impact of the Secularization in Southwest Germany," *Continuity and Change* 25 (2010): 313–344.

Making Knowledge in Early Modern Europe: Practices, Objects and Texts, 1400–1800. Ed. Pamela H. Smith and Benjamin Schmidt. Chicago: University of Chicago Press, 2007.

Malamani, Anita. "L'Ospedale degli Incurabili di Pavia dalle origini al suo assorbimento nel P. L. Pertusati," *Archivio storico lombardo* 100 (1974): 145–170.

Malena, Adelisa. *L'eresia dei perfetti. Inquisizione romana ed esperienze mistiche nel Seicento italiano.* Rome: Edizioni di storia e letteratura, 2003.

Martorelli Vico, Romana. "Madri, levatrici, balie e padre: Michele Savonarola, l'embriologia e la cura dei piccoli." In *Michele Savonarola. Medicina e cultura di corte.* Ed. Chiara Crisciani and Gabriella Zuccolin. Florence: SISMEL, 2011, 127–135.

Masetti Zannini, Gian Lodovico. "Una canonichessa erborista: Semidea Poggi (sec. XVI-XVII)," *Strenna storica bolognese* 45 (1995): 395–401.

Masetti Zannini, Gian Lodovico. *Motivi storici della educazione femmnile. Scienza, lavoro, giuochi.* Naples: D'Auria, 1981.

Matthew, Louisa. "'Vendecolori a Venezia': The Reconstruction of a Profession," *Burlington Magazine* 144 (2002): 680–686.

McClive, Cathy. "Blood and Expertise: The Trials of the Female Medical Expert in the Ancien-Régime Courtroom," *BHM* 82 (2008): 86–108.

McClure, George. "Healing Eloquence: Petrarch, Salutati, and the Physicians," *Journal of Medieval and Renaissance Studies* 15 (1985): 317–346.

McGough, Laura J. *Gender, Sexuality, and Syphilis in Early Modern Venice: The Disease that Came to Stay.* Basingstoke: Palgrave, 2011.

McVaugh, Michael R. "Smell and the Medieval Surgeon," *Micrologus* 10 (2002): 113–132.

Mikkeli, Heiki. *Hygiene in the Early Modern Medical Tradition.* Helsinki: Finnish Academy of Science and Letters, 1999.

The Mindful Hand: Inquiry and Invention from the Late Renaissance to Early Industrialization. Ed. Lissa Roberts, Simon Schaffer, and Peter Dear. Amsterdam: Edita, 2007.

Molà, Luca. "Inventors, Patents and the Market for Innovations in Renaissance Italy," *History of Technology* 32 (2014): 7–34.

Monachus Aromatarius. "Il contributo degli ordini monastici alla botanica in Italia." In *Di sana pianta. Erbaria e taccuini di sanità.* Ed. Rolando Bussi. Modena: Panini, 1988, 29–32.

Monahan, Erika. "Locating Rhubarb: Early Modernity's Relevant Obscurity." In *Early Modern Things: Objects and Their Histories, 1500–1800.* Ed. Paula Findlen. New York: Routledge, 2013, 227–251.

Monson, Craig. *Nuns Behaving Badly: Tales of Music, Magic, Art and Arson in the Convents of Italy.* Chicago: University of Chicago Press, 2010.

Moran, Bruce T. *Distilling Knowledge: Alchemy, Chemistry, and the Scientific Revolution.* Cambridge, MA: Harvard University Press, 2005.

Moran, Bruce T. "Prince-Practitioning and the Direction of Medical Roles at the German Court: Maurice of Hesse-Kassel and His Physicians." In *Medicine at the Courts of Europe, 1500–1837.* Ed. Vivian Nutton. London: Routledge, 1990, 95–116.

Munger, Robert S. "Guaiacum, the Holy Wood from the New World," *BHM* 4 (1949): 196–229.

Munkhoff, Richelle. "Poor Women and Parish Public Health in Sixteenth-Century London," *Renaissance Studies* 28 (2014): 579–596.

Munkhoff, Richelle. "Searchers of the Dead: Authority, Marginality, and the Interpretation of Plague in England, 1574–1665," *Gender & History* 11 (1999): 1–29.

Murphy, Caroline P. "In Praise of the Ladies of Bologna: The Image and Identity of the Sixteenth-Century Bolognese Female Patriciate," *Renaissance Studies* 31 (1999): 440–454.

Murphy, Caroline P. *Murder of a Medici Princess.* Oxford: Oxford University Press, 2009.

Musacchio, Jacqueline. *The Art and Ritual of Childbirth in Renaissance Italy.* New Haven, CT: Yale University Press, 1999.

Nannizzi, A. "L'arte degli speziali in Siena," *Bullettino senese di storia patria* 46 (1939): 93–131, 214–260.

Naso, Irma. "I Savoia e la cura del corpo. Medici a corte nel tardo medioevo." In *Être médicin à la cour (Italie, France, Espagne, XIIIe–XVIIIe siècle)*. Ed. Elisa Andretta and Marilyn Nicoud. Florence: SISMEL, 2013, 51–85.

Newton, Hannah. *Misery to Mirth: Recovery from Illness in Early Modern England.* Oxford: Oxford University Press, 2018.

Newton, Hannah. "'She Sleeps Well and Eats an Egg': Convalescent Care in Early Modern England." In *Conserving Health in Early Modern Culture: Bodies and Environments in Italy and England.* Ed. Sandra Cavallo and Tessa Storey. Manchester, UK: Manchester University Press, 2017, 104–132.

Nicoud, Marilyn. "Diététique et alimentation des élites princières dans l'Italie medieval." In *Pratiques et discours alimentaires en Méditerranée de l'Antiquité à la Renaissance.* Ed. Jean Leclant, André Vauchez, and Michel Sartre. Paris: PERSEE, 2008, 317–336.

Nolte, Cordula. "Domestic Care in the Sixteenth Century: Expectations, Experiences, and Practices from a Gendered Perspective." In *Gender, Health, and Healing, 1250–1550.* Ed. Sara Ritchey and Sharon Strocchia. Amsterdam: Amsterdam University Press, forthcoming.

Nordio, Andrea. "Presenze femminili nella nascita dell'ospedale degli Incurabili di Venezia," *Regnum dei* 120 (1994): 11–39.

Nutton, Vivian. "The Reception of Fracastoro's Theory of Contagion: The Seed that Fell Among Thorns?," *Osiris,* ser. 2, vol. 6 (1990): 196–234.

Nutton, Vivian. "The Seeds of Disease: An Explanation of Contagion and Infection from the Greeks to the Renaissance," *Medical History* 27 (1983): 1–34.

Olsan, Lea T. "Charms and Prayers in Medieval Medical Practice," *Social History of Medicine* 16 (2003): 343–366.

Palmer, Richard. "Pharmacy in the Republic of Venice in the Sixteenth Century." In *The Medical Renaissance of the Sixteenth Century.* Ed. Andrew Wear, Roger French, and I. M. Lonie. Cambridge, UK: Cambridge University Press, 1985, 100–117.

Paoli, Maria Pia. "Di madre in figlio: per una storia dell'educazione alla corte dei Medici," *Annali di storia di Firenze* 3 (2008): 65–145.

Parascandola, John. "From Mercury to Miracle Drugs: Syphilis Therapy over the Centuries," *Pharmacy in History* 51 (2009): 14–23.

Park, Katharine. *Doctors and Medicine in Early Renaissance Florence.* Princeton, NJ: Princeton University Press, 1985.

Park, Katharine. *Secrets of Women: Gender, Generation, and the Origins of Human Dissection.* New York: Zone Books, 2006.

Park, Katharine, and John Henderson, "'The First Hospital among Christians': The Ospedale di Santa Maria Nuova in Early Sixteenth-Century Florence," *Medical History* 35 (1991): 164–188.

Passerini, Luigi. *Storia degli stabilimenti di beneficenza e d'istruzione elementare gratuita della città di Firenze.* Florence: Le Monnier, 1853.

Pelling, Margaret. *The Common Lot: Sickness, Medical Occupations and the Urban Poor in Early Modern England.* London: Longman, 1998.

Pelling, Margaret. "'Thoroughly Resented?' Older Women and the Medical Role in Early Modern London." In *Women, Science and Medicine 1500–1700: Mothers and Sisters of the Royal Society.* Ed. Lynette Hunter and Sarah Hutton. Stroud: Sutton, 1997, 63–88.

Pennell, Sara. "Mundane Materiality, or, Should Small Things Still Be Forgotten? Material Culture, Micro-Histories and the Problem of Scale." In *History and Material Culture: A Student's Guide to Approaching Alternative Sources.* Ed. Karen Harvey. London: Routledge, 2009, 173–191.

Pennell, Sara. "Perfecting Practice? Women, Manuscript Recipes, and Knowledge in Early Modern England." In *Early Modern Women's Manuscript Writing.* Ed. Victoria E. Burke and Jonathan Gibson. Aldershot, UK: Ashgate, 2004, 237–258.

Perifano, Alfredo. *L'Alchimie à la cour de Côme Ier de Médicis: savoirs, culture et politique.* Paris: Honoré Champion, 1997.

Pesenti, Tiziana. *Fasiculo de medicina in volgare: Venezia, Giovanni e Gregorio De Gregori, 1494.* 2 vols. Treviso: Antilia, 2001.

Pieraccini, Gaetano. *La stirpe de' Medici di Cafaggiolo: Saggio di ricerche sulla trasmissione ereditaria dei caratteri biologici.* 3 vols. Florence: Vallecchi, 1924, reprinted 1986.

Pizzagalli, Daniela. *La signora del Rinascimento, vita e splendori di Isabella d'Este alla corte di Mantova.* Milan: BUR Biblioteca Univ. Rizzoli, 2013.

Polizzotto, Lorenzo. *The Elect Nation: The Savonarolan Movement in Florence.* Oxford: Oxford University Press, 1994.

Polizzotto, Lorenzo. "When Saints Fall Out: Women and the Savonarolan Movement in Early Sixteenth-Century Florence," *RQ* 46 (1993): 486–525.

Pomata, Gianna. *Contracting a Cure: Patients, Healers, and the Law in Early Modern Bologna.* Baltimore, MD: Johns Hopkins University Press, 1998.

Pomata, Gianna. "Malpighi and the Holy Body: Medical Experts and Miraculous Evidence in Seventeenth-Century Italy," *Renaissance Studies* 21 (2007): 568–586.

Pomata, Gianna. "Medicina delle monache. Pratiche terapeutiche nei monasteri femminili di Bologna in età moderna." In *I monasteri femminili come centri di cultura fra Rinascimento e Barocco.* Ed. Gianna Pomata and Gabriella Zarri. Rome: Edizioni di storia e letteratura, 2005, 331–363.

Pomata, Gianna. "Practicing between Earth and Heaven: Women Healers in Early Modern Bologna," *Dynamis* 19 (1999): 119–143.

Porter, Roy. "The Rise of Physical Examination." In *Medicine and the Five Senses.* Ed. W. F. Bynum and Roy Porter. Cambridge, UK: Cambridge University Press, 1993, 179–197.

Presciutti, Diana Bullen. "Picturing Institutional Wet-Nursing in Medicean Siena." In *Medieval and Renaissance Lactations: Images, Rhetorics, Practices.* Ed. Jutta Gisela Sperling. Burlington, VT: Ashgate, 2013, 129–146.

Primhak, Victoria. "Women in Religious Communities: The Benedictine Convents in Venice, 1400–1550," PhD dissertation, University of London, 1991.

Prosperi, Adriano. *Tribunali della coscienza. Inquisitori, confessori, missionari.* Torino: Einaudi, 1996.

Pullan, Brian. *Tolerance, Regulation and Rescue: Dishonoured Women and Abandoned Children in Italy, 1300–1800.* Manchester, UK: Manchester University Press, 2016.

Quétel, Claude. *History of Syphilis.* Trans. Judith Braddock and Brian Pike. Baltimore, MD: Johns Hopkins University Press, 1990.

Ragab, Ahmed. *The Medieval Islamic Hospital: Medicine, Religion, and Charity.* Cambridge, UK: Cambridge University Press, 2015.

Rankin, Alisha. "Becoming an Expert Practitioner: Court Experimentalism and the Medical Skills of Anna of Saxony (1532–1585)," *Isis* 98 (2007): 23–53.

Rankin, Alisha. "Exotic Materials and Treasured Knowledge: The Valuable Legacy of Noblewomen's Remedies in Early Modern Germany," *Renaissance Studies* 28 (2014): 533–555.

Rankin, Alisha. "How to Cure the Golden Vein: Medical Remedies as *Wissenschaft* in Early Modern Germany." In *Ways of Making and Knowing: The Material Culture of Empirical Knowledge.* Ed. Pamela H. Smith, Amy R. W. Meyers, and Harold J. Cook. Ann Arbor: University of Michigan Press, 2014, 113–137.

Rankin, Alisha. "On Anecdote and Antidotes: Poison Trials in Sixteenth-Century Europe," *BHM* 91 (2017): 274–302.

Rankin, Alisha. *Panaceia's Daughters: Noblewomen as Healers in Early Modern Germany.* Chicago: University of Chicago Press, 2013.

Rawcliffe, Carole. "Hospital Nurses and Their Work." In *Daily Life in the Late Middle Ages.* Ed. Richard Britnell. Gloucestershire: Sutton, 1998, 43–64.

Rawlings, F. H. "Two 17th Century Women Apothecaries," *Pharmaceutical Historian* 14 (1984): 7.

Ray, Meredith K. *Daughters of Alchemy: Women and Scientific Culture in Early Modern Italy.* Cambridge, MA: Harvard University Press, 2015.

Ray, Meredith K. "Letters and Lace: Arcangela Tarabotti and Convent Culture in *Seicento* Venice." In *Early Modern Women and Transnational Communities of Letters.* Ed. Julie D. Campbell and Anne R. Larsen. Aldershot, UK: Ashgate, 2009, 45–73.

Reinarz, Jonathan. "Learning to Use Their Senses: Visitors to Voluntary Hospitals in Eighteenth-Century England," *Journal for Eighteenth-Century Studies* 35 (2012): 505–520.

Rey Bueno, Mar. "The Health of Philip II, A Matter of State. Medicines and Medical Institutions in the Spanish Court (1556–1598)." In *Être médicin à la cour (Italie, France, Espagne, XIIIe-XVIIIe siècle).* Ed. Elisa Andretta and Marilyn Nicoud. Florence: SISMEL, 2013, 149–159.

Richa, Giuseppe. *Notizie istoriche delle chiese fiorentine.* 10 vols. Florence: P.G. Viviani, 1754–1762.

Ritchey, Sara. "Affective Medicine: Later Medieval Healing Communities and the Feminization of Health Care Practices in the Thirteenth-Century Low Countries," *Journal of Medieval Religious Cultures* 40 (2014): 113–143.

Rocke, Michael. *Forbidden Friendships: Homosexuality and Male Culture in Renaissance Florence.* New York: Oxford University Press, 1996.

Rodrigues, Lisbeth de Oliveira and Isabel dos Guimarães Sá, "Sugar and Spices in Portuguese Renaissance Medicine," *Journal of Medieval Iberian Studies* 7 (2015): 176–196.

Rose, Colin. "Plague and Violence in Early Modern Italy," *RQ* 71 (2018): 1000–1035.

Rotolo, Vittorio. "La storia medica dello zucchero," *Rivista di storia della medicina,* ser. 2, vol. 8 (1998): 15–25.

Ruberg, Willemijn. "The Letter as Medicine: Studying Health and Illness in Dutch Daily Correspondence, 1770–1850," *Social History of Medicine* 23 (2010): 492–508.

Rublack, Ulinka. "Fluxes: The Early Modern Body and the Emotions," *History Workshop Journal* 53 (2002): 1–16.

Sá, Isabel dos Guimarães. "Between Spiritual and Material Culture: Male and Female Objects at the Portuguese Court, 1480–1580." In *Domestic Institutional Interiors in Early Modern Europe.* Ed. Sandra Cavallo and Silvia Evangelisti. Burlington, VT: Ashgate, 2009, 181–199.

Sá, Isabel dos Guimarães. "Catholic Charity in Perspective: The Social Life of Devotion in Portugal and Its Empire (1450–1700)," *e-Journal of Portuguese History* 2 (2004): 1–20.

Saltini, Guglielmo Enrico. "Di una visita che fece in Genova nel 1548 il fanciullo Don Francesco di Cosimo I de' Medici al Principe Don Filippo di Spagna," *Archivio storico italiano,* ser. 4, vol. 4 (1879): 19–34.

Sandri, Lucia. "Fanciulle e balie. La presenza femminile nell'Ospedale." In *Madri, figlie, balie. Il coretto della chiesa e la comunità femminile degli Innocenti.* Ed. Stefano Filipponi and Eleonora Mazzocchi. Florence: Nardini, 2010, 12–23.

Sanger, Alice E. *Art, Gender, and Religious Devotion in Grand Ducal Tuscany.* Aldershot, UK: Ashgate, 2014.

Sanger, Alice E. "Maria Maddalena d'Austria's Pilgrimage to Loreto: Visuality, Liminality and Exchange." In *Medici Women as Cultural Mediators (1533–1743). Le donne di casa Medici e il loro ruolo di mediatrici culturali fra le corti d'Europa.* Ed. Christina Strunck. Milan: Silvana, 2011, 253–265.

Savoia, Paolo. "The *Book of the Sick* of Santa Maria della Morte in Bologna and the Medical Organization of a Hospital in the Sixteenth Century," *Nuncius* 31 (2016): 163–235.

Savoia, Paolo. "Skills, Knowledge, and Status: The Career of an Early Modern Italian Surgeon," *BHM* 93 (2019): 27–54.

Schleiner, Winfried. "Moral Attitudes towards Syphilis and Its Prevention in the Renaissance," *BHM* 68 (1994): 389–410.

Schutte, Anne Jacobsen. *Aspiring Saints: Pretense of Holiness, Inquisition, and Gender in the Republic of Venice, 1618–1750.* Baltimore, MD: Johns Hopkins University Press, 2001.

Shapin, Steven. *A Social History of Truth: Civility and Science in Seventeenth-Century England.* Chicago: University of Chicago Press, 1994.

Shaw, James, and Evelyn Welch, *Making and Marketing Medicines in Renaissance Florence.* Amsterdam: Rodopi, 2011.

Shemek, Deanna. "In Continuous Expectation: Isabella d'Este's Epistolary Desire." In *Phaethon's Children: The Este Court and Its Culture in Early Modern Ferrara.* Ed. Dennis Looney and Deanna Shemek. Tempe: Arizona Center for Medieval and Renaissance Studies, 2005, 269–300.

Siena, Kevin. "The Clean and the Foul: Paupers and the Pox in London Hospitals, c. 1550–c. 1700." In *Sins of the Flesh: Responding to Sexual Disease in Early Modern Europe.* Ed. Kevin Siena. Toronto: University of Toronto Press, 2005, 261–284.

Siena, Kevin, and Jonathan Reinarz, "Scratching the Surface: An Introduction." In *A Medical History of Skin: Scratching the Surface.* Ed. Jonathan Reinarz and Kevin Siena. London: Pickering & Chatto, 2013, 1–15.

Skemer, Don C. *Binding Words: Textual Amulets in the Middle Ages.* University Park: Penn State University Press, 2006.

Smith, Pamela H. *The Body of the Artisan: Art and Experience in the Scientific Revolution.* Chicago: University of Chicago Press, 2004.

Smith, Pamela H. "Making Things: Techniques and Books in Early Modern Europe." In *Early Modern Things: Objects and Their Histories, 1500–1800.* Ed. Paula Findlen. New York: Routledge, 2013, 173–203.

Sobel, Dava. *Galileo's Daughter: A Historical Memoir of Science, Faith and Love.* New York: Walker & Company, 1999.

Sperling, Jutta Gisela. *Convents and the Body Politic in Late Renaissance Venice.* Chicago: University of Chicago Press, 1999.

La spezieria di San Benedetto a Montefiascone. Dalle collezioni di Palazzo Venezia in Roma. Ed. Maria Selene Sconci and Romualdo Luzi. Ferrara: Belriguardo, 1994.

Stampino, Maria Galli. "Maria Maddalena, Archduchess of Austria and Grand Duchess of Florence: Negotiating Performance, Traditions, and Taste." In *Early Modern Hapsburg Women: Transnational Contexts, Cultural Conflicts, Dynastic Continuities.* Ed. Anne J. Cruz and Maria Galli Stampino. Burlington, VT: Ashgate, 2013, 41–56.

Steegman, Mary G. *Bianca Cappello.* London: Constable and Company, 1913.

Stein, Claudia. *Negotiating the French Pox in Early Modern Germany.* Aldershot, UK: Ashgate, 2009.

Stevens Crawshaw, Jane. "Families, Medical Secrets and Public Health in Early Modern Venice," *Renaissance Studies* 28 (2014): 597–618.

Stevens Crawshaw, Jane. *Plague Hospitals: Public Health for the City in Early Modern Venice.* Aldershot, UK: Ashgate, 2012.

Still, George Frederic. *The History of Paediatrics.* Oxford: Oxford University Press, 1931.

Stocchetti, Denise. "La fondazione del monastero fiorentino delle Murate e la pellegrina Eugenia," *Archivio italiano per la storia della pietà* 18 (2005): 177–247.

Stolberg, Michael. *Experiencing Illness and the Sick Body in Early Modern Europe.* New York: Palgrave, 2011.

Storey, Tessa. "Face Waters, Oils, Love Magic and Poison: Making and Selling Secrets in Early Modern Rome." In *Secrets and Knowledge in Medicine and Science, 1500–1800.* Ed. Elaine Leong and Alisha Rankin. Aldershot, UK: Ashgate, 2011, 143–163.

Strocchia, Sharon T. "Introduction: Women and Healthcare in Early Modern Europe," *Renaissance Studies* 28 (2014): 496–514.

Strocchia, Sharon T. *Nuns and Nunneries in Renaissance Florence.* Baltimore, MD: Johns Hopkins University Press, 2009.

Strocchia, Sharon T. "The Nun Apothecaries of Renaissance Florence: Marketing Medicines in the Convent," *Renaissance Studies* 25 (2011): 627–647.

Strocchia, Sharon, and Julia Rombough, "Women behind Walls: Tracking Nuns and Socio-Spatial Networks in Sixteenth-Century Florence." In *Mapping Space, Sense, and Movement in Florence: Historical GIS and the Early Modern City.* Ed. Nicholas Terpstra and Colin Rose. New York: Routledge, 2016, 87–106.

Stuart, Kathy. *Defiled Trades and Social Outcasts: Honor and Ritual Pollution in Early Modern Germany.* Cambridge, UK: Cambridge University Press, 1999.

Styles, John. "Product Innovation in Early Modern London," *Past and Present* 168 (2000): 124–169.

Tazzara, Corey. "Capricious Demands: Artisanal Goods, Business Strategies, and Consumer Behavior in Seventeenth-Century Florence." In *Early Modern Things: Objects and Their Histories, 1500–1800.* Ed. Paula Findlen. New York: Routledge, 2013, 204–224.

Tchikine, Anatole. "Horticultural Differences: The Florentine Garden of Don Luis de Toledo and the Nuns of S. Domenico del Maglio," *Studies in the History of Gardens and Designed Landscapes* 30 (2010): 224–240.

Terpstra, Nicholas. *Abandoned Children of the Italian Renaissance: Orphan Care in Florence and Bologna.* Baltimore, MD: Johns Hopkins University Press, 2005.

Terpstra, Nicholas. "Competing Visions of the State and Social Welfare: The Medici Dukes, the Bigallo Magistrates, and Local Hospitals in Sixteenth-Century Tuscany," *RQ* 54 (2001): 1319–1355.

Terpstra, Nicholas. *Cultures of Charity: Women, Politics and the Reform of Poor Relief in Renaissance Italy.* Cambridge, MA: Harvard University Press, 2013.

Terpstra, Nicholas. "Locating the Sex Trade in the Early Modern City: Space, Sense, and Regulation in Sixteenth-Century Florence." In *Mapping Space, Sense, and Movement in Florence: Historical GIS and the Early Modern City.* Ed. Nicholas Terpstra and Colin Rose. New York: Routledge, 2016, 107–124.

Terpstra, Nicholas. *Lost Girls: Sex and Death in Renaissance Florence.* Baltimore, MD: Johns Hopkins University Press, 2010.

Terpstra, Nicholas. "Making a Living, Making a Life: Work in the Orphanages of Florence and Bologna," *Sixteenth Century Journal* 31 (2000): 1063–1079.

Terpstra, Nicholas. "Mothers, Sisters, and Daughters: Girls and Conservatory Guardianship in Late Renaissance Florence," *Renaissance Studies* 17 (2003): 201–229.

Thomas, Keith. "Magical Healing: The King's Touch." In *The Book of Touch.* Ed. Constance Classen. Oxford: Berg, 2005, 354–362.

Tomas, Natalie. "Commemorating a Mortal Goddess: Maria Salviati de' Medici and the Cultural Politics of Duke Cosimo I." In *Practices of Gender in Late Medieval and Early Modern Europe.* Ed. Megan Cassidy-Welch and Peter Sherlock. Turnhout, Belgium: Brepols, 2008, 261–278.

Tomas, Natalie. "Eleonora di Toledo, Regency, and State Formation in Tuscany." In *Medici Women: The Making of a Dynasty in Grand Ducal Tuscany.* Ed. Giovanna Benadusi and Judith C. Brown. Toronto: University of Toronto Press, 2015, 59–89.

Tomas, Natalie. "'With His Authority She Used to Manage Much Business': The Career of Signora Maria Salviati and Duke Cosimo I de' Medici." In *Studies on Florence and the Italian Renaissance in Honour of F.W. Kent.* Ed. Peter Howard and Cecilia Hewlett. Turnhout, Belgium: Brepols, 2016, 133–148.

Trexler, Bernice J. "Hospital Patients in Florence: San Paolo, 1567–68," *BHM* 48 (1974): 41–59.

Trivellato, Francesca. "Guilds, Technology and Economic Change in Early Modern Venice." In *Guilds, Innovation and the European Economy, 1400–1800.* Ed. Steven R. Epstein and Maarten Prak. Cambridge, UK: Cambridge University Press, 2010, 199–231.

Turrill, Catherine. "*Compagnie* and *Discepole:* The Presence of Other Women Artists at Santa Caterina da Siena." In *Suor Plautilla Nelli (1523–1588): The First Woman Painter of Florence.* Ed. Jonathan Nelson. Florence: Cadmo, 2000, 83–102.

Vanden Berghe, Evelien. "The Pharmacy of the St. John's Hospital in Bruges." https://farmasihistorie.com/w/index.php?title=The_pharmacy_of_the_St._John%E2%80%99s_hospital_in_Bruges, accessed April 2019.

Vannini, Guido. "La spezieria: formazione e dotazione." In *Una farmacia preindustriale in Valdelsa. La spezieria e lo spedale di Santa Fina nella città di San Gimignano, secc. XIV–XVIII.* Ed. Gabriele Borghini. San Gimignano: Città di San Gimignano, 1981, 37–38.

Vanti, Mario. *S. Giacomo degl'Incurabili di Roma nel Cinquecento.* Rome: Pustet, 1938.

Vigarello, Georges. *Concepts of Cleanliness: Changing Attitudes in France since the Middle Ages.* Trans. Jean Birrell. Cambridge, UK: Cambridge University Press, 1988.

Vilares Cepeda, Isabel. "Os Livros da Rainha D. Leonor, segundo o inventario de 1537 do Convento da Madre de Deus," *Revista da Biblioteca Nacional* ser. 2, vol. 2 (1987): 51–81.

Viroli, Marco. *Caterina Sforza, Leonessa di Romagna*. Cesena: Il Ponte Vecchio, 2008.

Viviani della Robbia, Enrica. *Nei monasteri fiorentini*. Florence: Sansoni, 1946.

Weaver, Elissa. *Convent Theatre in Early Modern Italy: Spiritual Fun and Learning for Women*. Cambridge, UK: Cambridge University Press, 2002.

Weaver, Elissa. "Fiammetta Frescobaldi," http://www.lib.uchicago.edu/efts/IWW/BIOS/A0160.html, accessed April 2018.

Weddle, Saundra. "Identity and Alliance: Urban Presence, Spatial Privilege, and Florentine Renaissance Convents." In *Renaissance Florence: A Social History*. Ed. Roger Crum and John Paoletti. Cambridge, UK: Cambridge University Press, 2006, 394–412.

Weiner, Annette B. *Inalienable Possessions: The Paradox of Keeping-While-Giving*. Berkeley: University of California Press, 1992.

Weisser, Olivia. *Ill Composed: Sickness, Gender, and Belief in Early Modern England*. New Haven, CT: Yale University Press, 2015.

Welch, Evelyn. "Scented Buttons and Perfumed Gloves: Smelling Things in Renaissance Italy." In *Ornamentalism: The Art of Renaissance Accessories*. Ed. Bella Mirabella. Ann Arbor: University of Michigan Press, 2011, 13–39.

Welch, Evelyn. *Shopping in the Renaissance: Consumer Cultures in Italy, 1400–1600*. New Haven, CT: Yale University Press, 2005.

Welch, Evelyn. "Space and Spectacle in the Renaissance Pharmacy," *Medicina e storia* 15 (2008). 127–158.

Westwater, Lynn. "A Rediscovered Friendship in the Republic of Letters: The Unpublished Correspondence of Arcangela Tarabotti and Ismaël Bouilliau," *RQ* 65 (2012): 67–134.

Wheeler, Jo. *Renaissance Secrets: Recipes and Formulas*. London: V & A, 2009.

Wheeler, Jo. http://renaissancesecrets.blogspot.com/2014/09/stefano-rosselli-secrets-and-medici.html, accessed 15 April 2018.

Wiesner, Merry. *Working Women in Renaissance Germany*. New Brunswick, NJ: Rutgers University Press, 1986.

Wolfthal, Diane. "Household Help: Early Modern Portraits of Female Servants," *Early Modern Women* 8 (2013): 5–52.

Woods, May, and Arete Swartz Warren, *Glass Houses: A History of Greenhouses, Orangeries and Conservatories*. New York: Rizzoli, 1988.

Zardin, Danilo. "Libri e biblioteche negli ambienti monastici dell'Italia del primo Seicento." In *Donne, filosofia e cultura nel Seicento*. Ed. Pina Totaro. Rome: Consiglio nazionale delle ricerche, 1999, 347–383.

Zarri, Gabriella. "Matronage / maternage. Tipologie di rapporti tra corti femminili e istituzioni religiose." In *Le donne medici nel sistema europeo delle corti (XVI–XVIII secolo).* Ed. Giulia Calvi and Riccardo Spinelli. Florence: Polistampa, 2008, 67–74.

Zarri, Gabriella. "I monasteri femminili a Bologna tra il XIII e il XVII secolo," *Atti e memorie: Deputazione di storia patria per le provincie di Romagna,* n.s. 24 (1973): 133–224.

Zobi, Antonio. *Manuale storico degli ordinamenti economici vigenti in Toscana.* Florence: Accademia de'Georgofili, 1858.

Zuccolin, Gabriella. "Medici a corte e formazione del signore." In *Costumi educativi nelle corti europee (XIV-XVIII secoli).* Ed. Monica Ferrari. Pavia: Pavia University Press, 2010, 77–102.

Zuccolin, Gabriella. "Nascere in latino e in volgare. Tra la *Pratica Maior* e il *De regimine pregnantium.*" In *Michele Savonarola. Medicina e cultura di corte.* Ed. Chiara Crisciani e Gabriella Zuccolin. Florence: SISMEL, 2011, 137–209.

Zuccolin, Gabriella. "Sapere medico e istruzioni etico-politiche: Michele Savonarola alla corte estense," *Micrologus* 16 (2008): 313–326.

ACKNOWLEDGMENTS

The idea for this project first took flight in the summer of 2009, when I participated in the NEH Summer Seminar for College Teachers, "Disease in the Middle Ages," led by Monica Green and Walt Schalick. Venturing into the seminar with no training in medical history was a daunting but transformative experience. I am grateful to the National Endowment for the Humanities for sponsoring this vibrant exchange, and to the Wellcome Trust for hosting it. My thanks to the other participants for sharing insights into disease, disability, illness, healing, care, and the premodern body over five weeks of intensive reading and discussion.

Generous funding from other agencies allowed me to pursue preliminary research and to take it in new directions. A Solmsen Fellowship at the Institute for Research in the Humanities at the University of Wisconsin in spring 2011 let me delve into the history of pharmacy and nursing at an early stage. The Fox Center for Humanistic Inquiry at Emory University provided the time and space needed to advance the project in 2013–2014. My thanks to former director Martine Brownley for ensuring that fellows remained unencumbered by other duties, and to the other fellows for stimulating conversation. The book assumed its final shape at the National Humanities Center in 2015–2016, where I had the privilege of holding the Ruth W. and A. Morris Williams Fellowship. My intellectual debt to the Williams, who believe in the power of the humanities to enrich our common life, extends well beyond that academic year. Thanks too to my fellow fellows, especially Michelle O'Malley, for many rewarding exchanges, and to the incomparable library staff there. Finally, I am grateful to the Emory College of Arts and Sciences for a PERS travel grant that supported research in London in late summer 2016.

Portions of this book were presented as lectures or seminars at the University of Cologne, University of Toronto, Wake Forest University, Elon University, the Sixteenth Century Studies Conference, and several meetings of the Renaissance Society of America. Audiences at these venues shaped the project by asking hard questions and forcing me to articulate my viewpoints more clearly. I remain grateful for their feedback, even if I haven't satisfied all their concerns.

Librarians and archivists are vital contributors to the research process in matters great and small. My sincere thanks to staff at the following Florentine repositories for going beyond the call of duty: the Archivio di Stato; Biblioteca Nazionale Centrale; Biblioteca Laurenziana; Biblioteca Marucelliana; Biblioteca Riccardiana; and the Ospedale degli Innocenti. In addition, staff members at several collections outside Florence deserve great thanks: the Archivio di Stato, Bologna; Biblioteca Comunale dell'Archiginnasio, Bologna; Archivio di Stato, Rome; Wellcome Library, London; and the British Library, London. Closer to home, the Interlibrary Loan staff at Emory University tracked down obscure publications with lightning speed.

Friends and colleagues helped bring this book to fruition in countless ways. I am profoundly grateful to Judith Brown, Stanley Chojnacki, and the late Anne Schutte not only for supporting grant applications over the years, but also for setting such high standards for Italian historical scholarship. Sheila Barker deserves special thanks for sharing key documents and images; she has been a terrific sounding board as ideas evolved. Nicholas Terpstra has been a continual source of insight into the nature of early modern charity. Perceptive comments by Nicholas Baker about the early Medici court improved the first chapter significantly. Richard Goldthwaite asked provocative questions about the Florentine economy that proved seminal to my thinking. Saundra Weddle and Julia DeLancey offered unflagging support as the project progressed; their wide-ranging curiosity made this a better book. John Christopoulous, Nicholas Eckstein, Cynthia Klestinec, Elizabeth Mellyn, Julia Rombough, Natalie Tomas, Catherine Turrill, and Elissa Weaver shared food, friendship, archival references, and insight into all things Italian. John Henderson, Lisa Kaborycha, and Guy Geltner listened thoughtfully as I tested new ideas, while Alisha Rankin and Meredith Ray contributed important comparative perspectives. Shannon McHugh and Daniella

Cagliari kindly proffered information about the Bolognese nun Semidea Poggi at a pivotal point in the project. Suzanne Cusick and Jennifer Hariguchi generously shared transcriptions of archival materials with me. Megan Slemons, GIS specialist at Emory, provided valuable assistance in making the map of Florentine convent pharmacies. Kate Lowe ably shepherded this book through the early phases of publication. My thanks to the two anonymous reviewers for their constructive comments, and to Andrew Kinney and Olivia Woods at Harvard University Press for their help in bringing the book to completion. Of course, any remaining errors are my own responsibility. All translations are mine unless noted otherwise.

A host of friends sustained me through thick and thin over the course of writing and research. Heartfelt thanks to Robyn Fivush, Jim Roark, and Diane Jones Palm for their enduring friendship. Karen Stolley, Peggy Barlett, Judith Miller, Leslie Harris, and Laurie Patton buoyed my spirits, even from afar. My sister Susan Hall, my nieces Jamie and Stacey Hall, and my nephew Stephen Hall, remain my favorite cheerleaders. Other members of the Strocchia clan—Rachele, Domenico and Barbara, Flora, Francesco—welcomed me during research trips to Florence. Sadly, Felice Strocchia did not live to see the completion of this book, but I hope he would be proud to see another volume on the family bookshelf. Finally, this book is dedicated to the memory of my mother Stella, who embodied the very essence of care.

Portions of Chapters 3 and 4 were first published in "The Nun Apothecaries of Renaissance Florence: Marketing Medicines in the Convent," *Renaissance Studies* 25, no. 5 (2011): 627–647. Sections of Chapter 5 come from my essay, "Caring for the 'Incurable' in Renaissance Pox Hospitals," in *Hospital Life: Theory and Practice from the Medieval to the Modern,* ed. Laurinda Abreu and Sally Sheard (Oxford: Peter Berg, 2013), 67–92. I thank both publishers for the permission to reprint these text selections.

INDEX